It Runs in the Family: The Importance of the Lamiaceae Family Species

It Runs in the Family: The Importance of the Lamiaceae Family Species

Guest Editor

Antonios Chrysargyris

Basel • Beijing • Wuhan • Barcelona • Belgrade • Novi Sad • Cluj • Manchester

Guest Editor
Antonios Chrysargyris
Department of Agricultural
Sciences, Biotechnology and
Food Science
Cyprus University of
Technology
Limassol
Cyprus

Editorial Office
MDPI AG
Grosspeteranlage 5
4052 Basel, Switzerland

This is a reprint of the Special Issue, published open access by the journal *Agronomy* (ISSN 2073-4395), freely accessible at: https://www.mdpi.com/journal/agronomy/special_issues/lamiaceae_family_species.

For citation purposes, cite each article independently as indicated on the article page online and as indicated below:

Lastname, A.A.; Lastname, B.B. Article Title. *Journal Name* **Year**, *Volume Number*, Page Range.

ISBN 978-3-7258-4433-3 (Hbk)
ISBN 978-3-7258-4434-0 (PDF)
https://doi.org/10.3390/books978-3-7258-4434-0

Cover image courtesy of Antonios Chrysargyris
Spearmint (*Mentha spicata*)

Contents

About the Editor

Antonios Chrysargyris

Dr. Antonios Chrysargyris is a biologist, and he is currently a member of the Special Teaching Staff, at the Department of Agricultural Sciences, Biotechnology and Food Science, at the Cyprus University of Technology (CUT). He holds a master's degree in "Land Ecosystems/Biological Resources" from the University of Crete (Greece) and earned his PhD from the Cyprus University of Technology (CUT) in 2016. His research focuses on the complex interactions between medicinal and ornamental plants and environmental stressors such as salinity and mineral imbalances, and plant mineral requirements. He specializes in the cultivation of medicinal, aromatic, ornamental, and wild plant species, investigating their mineral requirements, nutritional value, and the biological activity of their extracts and essential oils, including antioxidant, antimicrobial, and insecticidal properties. He is also engaged in the introduction of unexplored and underutilized plant species into intensive cultivation systems as hydroponics and soilless cultures, where he assesses innovative organic and inorganic materials as alternative growing media. He has participated as a researcher in numerous National and European funded projects. To date, he has published 135 articles in peer-reviewed journals, contributed four chapters in peer-reviewed books, and delivered over 120 presentations at National and International conferences. With an H-index of 34, he has been ranked among the TOP 2% of most cited researchers worldwide in the field of plant biology and agronomy for the years 2021, 2022, and 2023. Dr. Chrysargyris also serves as an Associate or Guest Editor for several academic journals, including *Agronomy, Heliyon, Phyton, Frontiers in Plant Sciences, Horticulturae, Chemical, and Biological Technologies in Agriculture.*

agronomy

Editorial

It Runs in the Family: The Importance of the Lamiaceae Family Species

Antonios Chrysargyris

Department of Agricultural Sciences, Biotechnology and Food Science, Cyprus University of Technology, 3603 Lemesos, Cyprus; a.chrysargyris@cut.ac.cy

Medicinal and aromatic plants (MAPs) are becoming increasingly popular in industry, education, agriculture, and health disciplines due to their extensive collection of bioactive chemicals that enhance biological activity in vitro and in vivo. This trend is occurring because MAPs secondary metabolites contain significant antioxidant, anti-inflammatory, antibacterial, antiviral, and anticancer properties and biological activities that outperform many regularly used both synthetic and natural antioxidants [1,2]. These features are due to the abundance of various component groups (phenols, flavonols/flavonoids, alkaloids, polypeptides, vitamins, catechins, phytoestrogens, carotenoids, chlorophyll, minerals, etc.).

The Lamiaceae (Labiatae) family is one of the most significant groups of flowering plants, with a diverse range of species with ecological and medicinal applications. They are primarily herbs and shrubs with a fragrant perfume and a high level of beneficial substances used in natural therapy. Lamiaceae is a significant plant family with 250 genera and over 7000 species. The major genera in this family include *Salvia*, *Scutellaria*, *Stachys*, *Plectranthus*, *Hyptis*, *Teucrium*, *Thymus*, *Vitex*, and *Nepeta*. The most well-known examples include thyme, mint, oregano, basil, sage, savory, rosemary, hyssop, and lemon balm, all of which are aromatic spices, as well as a few others with more limited applications [3]. Lamiaceae species are diverse and widely distributed across several environments but are particularly abundant in the Mediterranean basin [4]. Species from this family are of enormous economic value on a global scale, having numerous applications, including food, cosmetics, flavoring, perfumery, pesticides, and pharmaceuticals [5].

The present Special Issue "It Runs in the Family: The Importance of the Lamiaceae Family Species" compiles 11 original research articles addressing the recent developments in growing plants in soilless culture, i.e., hydroponics and aquaponics, the changes on plants secondary metabolites under environmental stressors and cultivation practices, landrace maintenance and sustainable agriculture, the effective way of individual compounds identification and extraction methods, as well as the preservative role of essential oils (EOs) and natural compounds on fresh produce and stored grains. The present Special Issue contains scientific papers of high-quality standard coming from several prestigious and renown research groups.

The collection provides insights on the plant extracts and natural compound composition and their role in human health by inhibiting scavenging reactive oxygen species (ROS), stimulating the production of melanin, and absorbing UV radiation. Tsitsigianni et al. [6] investigated the biological activities of different extracts/infusions from ten Lamiaceae species with a rich phytochemical profile, exploring the cytotoxic, photoprotective, antioxidant, and wound-healing properties of different compounds. In that study, five compounds were isolated and identified through NMR spectra, namely salvianic acid A, rosmarinic acid, salvianolic acid K, luteolin-3-O-D-glucuronide, and hispidulin-7-O-D-glucuronide, introducing their photoprotective and non-cytotoxic roles as well as providing a better insight into *Salvia officinalis* and its bioactive constituents. In another study, Tomou et al. [7] investigated the metabolic profiles of the infusions of four *Stachys* members (*S. candida*, *S.*

chrysantha, *S. leucoglossa* subsp. *Leucoglossa*, and *S. spinulosa*) through NMR and HPLC-PDA-MS analyses, detecting 26 compounds belonging to flavonoids, phenylethanoid glycosides, and phenolic acids. Among them, chlorogenic acid was identified in all samples as one of their main metabolites. Innovative inputs of this study were reports for the first time on the metabolic characterization of *S. spinulosa* and discusses the chemotaxonomic significance of such findings.

Standard techniques for extracting EOs from plants include cold pressing, hydro-distillation, and steam distillation, but these technologies have limitations, including extended extraction durations, high energy requirements, and solvent usage. Gavrila et al. [8] investigated the combination of ultrasound and microwave techniques to enhance the thyme EO extraction. It was found that using ultrasound pre-treatment and microwave extraction processes, the extraction time was reduced by 72% compared to conventional hydrodistillation. The benefits of using ultrasound pre-treatment revealed a 23% increase in EO content when compared to extraction without pre-treatment.

Fresh produce preservation remains a challenge, garnering researchers' interest in the postharvest sector for alternative solutions due to uncontrolled spoilage and customer concerns about synthetic food safety. Natural preservative means, including Eos, are attracting interest nowadays, being easily accessible, environmentally safe, often less expensive, and less dangerous for non-target organisms than chemical treatments. In a study of Tzortzakis [9], *Origanum dictamnus* EOs were applied for the preservation of pepper fruits against a widespread postharvest fungus, *Botrytis cinerea*, indicating a 6-day lasting effect of the EOs vapor application on fresh produce preservation. In contrast to sanitary dips, vapor treatment proved to be more effective in pepper fruit preservation. However, on top of the antimicrobial properties of the EOs, fruit quality and safety are also fundamental. To that direction, Xylia et al. [10] evaluated the effectiveness of EOs of *Lavandula angustifolia* Mill. and *Rosmarinus officinalis*, their possible synergistic effects of the EO mixture, and the role of their common main EO component (eucalyptol) for the preservation of cucumber fruits. In that study, the appropriate EO levels were highlighted to avoid any oxidative stress to the fresh commodities, and the synergistic role of the different EO components was addressed in comparison to the main component, providing new insights in natural-based sanitation solutions.

Fresh produce preservation is undoubtedly a challenge, but equally challenging is the stored grain, including *Avena sativa* (L.), *Hordeum vulgare* (L.), *Sorghum bicolor* (L.), *Triticum aestivum* (L.), and *Zea mays* (L.), as several opportunistic pests cause significant losses. To that sense, Plata-Rueda et al. [11] evaluated the *Origanum vulgare* EOs against the stored product beetle, *Sitophilus granaries*, and revealed that *O. vulgare* EOs affect different biological functions in the insect, altering the behavioral pattern in terms of walking distance and resting time, displaying repellency and respiration rates, and insect survival. This represents a first step towards green pesticide innovation, opening new possibilities for pest management in storage.

Plant growth condition and the cultivation practices are important for a successful yield and quality of fresh commodities, with soilless systems being superior towards field conditions. In a study of Hazrati et al. [12], the optimal mineral levels and harvesting time for three *Mentha* species were determined, indicating that increased mineral levels affected positively yield but the increased nitrogen concentration in nutrient solution had a negative effect on specific quality parameters, such as higher NO_3^- content, especially at the third harvest time. Therefore, it is important to tailor the hydroponic nutrient solutions to specific plant species and environmental conditions for achieving optimal yields and quality in commercial mint cultivation. Another approach for sustainable crop cultivation is aquaponics, which is considered a system that can use the water-containing residues from fish production for producing plants. Albadwawi et al. [13] investigated the basil (*Ocimum basilicum* L.) growth in aquaponics compared to soil systems, indicating aquaponics as a sustainable system for both basil growth and increased antioxidant capacity of the plants when compared to conventional greenhouse cultivation, as plants were subjected to water stress in aquaponics but enhanced their biochemical profiles.

During crop production, not only environmental conditions (heat, drought, wind, etc.) and biotic conditions (pathogens, insects, etc.) but also cultivation practices (including fertilization, irrigation, variety, etc.) can affect the crop yield and performance of plants. In a study by Chrysargyris et al. [14], it was determined the impact of the cropping system (conventional *versus* organic) and irrigation regime (full *versus* deficit irrigation) on *Melissa officinalis*. Results of this work demonstrated that deficit irrigation is an environmentally friendly approach that might be used in both conventional and organic *Melissa officinalis* cropping systems, with the goal of reducing irrigation water use while compensating for reduced herb yields with higher essential oil and polyphenol content. In another study with water management efficiency, Yousefzadeh et al. [15] evaluated soil water availability and effects of developmental stages of *Thymus armeniacus* and *T. kotschyanus* under water shortage, concluding that cultivating *T. armeniacus* with adequate water availability leads to higher yields. Under water scarcity, *T. kotschyanus* is the preferred choice due to its drought tolerance, and selection of cultivars with resistance to abiotic or biotic stress factors is important for crop production.

Another important issue raised in this Special Issue is the environmental and socioeconomic dimensions of the MAPs, focusing on the Lamiaceae family and exploring the related local knowledge and cultural practices that influence their utilization for various purposes. Ivanova et al. [16] investigated the Lamiaceae diversity in home gardens, and semi-structured interviews focused on the cultivation, collection, and utilization practices common among elderly inhabitants in rural Bulgaria. It was highlighted that home gardens are important pools of plant genetic resources that should be preserved and further explored in the frame of the multitude of benefits provided by these plants, while traditional culinary practices were found to sustain the diversity of local forms (landraces).

This collection of high-level scientific publications aims to stimulate discussion and explore the potential of MAPs of the Lamiaceae family and their EOs in ecofriendly preservative means and uses in different industrial sectors with optimized growth conditions, extraction methods, and applications. Soilless cultivation methods can be introduced for other medicinal, aromatic, or culinary species of ethnobotanical interest. These species may have diverse bioactive capabilities that are not yet fully understood or utilized. Additionally, farmers need to optimize all elements that affect yield and produce quality in order to meet the high customer demand for high-quality food.

Funding: This work was funded by the Project "Opti-AromaQ" EXCELLENCE/0421/0299, which is co-financed by the European Union and the Republic of Cyprus through the Research and Innovation Foundation.

Conflicts of Interest: The author declares no conflicts of interest.

References

1. Seca, A.M.L.; Pinto, D.C.G.A. Biological Potential and Medical Use of Secondary Metabolites. *Medicines* **2019**, *6*, 66. [CrossRef] [PubMed]
2. Chrysargyris, A.; Petrovic, J.D.; Tomou, E.; Kyriakou, K.; Xylia, P.; Kotsoni, A.; Gkretsi, V.; Miltiadous, P.; Skaltsa, H.; Sokovi, M.D.; et al. Phytochemical Profiles and Biological Activities of Plant Extracts from Aromatic Plants Cultivated in Cyprus. *Biology* **2024**, *13*, 45. [CrossRef] [PubMed]
3. Bekut, M.; Brkić, S.; Kladar, N.; Dragović, G.; Gavarić, N.; Božin, B. Potential of selected Lamiaceae plants in anti(retro) viral therapy. *Pharmacol. Res.* **2020**, *133*, 301–314. [CrossRef]
4. Ramos Da Silva, L.R.; Ferreira, O.O.; Cruz, J.N.; De Jesus Pereira Franco, C.; Oliveira Dos Anjos, T.; Cascaes, M.M.; Almeida Da Costa, W.; Helena De Aguiar Andrade, E.; Santana De Oliveira, M. Lamiaceae Essential Oils, Phytochemical Profile, Antioxidant, and Biological Activities. *Evid.-Based Complement. Altern. Med.* **2021**, *2021*, 6748052. [CrossRef] [PubMed]
5. Stankovic, M. *Lamiaceae Species: Biology, Ecology and Practical Uses*; MDPI: Basel, Switzerland, 2020; ISBN 9783039284184.
6. Tsitsigianni, E.; Tomou, E.M.; Almpani, C.; Rallis, M.C.; Skaltsa, H. Biological Activities of Lamiaceae Species: Bio-Guided Isolation of Active Metabolites from *Salvia officinalis* L. *Agronomy* **2023**, *13*, 1224. [CrossRef]

7. Tomou, E.M.; Karioti, A.; Tsirogiannidis, G.; Krigas, N.; Skaltsa, H. Metabolic Characterization of Four Members of the Genus *Stachys* L. (Lamiaceae). *Agronomy* **2023**, *13*, 2624. [CrossRef]
8. Gavrila, A.I.; Chisega-Negrila, C.G.; Maholea, L.; Gavrila, M.L.; Parvulescu, O.C.; Popa, I. Enhancing the Extraction Process Efficiency of Thyme Essential Oil by Combined Ultrasound and Microwave Techniques. *Agronomy* **2023**, *13*, 2331. [CrossRef]
9. Tzortzakis, N. *Origanum dictamnus* Essential Oil in Vapour or Aqueous Solution Application for Pepper Fruit Preservation against *Botrytis cinerea*. *Agronomy* **2024**, *14*, 257. [CrossRef]
10. Xylia, P.; Goumenos, C.; Tzortzakis, N.; Chrysargyris, A. Application of Lavender and Rosemary Essential Oils (EOs), Their Mixture and Eucalyptol (EOs Main Compound) on Cucumber Fruit Quality Attributes and Microbial Load. *Agronomy* **2023**, *13*, 2493. [CrossRef]
11. Plata-Rueda, A.; Dos Santos, M.H.; Serrão, J.E.; Martínez, L.C. Chemical Composition and Insecticidal Properties of *Origanum vulgare* (Lamiaceae) Essential Oil against the Stored Product Beetle, *Sitophilus granarius*. *Agronomy* **2022**, *12*, 2204. [CrossRef]
12. Hazrati, S.; Pignata, G.; Casale, M.; Hosseini, S.J.; Nicola, S. Influence of Nutrient Solutions in an NGS® Soilless System on the Yield, Quality and Shelf Life of Fresh-Cut Commercial Mint at Different Harvest Times. *Agronomy* **2024**, *14*, 610. [CrossRef]
13. Albadwawi, M.A.O.K.; Ahmed, Z.F.R.; Kurup, S.S.; Alyafei, M.A.; Jaleel, A. A Comparative Evaluation of Aquaponic and Soil Systems on Yield and Antioxidant Levels in Basil, an Important Food Plant in Lamiaceae. *Agronomy* **2022**, *12*, 3007. [CrossRef]
14. Chrysargyris, A.; Petropoulos, S.A.; Tzortzakis, N. Essential Oil Composition and Bioactive Properties of Lemon Balm Aerial Parts as Affected by Cropping System and Irrigation Regime. *Agronomy* **2022**, *12*, 649. [CrossRef]
15. Yousefzadeh, K.; Houshmand, S.; Shiran, B.; Mousavi-Fard, S.; Zeinali, H.; Nikoloudakis, N.; Gheisari, M.M.; Fanourakis, D. Joint Effects of Developmental Stage and Water Deficit on Essential Oil Traits (Content, Yield, Composition) and Related Gene Expression: A Case Study in Two *Thymus* Species. *Agronomy* **2022**, *12*, 1008. [CrossRef]
16. Ivanova, T.; Bosseva, Y.; Chervenkov, M.; Dimitrova, D. Lamiaceae Plants in Bulgarian Rural Livelihoods—Diversity, Utilization, and Traditional Knowledge. *Agronomy* **2022**, *12*, 1631. [CrossRef]

agronomy

Article

Influence of Nutrient Solutions in an NGS® Soilless System on the Yield, Quality and Shelf Life of Fresh-Cut Commercial Mint at Different Harvest Times

Saeid Hazrati [1,2,*], Giuseppe Pignata [2], Manuela Casale [2], Seyyed Jaber Hosseini [3] and Silvana Nicola [2,*]

[1] Department of Agronomy and Plant Breeding, Faculty of Agriculture, Azarbaijan Shahid Madani University, Tabriz 53714-161, Iran
[2] Department of Agricultural, Forest, and Food Sciences—DISAFA, Horticultural Sciences—INHORTOSANITAS, Vegetable Crops and Medicinal and Aromatic Plants—VEGMAP, University of Torino, 10095 Grugliasco, Italy; giuseppe.pignata@unito.it (G.P.); manuela.casale@unito.it (M.C.)
[3] Department of Agronomy, Faculty of Agriculture, Tarbiat Modares University, Tehran 14115-336, Iran; jaberhosseini@yahoo.com
* Correspondence: saeid.hazrati@azaruniv.ac.ir (S.H.); silvana.nicola@unito.it (S.N.)

Abstract: The optimal fertilizer concentration for *Mentha* plants is contingent on the growing systems and harvest time, serving as operational solutions to control and enhance quality and yield. This study aimed to determine the effects of three macronutrients concentration in hydroponic nutrient solution (HNS) during three harvest times on the growth, quality, yield, and shelf life of three mint species (*M. spicata* L. var. *viridis* (MV); *M. piperita* L. (MP); *M. spicata* L. var. *rubra* (MR)) grown in a New Growing System (NGS®). Total dry matter (DM), nitrate (NO_3^-), phosphate (PO_4^{3-}), and calcium carbonate ($CaCO_3$) concentrations were increased with the addition of higher levels of nutrient fertilization in three species. When the ion concentration of the HNS was increased, total fresh yield decreased. The highest total bacterial count (TBC) was obtained in MR species in the three harvests in all the levels of HNS. The lowest browning potential (BP) and soluble o-quinone (So-Q) levels were observed at second harvest in the MR species with the application of one of the two HNS high in nitrogen (N). In conclusion, the combination of optimal HNS ion concentration and appropriate species is considered essential to obtain suitable yield, quality, and ensure shelf life of mint.

Keywords: biochemicals; fertilizer; *Mentha* sp.; post-harvest quality; soilless system

1. Introduction

The demand for medicinal and aromatic plants as sources of phytochemical composition in the food, medicine and cosmetic industries has recently increased [1]. In medicinal and aromatic plants, growth, yield, phytochemical composition, and shelf life are influenced by various environmental factors, growing practices, and nutrient supply [1,2].

Mint belongs to the Lamiaceae family, which are stoloniferous or rhizomatous, annual or perennial, medicinal plants that constitute one of the most popular aromatic plants in the world [1]. The genus *Mentha* is characterized by 62 taxa belonging to 18 species and 11 hybrids worldwide [3]. Also, an increasing number of mint cultivars are constantly being introduced. Mint species and cultivars differ widely in their characteristics, such as plant appearance, flavour, and architecture [2].

As described in the literature, the genus *Mentha* has been used in ethnomedicine for the treatment of bronchitis, flatulence, anorexia, nausea, ulcerative colitis, and liver problems due to its anti-emetic, anti-inflammatory, antispasmodic, sudorific, analgesic, and stimulating activities [2,3]. Moreover, it is well reported in the literature that some *Mentha* species, including *M. spicata* and *M. piperita*, possess antimicrobial and antioxidant activities [4,5]. Mint species have a wide global dispersal and are extensively cultivated in temperate or subtropical climates in the Mediterranean region, America, China, Europe,

India, and Brazil. Mint can be grown outdoors or in controlled environments, but in colder climates the year-round production is more possible in controlled environments [3,6].

The Piedmont Region (Italy) is famous for its production of the Lamiaceae genera. The main local species is black peppermint (*M. piperita* L. var. *officinalis* forma *rubescens* Camus), but other mints are also cultivated, such as red spearmint (*M. spicata* L. var. *rubra*) and *M. spicata* L. var. *viridis* (syn. *M. viridis* Auct.) [7].

Nowadays, the demand for fresh mint has increased, and it is necessary to find sustainable and efficient systems for its cultivation. One solution can be using the soilless culture systems (SCSs) which have several advantages compared to the traditional culture system (TCS) in soil which include an increase in yield, cost-efficiency of fertilizer application and water, lower use of products for plant protection, and lower environmental pollution [8]. Because of the water shortage and land availability, it is highly recommended to conduct more investigations on the SCSs and their potential [9].

SCSs represents an interesting approach to growing plants efficiently because they allow for the precise control of plant nutrition [10]. The concentration, composition, and interval of application of the HNS used in the SCS are among the main variables/factors in the success of crop production [11]. The regulations of these factors are of utmost importance because high concentrations of nutrients induce nutrient imbalance, osmotic stress, and ion toxicity, while low concentrations usually lead to nutrient insufficiency stress [12]. As SCSs are highly productive in high plant densities and short culture cycles, they are frequently used for the production of high-value-added crops [9,13]. Moreover, SCSs allow for a reduction in the use of products for the plant protection contributing to the potential decrement of the environmental pollution [14]. There are several settings of SCS which are typically used all over the world, mainly including floating systems, aeroponic systems, pot systems, the nutrient film technique (NFT) system, and others derivatives [8]. The New Growing System (NGS®) is a SCS similar to NFT based on the recirculation of water in a closed-cycle system of the HNS through a multi-channel film [9]. The multilevel system has been developed to favour HNS aeration and avoid the bung effect due to root growth. The HNS can be pumped into the system intermittently according to the growth conditions and the plant needs [15]. One advantage of the NGS® is the reduced volume of HNS necessary for plant production. This, consequently, reduces the energy required by the system both for the recirculation and the maintenance of the optimal water temperature [9,16]. Concerning the safety aspects, as in other SCSs [17,18], microbial contamination might be better controlled in NGS® than in TCS in the open field [8]. Therefore, the optimum fertilization regime is important for improving growth, yield, and quality, as well as reducing production costs of aromatic plants [11,12].

Aromatic plants grown in SCSs require an accurate setting of water and nutrient management because the buffer effect of the substrate is minimum while the plant density is high. Hence, the optimization of the nutrient supplementations is necessary for the farmers to quantitatively and qualitatively maximize their production. Also, the nutrient concentration in the HNS used in SCSs is one of the most important factors for improving crop production [19].

Similar to other aromatic plants, in the cultivation of mint, the HNS can affect the plants appearance, growth, nutritional value, and shelf life. Several authors have presented experiments on mint with the aim of determining the best relation between the ratio of essential nutrients with plant growth and yield [20,21].

It is well established that in SCSs, the commercial production of aromatic plants requires adequate and regular levels of macro-, meso- and micro-nutrients to provide high-quality post-harvest characteristics required for long shelf life [11]. For instance, supplementation of high levels of N usually leads to storage disorders quality loss in post -harvest [22,23] and low contents of vitamin C, sugars, and dry weight [24]. Conversely, a low level of N leads to a decrease in the growth of the aerial part. Also, some studies indicated that utilizing N at higher values led to increased growth, yield, and essential oil content. Consequently, this resulted in higher essential oil content, specifically an increased menthol content in the essential oil of *M. piperita* L. and *M. spicata* L., conversely, a deficit in N application reduces mint growth [25,26]. On the other hand, potassium (K) is involved in stomata closure and K deficit increases susceptibility to pathogens and may also interfere with the absorption of calcium (Ca^{2+}) [27].

In Italy, mint is usually cultivated in soil and the demand for the fresh-cut product is increasing in processing industry, cooking, and home consumption. The cultivation of mint in open fields gives yield only in a specific and limited period of the year (mainly in summer). The setting of NGS® in the greenhouse may be considered as a solution to alleviate the seasonal shortages of this crop. To the best of our knowledge, SCSs have not been adequately considered in the cultivation of mint. An innovative growing system such as NGS® can become an alternative to TCS in soil for mint cultivation to increase health, quality (without soil pollutants), and yield of this plant. This may also help to standardize the cultural practices and decrease the growing period and the production costs. Moreover, NGS® may be suitable for continuous harvests, regrowth, and exploiting the same root system [9]. The timing of harvesting significantly impacts the quantity and quality of mint. The yield, quality characteristics, and shelf life in mint plants exhibit a species-specific pattern, influenced by the expression of genes at various harvest stages. Consequently, research has delved into the impact of growth stages on mint [25,26,28,29], shedding light on their growth, yield, quality characteristics, shelf life, metabolites, and related biological activities, including antimicrobial and antioxidant properties. Research indicates that the timing of harvest significantly impacts both the yield and quality of *M. piperita*. The peak yields are achieved at the commencement of flowering, particularly concerning drug leaf yield, while optimal conditions for all other characteristics occur during the 100% flowering period [29,30].

Another critical issue in the cultivation of leafy vegetables is the frequency distribution of NO_3^- concentrations [31]. Due to the adverse effects of NO_3^- on human health, much attention has always been paid to the accumulation of this ion in vegetables [32]. Therefore, finding the best nutrition treatment to achieve a safe concentration of NO_3^- (not harmful to human health) is also considered in a part of our study.

The study aims to investigate the impact of varying N concentrations, specifically at doses of 6 and 12 mmol·L^{-1}, while maintaining constant levels of phosphor (P) and K (2 and 6 mmol·L^{-1}, respectively), or by doubling their concentrations in the HNS, on the yield and quality traits of three mint species grown in NGS® in a greenhouse. The evaluation also included assessing the shelf life of the fresh-cut product obtained from these mint species. The hypothesis posits that manipulating macronutrient concentrations in the nutrient solution will significantly influence the growth, yield, and quality characteristics of the mint plants, ultimately affecting the shelf life of the fresh-cut product obtained. Additionally, the study aims to address the issue of NO_3^- concentration in leafy vegetables, emphasizing the importance of finding optimal nutrition treatments that achieve safe NO_3^- levels, ensuring the produced mint is not harmful to human health. The results of the research are expected to contribute valuable insights into the potential benefits of using SCSs, particularly NGS®, for mint cultivation, offering a sustainable and efficient alternative to traditional soil-based methods or other soilless systems.

2. Materials and Methods

2.1. Plant Material and Growing Conditions

The study was conducted at the Experimental Center of the Department of Agricultural, Forest and Food Sciences (DISAFA) (44°53′11.67″ N; 7°41′7.00″ E—231 m a.s.l.), in Tetti Frati, Carmagnola (TO), Italy, from January to July in an automatically controlled greenhouse. A factorial in time experiment was carried out in a randomized complete block design (RCBD) with three replications. The factorial design included two levels of N (6 mmol·L^{-1} (N6) and 12 mmol·L^{-1} (N12)), and two ratios of P and K (constant K-P and doubled, with levels of 2 and 6 mmol·L^{-1} (P2K6) and 4 and 12 mmol·L^{-1} (P4K12)). The plant materials used were commercial mints (*M. spicata* L. var. *viridis* (MV);

M. piperita L. (MP); *M. spicata* L. var. *rubra* (MR) (S.A.I.S. S.p.A., Cesena (FC), Italy) propagated from mother plant cuttings in 60-cell Styrofoam trays; then, the plants were subsequently transferred into pots utilizing the Neuhaus Huminsubstrat N17, which is a peat-based horticultural medium. Plant lets were kept at 20 °C day and night and were overhead irrigated twice daily for 1 min (until transplanting). The nursery phase was conducted following standard cultural practices. When the plants were at an appropriate development stage, they were moved into a Lab-scale Pilot Plant (LSPP) based on the NGS® technology [9]. Max, min, and mean temperatures during the growing period in the greenhouse were 43 °C, 2 °C, and 17.3 °C, respectively.

During the cultivation in NGS®, four levels of HNS were tested (Table 1). Salts (purity > 98%) were dissolved in tap water with a known salt content to create the HNS. The tap water has an electrical conductivity (EC) of 440 $\mu S \cdot cm^{-1}$, with a pH of 7.5 and 24 °F of hardness. During the cultivation in NGS®, the nutrient solution was used in this protocol, which contained the following macroelements: $(NH_4)\ H_2PO_4$, $(NH_4)_2SO_4$, K_2SO_4, KH_2PO_4, $Ca(OH)_2$, $MgSO_4$, $7H_2O$ NH_4NO_3, and microelements (Oligogreen 30 mg L^{-1} with the following composition: Fe EDTA: 2%, Mn EDTA: 4%, Zn EDTA: 3%, Cu EDTA: 1%, B: 0.05%, and Mo 0.05% and Kelagreen 10 mg L^{-1} with the following composition: Fe 11%, Mn 13%, Zn 14% and Cu 14%; Green Has Italia S.p.a., Canale d'Alba (CN), Italy) were added to the HNS. All the HNS prepared had a ratio of 40/60 $N\text{-}NO_3^-/N\text{-}NH_4^+$. The EC, pH, and temperature were measured in the HNS by means of a Waterproof CyberScan PC 650 (Eutech Instruments Pte Ltd., Singapore), equipped with an EC/temperature probe (CONSEN9203J) and a submersible pH electrode (ECFC7252203B). The pH in the HNS was monitored continuously and kept close to 5.5, while the EC was between 2.0 and 2.5 dS m^{-1}. The temperature of the HNS was 20–25 °C, and the dissolved oxygen was measured by using an oximeter (YSI 550A; YSI, Inc., Yellow Springs, OH, USA), and was between ca. 7 and 9 ppm throughout the growing cycle. The total number of plants used was ca. 360 (ca. 40 plants/m^2).

Table 1. Hydroponic nutrient solutions (HNS) composition used in the experiment ($mmol \cdot L^{-1}$). N6: 6 $mmol \cdot L^{-1}$; N12: 12 $mmol \cdot L^{-1}$; P2: 2 $mmol \cdot L^{-1}$; P4: 4 $mmol \cdot L^{-1}$; K6: 6 $mmol \cdot L^{-1}$, and K12: 12 $mmol \cdot L^{-1}$.

	N6P2K6	N12P2K6	N6P4K12	N12P4K12
N	6.0	12.0	6.0	12.0
P	2.0	2.0	4.0	4.0
K	6.0	6.0	12.0	12.0
Ca	2.5	2.5	2.5	2.5
Mg	2.0	2.0	2.0	2.0

The experiment consisted of harvesting mint plants three times when the plants reached a suitable maturity stage. After the first canopy harvest, plants were allowed to regrow twice with the aim of evaluating the system efficiency and having an alternative to the common agronomic practice of the mint in the surrounding area [33]. Normal/conventional cultivation practices consist of two annual harvests, one in early and one in late summer.

The first harvesting took place 37, 42, and 49 days after transplanting for MV, MR, and MP, respectively, while the second harvesting took place 46, 37, and 43 days after the first harvest, and the third harvesting took place 30, 37, and 30 days after the second harvest. All the procedures followed the standard practices to obtain replicable and comparable data, and the timing efficiency of the sampling procedures [9,34]. Upon the completion of the harvest, the raw material was promptly conveyed to the post-harvest laboratory for further analysis and the preparation of the fresh-cut products.

2.2. Raw Material Analysis

The biometrical measurements recorded were the leaf fresh weight (LFW) per m^2 and LFW per plant. The dry matter (DM) was determined through a process of drying at a temperature of 40 °C.

2.3. Weight Loss

Weight loss (WL) of fresh-cut products was determined by weighing the bags daily during storage and it was calculated progressively based on the comparison with the at-harvest (d0) value as an index of freshness decay.

2.4. Fresh-Cut Processing

Samples (50 g) were packaged in thermo-sealed bags (0.25 m × 0.35 m) that had previously been prepared with polypropylene film (Alvapack S.r.l., Bologna, Italy). The packaged specimens were maintained at a temperature of 4 °C for a duration of nine days in refrigerated chambers, which were completely shielded from any form of light within the display cabinet.

2.5. Headspace Analysis

Individual packages' headspace gas composition was monitored using an O_2 analyzer with an electrochemical ceramic oxide–zirconia detector (CG-1000, Ametek, Thermox Instruments Co., Pittsburgh, PA, USA) and infrared CO_2 detectors (Via 510, Horiba Instruments Co., Irvine, CA, USA). Overall, 1 mL samples (per package) gas were measured, and the results were reported as the mean of the three packages.

2.6. Leaf Colour

The measurement of leaf colour was conducted utilizing a CR10 colorimeter (Konica-Minolta Sensing Inc., Osaka, Japan). The L^* component signifies lightness, the a^* component represents values from green (−) to red (+), and the b^* component represents values from blue (−) to yellow (+). Prior to sampling mint leaves, the instrument underwent calibration using a Minolta standard white reflector plate.

2.7. Tissue Ion and Salt Content

Nitrate (NO_3^-), phosphate (PO_4^{3-}) and calcium carbonate ($CaCO_3$) contents were determined using a refractometric kit (Merck Reflectoquant RQflex2©; Darmstadt, Germany) following the manufacturer's instructions. A frozen tissue sample (10.0 g) was stomached with distilled water for 2 min and then filtered for refractometric measurement and expressed as mg kg^{-1} LFW for NO_3^- and mg g^{-1} LFW for PO_4^{3-}, and $CaCO_3$.

2.8. Browning Potential and Soluble o-Quinone Content

The browning potential (BP) and soluble o-quinone (So-Q) content were determined from 5 g of frozen tissue according to the methodologies of Couture et al. [35] and Loaiza-Velarde and Saltveit [36]. The results were stated as raw absorbance units (Abs_{340} and Abs_{437} for BP and So-Q, respectively).

2.9. Microbial Analysis

The total bacterial count (TBC) was determined using the Plate Count Agar substrate and the mould and yeast count (MC and YC, respectively) were determined using the Yeast Extract Glucose Chloramphenicol Agar substrate. A total of 25.0 g of fresh tissue from each sample was subjected to stomaching for a duration of 2 min at normal speed, using 225.0 mL of Ringer's buffer. The resulting mixture was then diluted and subsequently transferred into Petri dishes containing the selective substrate. The TBC was performed after incubation for 48 h at 30 °C, while both MC and YC were performed after incubation for 5 days at 30 °C. The results were expressed as colony-forming units (CFU) g^{-1} LFW.

2.10. Experimental Design and Statistical Analysis

During the data analysis, the effects of various factors were determined through the utilization of analysis of variance (ANOVA) employing the general linear model (GLM) procedure within the Statistical Analysis System (SAS 9.4) software. The PROC UNIVARIATE function within SAS was employed to assess the assumptions of ANOVA, and it was determined that the residuals exhibited a normal distribution. To check significant differences between the means, the least significant difference (LSD) test with p values < 0.05 was used. In the case of significant interaction, the LS means procedure was used to compare significant interactions. When an F-test indicated statistical significance at $p < 0.05$, the protected least significant difference was used to separate the means of main effect and the significant interactions were separated by the slicing method. When the interactions were not significant, we only discussed the main effects. When the main effects, two-way, and the three-way interaction effects of the traits were significant, we only discussed the three-way interaction effects or when the main effects, or two-way interaction traits were significant, we only discussed the two-way interaction effects.

3. Results

3.1. Quantitative Characteristics

The results of analysis of variance of DM, LFW, yield, and LDG are shown in Table S1. The highest amount of LFW and yield in all HNS treatments was related to the third harvest time. The results of HNS and harvest time interaction showed that the highest amount of LFW was related to the application of N6P2K6 and N12P2K6 treatments at the second harvest (29.99 and 30.04 g/plant, respectively), and the lowest amount of LFW was observed in the application of N12P4K12 treatment at the third harvest time (11.76 g/plant).

The results of HNS and harvest time interaction showed that the highest yield was observed in the application of N6P2K6 and N12P2K6 at the second harvest with 1199 and 1201 g·m^{-2}, respectively (Table 2). In contrast, the lowest yield was observed in the application of N12P4K12 at the third harvest with 470 g·m^{-2}.

Table 2. Influence hydroponic nutrient solutions (HNS) and harvest time (first harvest: I harvest, second harvest: II harvest, regrowth and third harvest: III harvest, regrowth) on the leaf fresh weight per plant (LFW), leaf daily growth (LDG) and yield of mint species in a new growing system (NGS®). N6: 6 mmol·L^{-1}; N12: 12 mmol·L^{-1}; P2: 2 mmol L^{-1}; P4: 4 mmol L^{-1}; K6: 6 mmol L^{-1} and K12: 12 mmol L^{-1}.

HNS	Harvest Time	LFW (g/plant)	Yield (g m^{-2})	LDG (g/plant/d)
N6P2K6	I harvest	17.9 ± 1.3 cd	719.4 ± 52.9 cd	0.43 ± 0.03 c–e
	II harvest, regrowth	29.9 ± 2.6 a	1199.5 ± 105.7 a	0.72 ± 0.07 a
	III harvest, regrowth	21.9 ± 3.5 b	877.6 ± 138.3 b	0.67 ± 0.09 a
N12P2K6	I harvest	19.2 ± 0.9 bc	766.7 ± 36.4 bc	0.45 ± 0.02 cd
	II harvest, regrowth	30.1 ± 2.0 a	1201.8 ± 80.2 a	0.72 ± 0.05 a
	III harvest, regrowth	18.2 ± 2.1 cd	729.4 ± 85.1 cd	0.56 ± 0.06 b
N6P4K12	I harvest	13.6 ± 1.2 ef	544.7 ± 46.4 ef	0.32 ± 0.03 f
	II harvest, regrowth	20.3 ± 1.9 bc	812.9 ± 79.2 bc	0.49 ± 0.05 bc
	III harvest, regrowth	15.1 ± 1.7 de	605.9 ± 66.2 de	0.46 ± 0.04 c
N12P4K12	I harvest	14.3 ± 0.5 ef	571.5 ± 20.0 ef	0.34 ± 0.01 ef
	II harvest, regrowth	20.1 ± 1.7 bc	802.8 ± 66.9 bc	0.49 ± 0.03 bc
	III harvest, regrowth	11.8 ± 1.6 f	470.2 ± 65.8 f	0.36 ± 0.05 d–f

Means followed by different letters in the same column for the same factor are significantly different ($p \leq 0.05$) according to the LSD test.

The results of LDG also showed that LDG was affected by the interaction of HNS and harvest time and its highest value was obtained in the treatment of N12P2K6 at the second harvest (0.72 g/plant/d) and N6P2K6 at the second and third harvests (0.72 and 0.67 g/plant/d, respectively). In contrast, the lowest amount of LDG was observed in

N6P4K12 treatment at the first harvest (0.32 g/plant/d). The results showed that the amount of LDG at the first and third harvests was the lowest, and the amount at the second harvest was the highest in all HNS treatments.

The interaction of harvest time and species showed that the highest amount of DM was related to MP species at the third harvest (22.75%), whereas the lowest amount of DM was observed in MR species in the second harvest (16.06%) (Table 3). The results showed that in all species, the highest amount of DM was related to the third harvest and the lowest amount was observed at the second harvest.

Table 3. Influence of harvest time (first harvest: I harvest, second harvest: II harvest, regrowth and thrid harvest: III harvest, regrowth) on the dry matter (DM), leaf fresh weight per plant (LFW), yield and leaf daily growth (LDG) of three commercial mint (*M. spicata* L. var. *viridis* (MV); *M. piperita* L. (MP); *M. spicata* L. var. *rubra* (MR) cultivated in a new growing system (NGS®).

Species	Harvest Time	DM (%)	LFW (g/plant)	Yield (g m^{-2})	LDG (g/plant/d)
MW	I harvest	18.2 ± 0.3 c	14.1 ± 0.9 e	562.1 ± 37.6 e	0.38 ± 0.0 de
	II harvest, regrowth	16.0 ± 0.4 de	30.8 ± 1.6 a	1233.9 ± 63.9 a	0.67 ± 0.04 ab
	III harvest, regrowth	19.7 ± 0.3 b	17.4 ± 1.6 d	695.5 ± 65.3 d	0.58 ± 0.05 c
MP	I harvest	17.2 ± 0.3 cd	16.6 ± 0.9 de	662.4 ± 37.1 de	0.34 ± 0.02 e
	II harvest, regrowth	17.2 ± 0.3 cd	18.4 ± 1.4 d	734.8 ± 57.4 d	0.44 ± 0.03 d
	III harvest, regrowth	22.7 ± 0.6 a	10.4 ± 1.5 f	417.1 ± 58.4 f	0.35 ± 0.05 e
MR	I harvest	17.8 ± 0.3 cd	18.2 ± 1.1 d	727.2 ± 45.1 d	0.43 ± 0.03 d
	II harvest, regrowth	16.1 ± 0.3 e	26.1 ± 2.1 b	1044.1 ± 84.2 b	0.71 ± 0.06 a
	III harvest, regrowth	19.3 ± 0.4 b	22.5 ± 2.1 c	899.8 ± 83.5 c	0.61 ± 0.06 bc

Means followed by different letters in the same column for the same factor are significantly different ($p \leq 0.05$) according to the LSD test.

Among the three studied species, MP had the highest amount of DM and MW and MR species did not differ significantly in terms of DM (Table 3). In addition, the highest amount of LFW was related to MW species at the second harvest (30.85 g/plant), while the lowest amount was observed in MP species in the third harvest (10.43 g/plant). Among the three studied species, MW and MR species had the highest amount of LFW and, while MP species had the lowest amount of LFW among all studied species, no significant differences were found between the species. The highest and the lowest yield in the interaction effect of harvest time and species were related to MW at the second harvest (30.85 g/plant) and MP at the third harvest (10.43 g/plant) (Table 3). Among the three studied species, MP had the lowest yield and MW and MR had the highest yield. Also, in all the studied species, the second harvest had the highest yield compared to the time of the first and third harvests.

The amount of LDG was affected by the interaction effect of harvest time and species. The lowest amount of LDG was related to MP species. On the other hand, the highest amount of LDG were observed in the MW and MR species. The highest level of LDG in MR species was related to the second harvest (0.71 g/plant/d). In MP species, the first and third harvests had the lowest LDG, with 0.34 and 0.35 g/plant/d, respectively (Table 3). Furthermore, LDG was affected by the osmotic stress of high-concentration treatments; however, the amount of N did not induce a significant difference in LDG.

LFW and yield were affected by the interaction of species and HNS. The highest LFW and yield were related to MR species and N6P2K6 (Figure 1a,b). In contrast, the lowest amount of LFW was observed in MP species in N6P4K12 and N12P4K12 treatments. Among different HNS treatments, N6P2K6 and N12P2K6 had the highest amount of LFW and yield and no significant difference was observed between them. In contrast, the lowest amount of LFW and yield were related to N6P4K12 and N12P4K12 treatments.

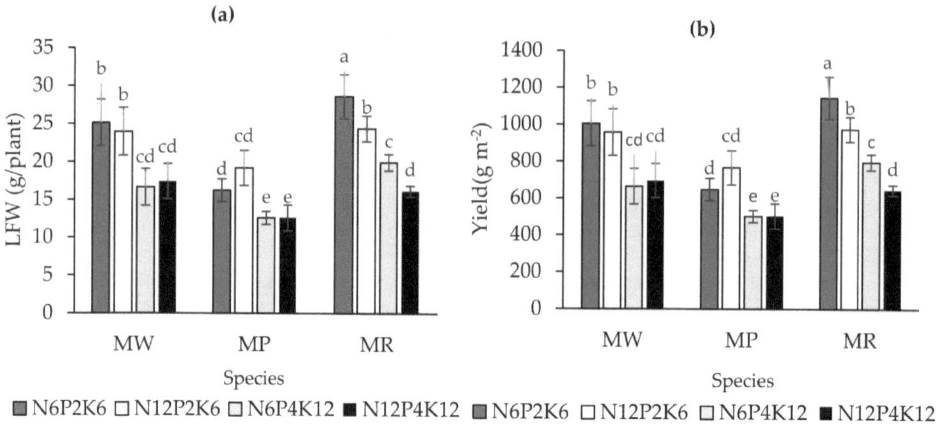

Figure 1. Influence hydroponic nutrient solutions (HNS) on the leaf fresh weight per plant (LFW) (**a**) and yield (**b**) of commercial mint (*M. spicata* L. var. *viridis* (MV); *M. piperita* L. (MP); *M. spicata* L. var. *rubra* (MR) in a new growing system (NGS®). N6: 6 mmol·L^{-1}; N12: 12 mmol L^{-1}; P2: 2 mmol·L^{-1}; P4: 4 mmol L^{-1}; K6: 6 mmol L^{-1} and K12: 12 mmol L^{-1}. Means followed by different letters in the same column for the same factor are significantly different ($p \leq 0.05$), according to LSD test.

3.2. Qualitative Characteristics

3.2.1. Colour Parameters

Red-green (a^*) colouring was also affected by species and harvest time interaction (Table S1). The highest amount of a^* was related to MR species at the second harvest time (11.93). In contrast, the lowest amount of a^* was observed in MP species at the second harvest time (7.83) (Figure 2).

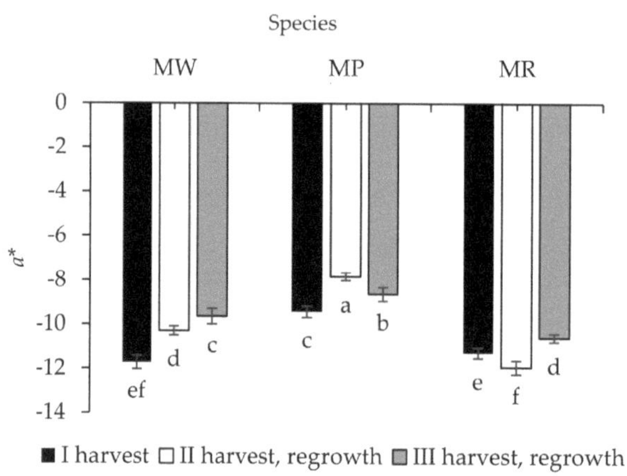

Figure 2. Influence of harvest time (first harvest: I harvest, second harvest: II harvest, regrowth and third harvest: III harvest, regrowth) on the colour parameter (a^*) of commercial mint (*M. spicata* L. var. *viridis* (MV); *M. piperita* L. (MP); *M. spicata* L. var. *rubra* (MR) in a new growing system (NGS®). Means followed by different letters in the same column for the same factor are significantly different ($p \leq 0.05$) according to LSD test.

Among the species, MR had the highest lightness (L^*) and yellow-blue colouring (b^*). In addition, the comparison between different HNS indicated that N6P2K6 treatment had the highest levels of L^* and b^*. Also, the comparison between different harvest times showed that the first and third harvest times had the highest amount of L^* and b^*, while the second harvest time had the lowest amount of L^* and b^* (Table 4). The effect of harvest time, species and HNS interaction showed that the amount of L^* in N6P2K6 treatment among MR species and N6P4K12 treatment in MR species had the highest values out of all harvest times. On the other hand, the highest amount of b^* was observed in N6P2K6 and MW species at the first and third harvest times. However, at the second harvest, N6P2K6 treatment in MR species had the highest amount of b^* (Table 4).

Table 4. Influence of harvest time (first harvest: I harvest, second harvest: II harvest, regrowth and third harvest: III harvest, regrowth) and hydroponic nutrient solutions (HNS) on the colour parameters value of three commercial mint (*M. spicata* L. var. *viridis* (MV); *M. piperita* L. (MP); *M. spicata* L. var. *rubra* (MR) cultivated in a new growing system (NGS®). N6: 6 mmol·L^{-1}; N12: 12 mmol L^{-1}; P2: 2 mmol L^{-1}; P4: 4 mmol L^{-1}; K6: 6 mmol L^{-1} and K12: 12 mmol L^{-1}.

HNS	Species	Harvest Time					
		I Harvest		II Harvest, Regrowth		III Harvest, Regrowth	
		L^*	b^*	L^*	b^*	L^*	b^*
N6P2K6	MW	43.7 ± 0.3 a–c	27.0 ± 0.1 a	39.5 ± 0.9 cd	19.8 ± 0.8 cd	44.8 ± 0.4 a	23.4 ± 1.1 a
	MP	42.8 ± 0.8 b–d	21.5 ± 0.9 cd	38.3 ± 0.7 de	14.2 ± 0.7 fg	42.9 ± 0.9 bc	20.1 ± 1.1 b
	MR	44.1 ± 0.4 ab	24.6 ± 0.5 b	43.9 ± 0.8 a	23.6 ± 1.6 a	44.6 ± 0.8 ab	22.6 ± 0.5 a
N12P2K6	MW	39.7 ± 0.1 e	19.2 ± 1.0 e	38.7 ± 0.4 de	15.8 ± 0.3 ef	37.8 ± 0.3 g	13.5 ± 0.3 e
	MP	38.9 ± 0.3 ef	15.1 ± 0.4 f	37.5 ± 0.2 e	12.3 ± 0.3 g	39.2 ± 1.1 fg	13.9 ± 1.1 de
	MR	39.9 ± 0.1 e	17.7 ± 0.1 e	40.9 ± 0.4 bc	18.8 ± 1.3 cd	40.9 ± 0.2 de	16.5 ± 0.1 c
N6P4K12	MW	42.7 ± 0.3 cd	25.3 ± 0.6 ab	41.4 ± 0.1 b	21.0 ± 0.1 bc	41.6 ± 0.5 cd	18.9 ± 0.2 b
	MP	43.6 ± 0.2 a–c	22.9 ± 0.2 c	38.1 ± 0.5 de	13.3 ± 0.6 fg	40.7 ± 0.6 def	16.6 ± 1.0 c
	MR	44.3 ± 0.3 a	24.6 ± 0.5 b	44.3 ± 0.5 a	23.6 ± 1.4 ab	43.4 ± 0.4 ab	19.8 ± 0.5 b
N12P4K12	MW	38.3 ± 0.3 f	18.9 ± 0.6 e	39.3 ± 0.7 cd	17.7 ± 0.7 de	39.2 ± 0.4 efg	16.0 ± 0.3 c
	MP	39.9 ± 0.2 e	17.6 ± 0.4 e	38.1 ± 0.2 de	12.4 ± 0.5 g	39.5 ± 0.4 ef	15.5 ± 0.6 cd
	MR	41.6 ± 0.1 d	21.1 ± 0.7 d	40.9 ± 0.5 bc	18.6 ± 0.8 cd	40.7 ± 0.2 def	16.2 ± 0.1 c

Means followed by different letters in the same column for the same factor are significantly different ($p \leq 0.05$) according to the LSD test.

3.2.2. Weight Loss

Different concentrations of HNS did not significantly affect the WL in different harvest time; however, in different species with different harvesting times, we observed a significant difference in WL (Table S2). After nine days of storage, the variation in WL was significantly different. In the first days of storage, weight gain was very low. This increase in weight was accelerated at the third harvest, especially in MR. The weight gain in the first days of storage, especially at the third harvest, can be due to the fact that plants were kept at warehousing bags with holes that allow for the exchange of carbon dioxide (CO_2), oxygen (O_2), and water vapor at the beginning of the storage period, which results in more available water in this period. However, the highest WL occurred at the first (1.07%) and the third (0.27%) harvest times in the MV species, and at the second harvest WL occurred in the MR species on the ninth day (1.26%). Among the three species, the lowest WL was related to MP (Figure 3). Among the harvesting times, the highest WL was observed at the second harvest time (0.11%), and the lowest WL belonged to the third harvest time. Among the HNS treatments, the highest rate of WL occurred in the N12P4K12; however, the WL rate was not significantly different from the HNS used. Nevertheless, at the first harvest time, the highest reduction in the MV species was found in the N12P4K12 and at the second harvest time treatment (0.61%). Furthermore, the highest reduction in MR species was observed in the N12P4K12 treatment (0.44%) (Figure 4).

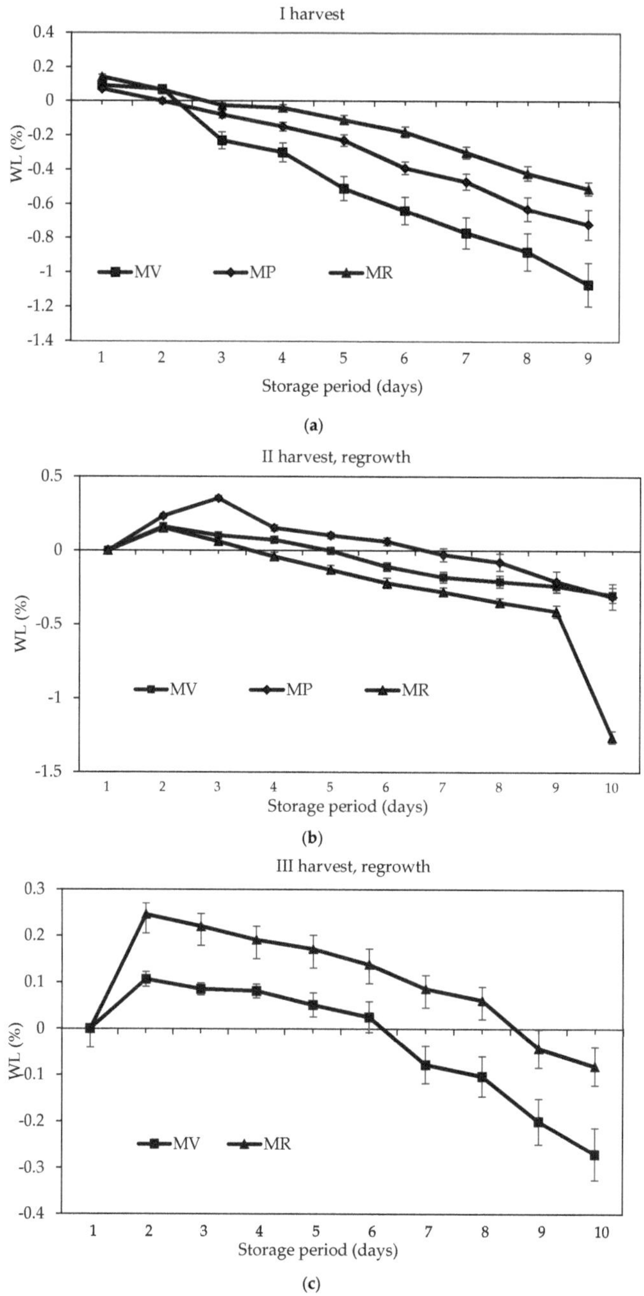

Figure 3. Influence of harvest time (first harvest: I harvest (**a**), second harvest: II harvest (**b**), regrowth and third harvest: III harvest (**c**), regrowth) on weight loss (WL) the headspace partial pressure of commercial mint (*M. spicata* L. var. *viridis* (MV); *M. piperita* L. (MP); *M. spicata* L. var. *rubra* (MR) daily during storage at 4 °C. Values are the mean of replicates ± SE.

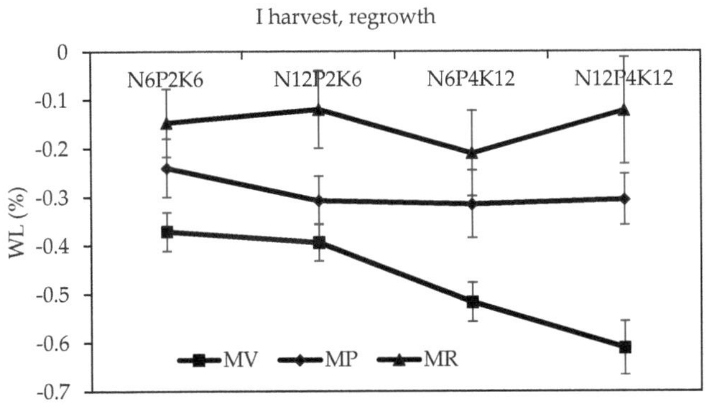

(a)

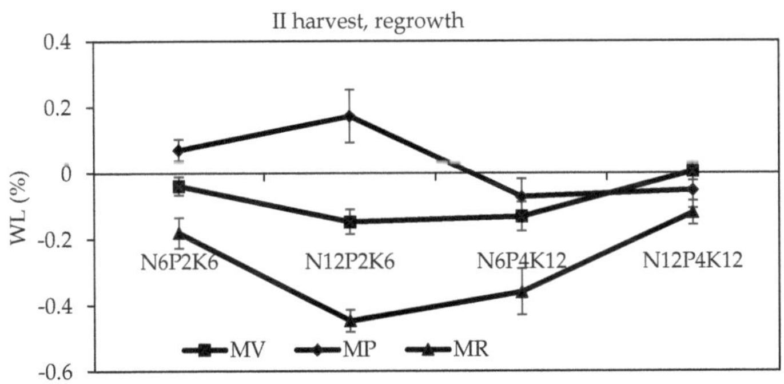

(b)

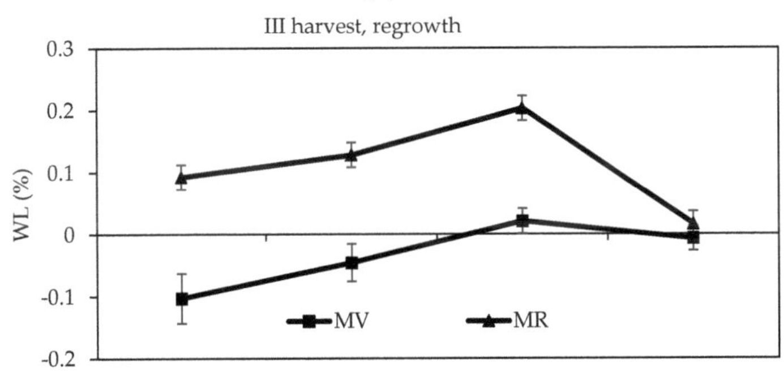

(c)

Figure 4. Influence of hydroponic nutrient solutions (HNS) and harvest time (first harvest: I harvest (**a**), second harvest: II harvest (**b**), regrowth and third harvest: III harvest (**c**), regrowth) on weight loss (WL) the headspace partial pressure of commercial mint (*M. spicata* L. var. *viridis* (MV); *M. piperita* L. (MP); *M. spicata* L. var. *rubra* (MR) stored at 4 °C. N6: 6 mmol·L^{-1}; N12: 12 mmol L^{-1}; P2: 2 mmol L^{-1}; P4: 4 mmol L^{-1}; K6: 6 mmol L^{-1} and K12: 12 mmol L^{-1}. Values are the mean of replicates ± SE.

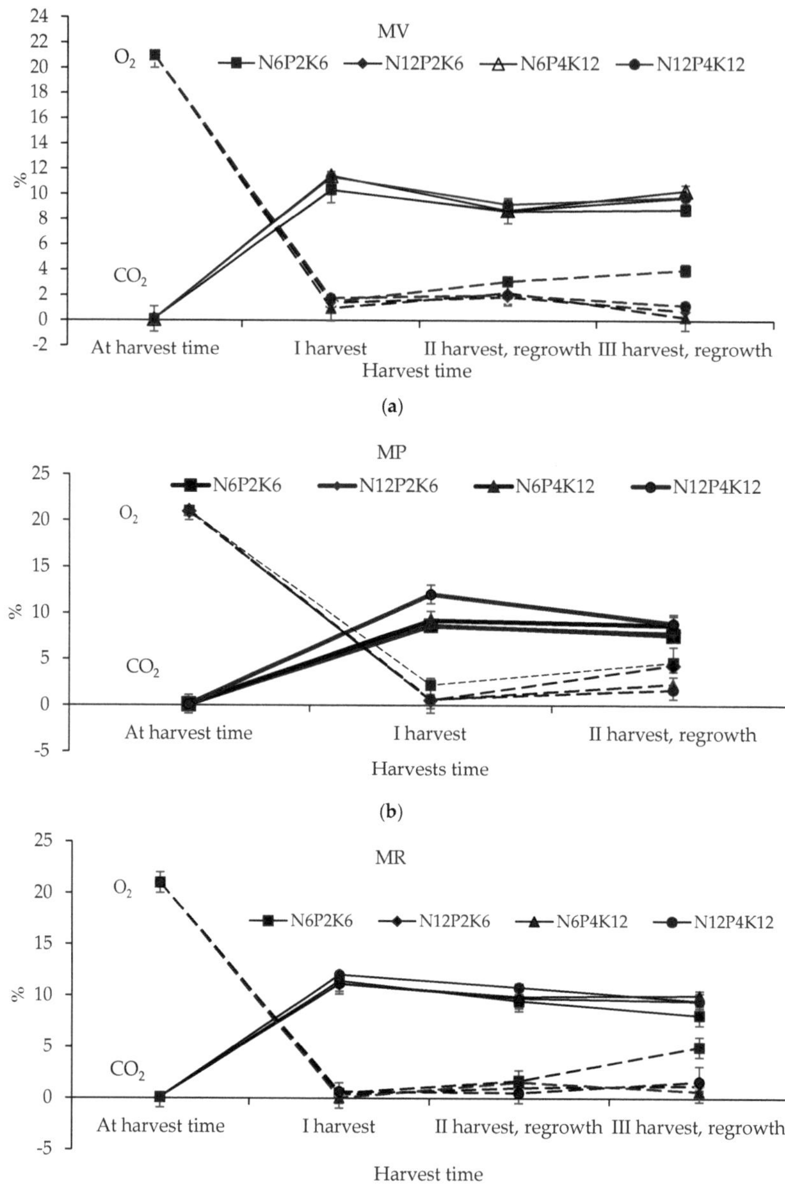

Figure 5. Influence of hydroponic nutrient solutions (HNS) and harvest time (first harvest: I harvest, second harvest: II harvest, regrowth and third harvest: III harvest, regrowth) on the headspace partial pressure of O_2 and CO_2 (%) of fresh-cut, commercial mint (*M. spicata* L. var. *viridis* (MV) (**a**); *M. piperita* L. (MP) (**b**); *M. spicata* L. var. *rubra* (MR) (**c**)© stored for 9 days at 4 °C. N6: 6 mmol L^{-1}; N12: 12 mmol L^{-1}; P2: 2 mmol L^{-1}; P4: 4 mmol L^{-1}; K6: 6 mmol L^{-1} and K12: 12 mmol L^{-1}. Values are the mean of replicates ± SE.

3.2.3. Respiration Rate and Headspace Analysis

Analyzing O_2 and CO_2 levels in the fresh-cut bags showed that HNS, species, and harvesting time significantly affected the parameters during the storage period (Table S2). In most tests, during the storage period, the O_2 rate usually decreased, and the rate of the produced CO_2 increased as a result of breathing. The initial concentrations of O_2 and CO_2 increased and decreased, respectively, at the same rate nine days after storage. As shown in Figure 5, after nine days of storage, the highest rate of CO_2 was obtained in the treatments with the highest concentrations of nutrients (N12P4K12), but the lowest rate was related to N6P2K6 and N12P2K6 treatments. In contrast, the N6P2K6 treatment resulted in the highest amount of O_2, while the lowest amount was associated with the N12P4K12 treatment (Figure 5). The highest rate of the produced CO_2 was related to the MR species and the lowest rate of produced CO_2 was recorded in the MP species (10.23 and 8.65%, respectively). In contrast, the highest rate of the produced O_2 was in the MP species and the least rate was related to the MR species (2.09 and 1.24%, respectively) at the first harvest time for the N6P2K6 treatment. The highest and lowest rates of CO_2 were obtained at the first and second harvest times, respectively (9.03 and 10.45%, respectively). On the other hand, the highest rate of O_2 was obtained at the second harvest time (2.25%), while the lowest rate was achieved at the first harvest time (0.91%) (Figure 5).

3.2.4. Mineral Ion Determinations

The results showed that HNS and species at different harvest times had a significant impact on the accumulation of NO_3^-, PO_4^{3-} and $CaCO_3$ in the tissue; however, the interaction between HNS and species was not significant (Table S2). The highest accumulation of NO_3^-, PO_4^{3-} and $CaCO_3$ was obtained at the third harvest time. The highest amount of NO_3^- was obtained in the N12P2K6 treatment on the first day of the third harvest time (2227.33 mg kg^{-1} LFW). Generally, the highest NO_3^- accumulation was observed at the third harvest, and the lowest accumulation rate was related to the first harvest time. Concerning the HNS, the highest accumulation of NO_3^- was obtained in N12P2K6 and the lowest accumulation values were observed in N6P2K6 and N6P4K12 treatments. Among the species, the highest NO_3^- accumulation in all harvesting times was in the MP species; however, MV and MR species did not significantly differ in NO_3^- accumulation (Table 5). There was a significant difference between the rates of PO_4^{3-} accumulation in the mint plant at different harvesting times. At all harvesting times, the highest accumulation of PO_4^{3-} was achieved in N6P4K12 and N12P4K12 treatments, and the lowest accumulation was obtained in N12P2K6. The results of the present experiment indicated that the high levels of P and K in the HNS led to an increase in PO_4^{3-} in the plant. Among the species, the highest accumulation of PO_4^{3-} was observed in MP at all harvesting times. Among the harvesting time, the highest accumulation of PO_4^{3-} was achieved at the third harvest time. The accumulation of $CaCO_3$ decreased by increasing the amount of K. According to the results, the highest accumulation of $CaCO_3$ was in N12P2K6 at all harvesting times; among the harvesting times, the highest accumulation was obtained at the third harvest. There was a significant difference among species in $CaCO_3$ at different harvest times, so the highest accumulation rate was observed in MP species.

Table 5. Influence hydroponic nutrient solutions (HNS) and harvest time (first harvest: I harvest, second harvest: II harvest, regrowth and third harvest: III harvest, regrowth) on the nutrients content (Nitrate (NO_3^-), phosphate (PO_4^{3-}) and calcium carbonate ($CaCO_3$) concentrations) of three commercial mint (*M. spicata* L. var. *viridis* (MV); *M. piperita* L. (MP); *M. spicata* L. var. *rubra* (MR) in the fresh-cut during its shelf life (at harvest (0) and nine days (9)) cultivated in a new growing system (NGS®). N6: 6 mmol L^{-1}; N12: 12 mmol·L^{-1}; P2: 2 mmol L^{-1}; P4: 4 mmol L^{-1}; K6: 6 mmol L^{-1} and K12: 12 mmol L^{-1}.

Species	Storage (Days at 4 °C)	I Harvest			II Harvest, Regrowth			III Harvest, Regrowth		
		NO_3^- mg kg^{-1}	$CaCO_3$ mg g^{-1}	PO_4^{3-} mg g^{-1}	NO_3^- mg kg^{-1}	$CaCO_3$ mg g^{-1}	PO_4^{3-} mg g^{-1}	NO_3^- mg kg^{-1}	$CaCO_3$ mg g^{-1}	PO_4^{3-} mg g^{-1}
MV	0	251.5 ± 67.5 b	0.14 ± 0.02 c	0.40 ± 0.034 c	743.3 ± 161.3 c	0.36 ± 0.02 ab	0.52 ± 0.05 ab	1145.5 ± 205.9 b	0.49 ± 0.05 b	0.50 ± 0.05 b
	9	129.7 ± 33.6 b	1.08 ± 0.15 a	0.27 ± 0.04 d	735.0 ± 168.8 c	0.27 ± 0.02 b	0.57 ± 0.02 a	1077.3 ± 152.7 b	0.45 ± 0.06 b	0.64 ± 0.10 b
MP	0	503.6 ± 137.7 a	0.37 ± 0.05 c	0.67 ± 0.05 b	1120.8 ± 146.3 b	0.32 ± 0.04 b	0.45 ± 0.01 bc	2024.3 ± 435.1 a	1.92 ± 0.23 a	1.12 ± 0.14 a
	9	626.3 ± 88.0 a	0.64 ± 0.06 b	0.87 ± 0.08 a	1504.3 ± 110.6 a	0.43 ± 0.02 a	0.44 ± 0.04 cd	-	-	-
MR	0	135.8 ± 40.9 b	0.61 ± 0.06 b	0.39 ± 0.03 c	578.3 ± 89.9 c	0.45 ± 0.07 a	0.42 ± 0.03 de	1137.0 ± 173.6 b	0.50 ± 0.07 b	0.62 ± 0.06 b
	9	144.2 ± 35.1 b	0.44 ± 0.07 bc	0.47 ± 0.06 c	649.2 ± 93.7 c	0.30 ± 0.04 b	0.36 ± 0.04 e	1354.3 ± 204.3 b	0.51 ± 0.07 b	0.57 ± 0.08 b
HNS										
N6P2K6	0	106.7 ± 29.2 c	0.53 ± 0.07 bc	0.45 ± 0.06 cd	391.1 ± 49.0 e	0.38 ± 0.05 bc	0.39 ± 0.04 cb	814.4 ± 109.9 de	1.10 ± 0.24 bc	0.66 ± 0.16 bc
	9	303.8 ± 125.2 b	0.59 ± 0.08 b	0.46 ± 0.12 bcd	548.7 ± 147.5 de	0.31 ± 0.04 cd	0.35 ± 0.05 c	756.0 ± 58.1 e	0.67 ± 0.10 cd	0.37 ± 0.04 c
N12P2K6	0	663.9 ± 163.9 a	0.57 ± 0.05 b	0.36 ± 0.03 de	1343.3 ± 146.1 a	0.50 ± 0.08 a	0.41 ± 0.04 ab	2227.3 ± 261.4 a	1.02 ± 0.24 a	0.69 ± 0.09 ab
	9	497.8 ± 93.8 a	0.91 ± 0.18 a	0.31 ± 0.07 e	1477.8 ± 134.7 a	0.45 ± 0.04 ab	0.46 ± 0.03 cb	1479.3 ± 304.4 a	0.55 ± 0.07 ab	0.35 ± 0.10 bc
N6P4K12	0	102.9 ± 38.0 c	0.40 ± 0.07 bc	0.58 ± 0.07 bc	537.8 ± 120.0 e	0.30 ± 0.02 cd	0.38 ± 0.05 cb	641.4 ± 125.5 e	0.39 ± 0.06 cd	0.58 ± 0.08 bc
	9	102.0 ± 4.0 c	0.53 ± 0.07 bc	0.76 ± 0.11 a	764.4 ± 193.6 cd	0.30 ± 0.04 cd	0.52 ± 0.05 a	943.3 ± 263.1 dc	0.32 ± 0.05 cd	0.95 ± 0.11 a
N12P4K12	0	314.3± 57.9 b	0.34 ± 0.03 c	0.51 ± 0.07 bc	984.4 ± 136.8 bc	0.38 ± 0.04 cd	0.51 ± 0.05 a	1385.0 ± 89.6 bc	0.33 ± 0.03 d	0.74 ± 0.09 a
	9	296.0 ± 95.0 b	0.83 ± 0.19 a	0.59 ± 0.09 b	1060.4 ± 162.9 b	0.27 ± 0.02 d	0.48 ± 0.05 ab	1610.2 ± 126.2 b	0.44 ± 0.09 cd	0.60 ± 0.07 ab
Harvest time		298.5 ± 38.1 b	0.59 ± 0.04 b	0.50 ± 0.03 ab	888.5 ± 64.7 a	0.35 ± 0.02 c	0.44 ± 0.02 b	1272.5 ± 99.2 a	0.64 ± 0.07 b	0.64 ± 0.04 b

Means followed by different letters in the same column for the same factor are significantly different ($p \leq 0.05$) according to the LSD test.

3.2.5. Browning Potential (BP) and Soluble o-Quinone (So-Q) Content

Browning potential and soluble o-quinone content were evaluated over a period of nine days. BP and So-Q content were higher on the first day in comparison with the results obtained at the end of the storage time. The results showed that HNS, species, and harvest time had a significant impact on BP and So-Q content (Table S2). The highest BP was achieved at the third harvest time in N6P2K6 and MP species that did not differ from the MV species; however, the values in MR species were significantly lower than other two species at all harvest times. The amount of BP in N12P2K6 was lower than other HNS treatments but it did not significantly differ from N12P4K12 treatment. In addition, the highest content was obtained in N6P2K6 treatment, and it did not differ from N6P4K12 treatment (Table 6).

Table 6. Influence hydroponic nutrient solutions (HNS) and harvest time (first harvest: I harvest, second harvest: II harvest, regrowth and third harvest: III harvest, regrowth) on the enzymatic browning parameters of three commercial mint (*M. spicata* L. var. *viridis* (MV); *M. piperita* L. (MP); *M. spicata* L. var. *rubra* (MR) in the fresh-cut during its shelf life (at harvest (0) and nine days (9)) cultivated in a new growing system (NGS®). N6: 6 mmol L^{-1}; N12: 12 mmol L^{-1}; P2: 2 mmol L^{-1}; P4: 4 mmol L^{-1}; K6: 6 mmol L^{-1} and K12: 12 mmol L^{-1}.

		I Harvest		II Harvest, Regrowth		III Harvest, Regrowth	
Species	**Storage (Days at 4 °C)**	**BP (A_{340} LFW)**	**So-Q (A_{437} LFW)**	**BP (A_{340} LFW)**	**So-Q (A_{437} LFW)**	**BP (A_{340} LFW)**	**So-Q (A_{437} LFW)**
MV	0	29.3 ± 3.4 a	1.8 ± 0.16 ab	16.3 ± 2.5 b	2.5 ± 0.2 b	27.2 ± 4.5 a	2.9 ± 0.2 a
	9	23.9 ± 5.3 ab	1.7 ± 0.25 b	9.2 ± 0.6 c	2.1 ± 0.1 bc	28.9 ± 4.4 a	2.8 ± 0.1 a
MP	0	33.8 ± 3.1 a	2.2 ± 0.17 a	16.6 ± 1.6 b	1.9 ± 0.2 c	26.0 ± 3.4 a	3.1 ± 0.5 a
	9	24.4 ± 2.6 ab	2.3 ± 0.22 a	21.2 ± 2.0 a	1.4 ± 0.1 d		
MR	0	16.6 ± 1.3 bc	1.1 ± 0.15 b	10.8 ± 0.8 c	3.2 ± 0.2 a	9.8 ± 1.2 b	2.1 ± 0.3 b
	9	12.3 ± 1.7 c	1.6 ± 0.15 b	8.3 ± 0.5 c	2.4 ± 0.2 bc	26.2 ± 4.4 a	2.8 ± 0.3 a
HNS							
N6P2K6	0	31.2 ± 4.6 a	2.3 ± 0.23 a	19.1 ± 3.0 a	2.9 ± 0.3 a	22.0 ± 5.6 b	2.1 ± 0.2 b
	9	28.6 ± 6.8 ab	1.9 ± 0.21 ab	13.0 ± 2.2 ab	1.7 ± 0.2 c	38.9 ± 6.2 a	3.4 ± 0.2 a
N12P2K6	0	16.4 ± 1.7 c	1.5 ± 0.22 b	10.7 ± 1.0 b	2.6 ± 0.3 ab	17.1 ± 2.8 b	2.7 ± 0.4 ab
	9	15.1 ± 2.5 c	2.0 ± 0.28 ab	10.6 ± 2.5 b	1.8 ± 0.2 c	22.8 ± 5.7 b	2.7 ± 0.4 ab
N6P4K12	0	32.1 ± 3.8 a	1.7 ± 0.15 ab	16.2 ± 1.9 ab	2.2 ± 0.3 bc	28.6 ± 7.3 ab	3.2 ± 0.3 a
	9	18.6 ± 3.9 bc	1.8 ± 0.29 ab	14.7 ± 3.0 ab	2.0 ± 0.2 bc	26.4 ± 9.1 ab	2.5 ± 0.3 ab
N12P4K12	0	26.8 ± 3.4 abc	1.8 ± 0.20 ab	12.1 ± 1.2 b	2.4 ± 0.3 abc	12.6 ± 1.8 b	2.6 ± 0.2 ab
	9	18.3 ± 5.3 bc	1.7 ± 0.29 ab	13.4 ± 2.2 ab	2.2 ± 0.3 abc	22.3 ± 2.2 b	2.7 ± 0.2 ab
Harvest time		23.4 ± 1.6 a	1.8 ± 0.08 c	13.7 ± 0.8 b	2.2 ± 0.1 b	23.3 ± 2.0 a	2.7 ± 0.1 a

Means followed by different letters in the same column for the same factor are significantly different ($p \leq 0.05$) according to the LSD test.

The So-Q content was affected by HNS, species, and harvesting time. The So-Q content at the first day of storage was higher than the ninth day of storage. In addition, among different harvesting times, the highest content was achieved at the third harvest time. Among the HNS, the highest amount of So-Q was obtained in N6P2K6 treatment, and there was no significant difference between the species (Table 6). The results of the present experiment indicated that the amounts of So-Q at the first and third harvest times were the highest in MP species, while for MR species the highest content was found at the second harvest time. The results of harvest time and HNS interaction showed that the highest amount of So-Q was related to N6P2K6 treatment at all harvest times. Also, N6P4K12 treatment represented a high So-Q content in all harvest times.

3.2.6. Microbial Evaluation

The results showed that the interaction of species and harvest time were significant for TBC, MC and TYC (Table S2). At the first and second harvest time among MR species, the highest amount of TBC was recorded (5424.23 and 5133 CFU·g^{-1} LFW, respectively). Also, the highest MC was observed at the first harvest time of MP species (1333.30 CFU·g^{-1} LFW). Furthermore, our results indicated that the amount of TYC at the first harvest time in MP species had the highest rate (250.23 CFU·g^{-1} LFW) (Table 7). Our findings showed that the interaction of harvest time and HNS was significant for TYC. The results indicated that the highest and the lowest TYC were observed at the first and third harvest time, respectively. Also, at the first harvest time of N6P2K6 treatment (136.44 CFU·g^{-1} LFW), the second and third harvests of N12P2K6 treatment, the highest TYC were 94.67 and 68.33 CFU·g^{-1} LFW, respectively (Table 7).

Among the species, the highest MC belonged to the MP and the lowest rate was found in MV. The highest MC was obtained at the first harvest time while the lowest rate was found at the third harvest time. In general, the rate of microbial contamination in the NGS® was significantly higher at the first harvest time in comparison to the second and third harvest time.

Regarding TYC, at the first and second harvest time, the highest rate of contamination was observed in N6P2K6 and N12P4K12 in MP species. At the third harvest time, the highest TYC was obtained in the MV species in N12P2K6 treatment. Among the species at the first and second harvest time, the highest TYC was recorded for MP species. The highest TYC belonged to the MR species, and the lowest rate was obtained in MV species at all harvest times (Table 7). It should be noted that there was no sample of MP species at the third harvest time.

Table 7. Influence hydroponic nutrient solutions (HNS) and harvest time (first harvest: I harvest, second harvest: II harvest, regrowth and third harvest: III harvest, regrowth) on the total bacterial count (TBC), the mould count (MC) and yeast count (YC) (CFU·g^{-1} LFW) of three commercial mint (*M. spicata* L. var. *viridis* (MV); *M. piperita* L. (MP); *M. spicata* L. var. *rubra* (MR) cultivated in a new growing system (NGS®). N6: 6 mmol L^{-1}; N12: 12 mmol L^{-1}; P2: 2 mmol L^{-1}; P4: 4 mmol L^{-1}; K6: 6 mmol L^{-1} and K12: 12 mmol L^{-1}.

		I Harvest			II Harvest, Regrowth			III Harvest, Regrowth		
HNS	Species	TBC	MC	TYC	TBC	MC	TYC	TBC	MC	TYC
MV	N6P2K6	1360.7 ± 46.7 c	481 ± 53.1 cd	19.3 ± 0.6 c	1581 ± 70.6 b	513.3 ± 81.9 abcd	6 ± 1.1 bc	860.3 ± 40.9 ab	412 ± 115.9 a	4.6 ± 1.8 d
	N12P2K6	1418 ± 68.1 c	474 ± 57.8 d	24 ± 3.1 c	425 ± 8.3 b	733.3 ± 70.5 ab	4 ± 0.6 c	249.3 ± 71.5 cd	227.3 ± 34.7 a	0.0 ± d
	N6P4K12	1123.7 ± 28.4 c	580.3 ± 60.1 bcd	44.7 ± 6.7	697 ± 31.2 b	693.3 ± 70.5 ab	12 ± 2.3 bc	142.7 ± 24.8 d	238.7 ± 67.5 a	0.0 ± d
	N12P4K12	843 ± 63.4 c	482 ± 56.6 cd	31.3 ± 1.7 c	530 ± 88.1 b	746.7 ± 89.7 ab	8 ± 2 bc	262.7 ± 19.1 cd	248.3 ± 19.2	6 ± 1.1 d
MP	N6P2K6	1706 ± 13.1 c	1413.3 ± 420.5 a	346.7 ± 17.6 a	221 ± 65.6 b	384 ± 66.2 bcd	20 ± 3.1 bc			
	N12P2K6	316 ± 38.1 c	1133.3 ± 209.5 ab	240 ± 23.0 ab	196 ± 29.9 b	100 ± 22.5 d	18.7 ± 1.1 bc			
	N6P4K12	518 ± 62.4 c	1120 ± 105.8 abc	253.3 ± 40.5 ab	245 ± 44.3 b	176 ± 29.0 cd	112.7 ± 5.3 abc			
	N12P4K12	123 ± 24.9 c	1666.7 ± 357.9 a	160 ± 26.4 bc	279 ± 58.1 b	127 ± 14.6 d	177.3 ± 12.3 ab			
MR	N6P2K6	5816.7 ± 214.3 ab	420 ± 23.1 d	43.3 ± 8.8 c	8460 ± 1075.6 a	800 ± 190.5 ab	2 ± 1.1 c	891 ± 66.4 b	303 ± 8.9 a	
	N12P2K6	4243 ± 427.2 b	336 ± 72.7 d	15.3 ± 2.4 c	5223 ± 756.7 ab	673.3 ± 33.3 abc	261.3 ± 2 a	581.7 ± 63.3 cbd	319.3 ± 93.3 a	
	N6P4K12	4770 ± 840.1 ab	439.3 ± 100.9 d	16.7 ± 4.1 c	4290 ± 342.1 ab	388.3 ± 81.8 bcd	0.00 ± 00 c	1270.7 ± 115.5 a	214.7 ± 31.5 a	
	N12P4K12	6867 ± 124.7 a	362.7 ± 81.4 d	78.6 ± 10.7 c	2560 ± 296.9 ab	940 ± 161.6 a	24 ± 2.30 bc	639.3 ± 44.9 ab	1120.7 ± 21.9 a	110 ± 8.1 b
MV		1235.6 ± 77.5 b	1235.6 ± 77.5 b	504.3 ± 27.6 b	29.8 ± 3.3 b	808 ± 138.1 b	671.7 ± 41.8 a	7.5 ± 1.1 a	378.8 ± 87.1 b	281.6 ± 37.3 a
MP		281.8 ± 49.3 b	281.8 ± 49.3 b	1333.3 ± 144.8 a	250 ± 23.3 a	235 ± 23.8 b	196.8 ± 37.4 b	82.2 ± 20.4 a		
MR		5424.2 ± 367.9 a	5424.2 ± 367.9 a	389.5 ± 34.5 b	38.5 ± 8.4 b	5133 ± 711.5 a	700.4 ± 84.4 a	71.8 ± 29.2 a	845.7 ± 88.2 a	489.5 ± 112.6 a
N6P2K6		2449 ± 861.7 a	2449 ± 861.7 a	771.4 ± 22.0 a	136.4 ± 59.9 a	3421 ± 1312.6 a	565.8 ± 91.7 a	9.3 ± 2.9 a	875.7 ± 35.5 a	357.7 ± 57.1 a
N12P2K6		2014 ± 596.6 a	2014 ± 595.6 a	647.8 ± 139.7 a	93.1 ± 37.4 a	1948 ± 848.2 a	502.2 ± 103.6 a	94.7 ± 36.5 a	415.5 ± 85.7 a	273.3 ± 49.0 a
N6P4K12		2182 ± 707.1 a	2182 ± 707.1 a	713.2 ± 113.3 a	104.9 ± 39.2 a	1744 ± 647.6 a	419.2 ± 81.7 a	41.6 ± 19.3 a	706.7 ± 257.7 a	226.7 ± 33.7 a
N12P4K12		2611 ± 1069.8 a	2611 ± 1069.8 a	837.1 ± 233.9 a	90 ± 20.5 a	1123 ± 368.1 a	604.7 ± 130.3 a	69.9 ± 27.3 a	451 ± 87 a	684.5 ± 195.5 a

Means followed by different letters in the same column for the same factor are significantly different ($p \leq 0.05$) according to the LSD test.

4. Discussion

Effective nutrient management in SCSs is crucial to balance robust biomass growth and the production of high-quality essential raw materials, ensuring profitable outcomes. The results indicated that optimal HNS concentrations varied among mint species. Enhancements occurred in various growth traits, primarily attributed to the roles of HNS in plants, especially N supplementation. N is a vital element for plant growth and yield as it constitutes several metabolic compounds, including chlorophyll, cell walls, enzymes, hormones, nucleic acids, proteins, and vitamins [37]. These compounds play essential roles in cell division, cell enlargement, tissue and organ formation, among other critical processes. The improvement in parameters related to these compounds likely contributed to more growth in shoot fresh weight, subsequently leading to higher dry weight [38].

The results showed that in all species, DM was maximum at the third harvest and minimum at the second harvest. Also, in the third harvest, plants had better developed roots, grew enough branches, and had spent more energy on DM production, which agrees with similar findings in the literature [39]. The nutritional needs of plants differ across developmental stages, and numerous studies have demonstrated that alterations in HNS can significantly change plant growth and yield characteristics [25,26,28,29]. The explanation for the increase in DM in the third harvest may be the accumulation of NO_3^-, PO_4^{3-} and $CaCO_3$ in tissues at this stage in addition to the stages before. A positive correlation occurred between nutrient concentration and DM.

Plants can decipher nutrient doses and adjust biomass partitioning accordingly [40]. Growth and yield are strongly influenced by genotype and environmental factors [41]. According to the available literature, these two variables are negatively affected by high and low nutrient concentrations [42,43]. Thus, the lowest LFW per plant and per m^2 was achieved in the most concentrated HNS, resulting from osmotic stress. LFW increased by the heightened presence of N; however, high concentrations of P and K decreased the LFW and yield due to osmotic stress. In the second harvest, none of the plants survived the treatment with the highest HNS concentration. In contrast, the highest LFW was observed in the least concentrated HNS. This result is in line with the findings of Roosta [44] who reported that spraying K solution increased the percentage of DM in mint. However, the current result is inconsistent with the increase in plant fresh weight. At all harvesting times, N12P4K6 and N12P4K12 resulted in the lowest fresh yields and the highest percentage of DM. By using N12P4K12, which had the highest concentration of nutrients, the lowest fresh yield was obtained. Therefore, HNS-dependent partitioning of biomass may differ among plant species in NGS® in the greenhouse. Decreasing in yield by increasing the concentration of nutrients due to osmotic stress has previously been reported in leafy vegetables [2,11,12,19].

Furthermore, osmotic stress at high concentrations appears to reduce LDG. The amount of N did not induce a significant difference in LDG. Previous research has shown that osmotic stress occurs due to high-concentrations of K and P, and reduces the LDG, which is consistent with the results of the current experiment [45]. Also, differences in LDG among various species may emanate from their genetic potential [46].

Colour is one of the most important traits characterizing the quality of vegetables since its change during storage affects marketing. Colour indirectly indicates freshness and microbial deterioration/contamination in stored plant products [47]. The values of *L**, *a**, and *b** significantly decreased by increasing the HNS concentration, especially by increasing the N concentration. In treatments with the N concentration of 6 $mmol \cdot L^{-1}$, compared to the two other HNS with the N value of 12 $mmol \cdot L^{-1}$, the values of colour parameters were higher in all harvest times and species. One of the possible reasons for the decrease in *b** in N12P2K6 and N12P4K12 treatments is the chlorophyll decomposition due to high concentration of salt and osmotic stress that leads to decreased absorption of nutrients, as well as plant metabolites biosynthesis. Similar results were observed in other leafy plants such as lettuce, where higher nutrient concentrations induced a decrease in *L** and *b** [48].

A vast majority of mints is sold as fresh-cut leaves or cuttings, where preservation of freshness, texture, shelf life, and aroma are essential [49]. WL is a physiological event in fresh-cut products and in fresh aromatic and medicinal herbs. It can be decelerated by controlling temperature and humidity and using suitable and efficient packaging [50,51]. Plastic film performance was significantly maintained in all species. The permeability characteristics of the film on the bags might have induced a low respiratory rate. In the current study, WL was very low at the beginning of storage. The highest WL occurred in bags prepared from the mint of the first harvest time. This accelerated WL in the first harvest time may be due to the crispness and wateriness of the tissues.

The responses of species were studied in terms of respiration rate and headspace analysis. Data analysis showed that the HNS did not affect the rate of CO_2 and O_2 production in any species, which is consistent with the findings of Luna et al. [11] on several varieties of lettuce. One of the reasons for justifying and explaining the reduced respiration rate in the third harvest in MP species compared to the first harvest time is that MP species had a higher DM, which means they had less water and bioactivity. This hypothesis has already been reported by Scuderi et al. [52] who highlighted that the reduced respiration of lettuce was related to the increased amount of DM and less water in plant tissues. The highest respiration rate was observed in response to the N12P4K12 treatment, with the highest amount of NS, which agrees with other reports that showed higher respiration rates in response to higher NS and EC concentrations [52].

The NO_3^- content in edible fruits and vegetables is an important quality characteristic, and the EU encourages good agricultural practices to reduce NO_3^- contents in leafy herbal plants [53]. The NO_3^- content in the three species during the first harvest time was very low, well below the maximum NO_3^- content allowed for vegetables by the EU (Commission Regulation No. 1258/2011). The highest accumulation of NO_3^- occurred in the treatment with the highest amount of N supplementation and the lowest amounts of P and K. Several authors showed that P and K deficiency reduced the activity of nitrate reductase enzyme and caused an increase in the NO_3^- content [54,55]. Therefore, it seems that in treatments with lower P and K, the activity of nitrate reductase enzyme decreased, which may induce an increase in NO_3^- [56], as observed by Miceli et al. [47] on Swiss Chard plants. Nevertheless, a significant challenge with leafy herbal plants, such as mint, is the substantial accumulation of NO_3^- in their tissues, often surpassing 2500 mg NO_3^- kg^{-1}. European Commission regulations No1881/2006 and 1258/2011 have set NO_3^- thresholds for herbal plants at 1000–2500 mg NO_3^- kg^{-1} [57]. Another item that is affected by the accumulation rate of NO_3^- in plant tissues is shelf life. The rate of NO_3^- accumulation in the first day was highest, which is consistent with the results of Miceli et al. [47] (Table 5). Some plants reduce NO_3^- to ammonium ions in the aerial parts, which is mainly due to the different processes of NO_3^- reductase in different parts of the plants. NO_3^- reductase is known as a regulatory enzyme that limits the amount of plant NO_3^-. The activity of this enzyme shows a large variation in various plant species, depending on HNS and environmental conditions. One of the main reasons accounting for this variation could be related to differences in genetic origins of plant species or maybe indirectly with the concentrations of Mo and Fe in plants [1,58–60].

NO_3^- in the plant cell cytosol is converted to nitrite by the NO_3^- reductase enzyme. Then, NO_3^- is reduced to ammonium or ammonia by the nitrite reductase enzyme. While these compounds enter the structure of organic compounds, the resultant NADPH and NADH act as electron donors for the activity of the nitrate reductase enzyme. Therefore, the synthesis of the nitrate reductase enzyme is controlled by NO_3^- presence, and this ion induces the relevant enzymatic synthesis pathway [58]. Thus, any factor that leads to a decrease in the activity of nitrate reductase enzyme in plant cells is potentially associated with the accumulation of NO_3^- in plant shoots. In a relevant experiment, Atkin and Cummins et al. [61] showed that the increase in temperature reduces the activity of the nitrate reductase enzyme. Therefore, it seems that in the present experiment, due to the high temperature at the third harvest, the activity of nitrate reductase enzyme decreased, thus increasing foliar NO_3^- accumulation. In addition, the previous study which was conducted in summer at high temperatures showed a higher accumulation of NO_3^-. Similarly, in our results, the highest accumulation occurred in the third harvest that coincided with high temperatures [56,62,63].

Under management conditions with low N intake, plants reduce the activity of nitrate reductase enzyme, but with exposure to NO_3^- ions, the rate of activity of this enzyme increases [64]. The concentration of chemical fertilizers is known as one of the factors affecting NO_3^- accumulation. In vegetables, this accumulation often depends on the amount and type of nutrients around the roots, which is closely related to the amount and timing of chemical fertilizer consumption [65].

The HNS with the highest P concentration determined the highest PO_4^{3-} accumulation in plants. In a study by Fallovo et al. [56] on lettuce plants, the increase in nutrient supplementation increased the accumulation of PO_4^{3-}, which is consistent with the results of the current study.

$CaCO_3$ uptake appears to be directly related to high N concentrations, while it is inversely related to high K and P concentrations. The rate of $CaCO_3$ accumulation significantly decreased during the shelf life for all three harvests. In a study on mint plants, spraying K reduced the concentration of $CaCO_3$ in both leaves and roots [44], which is consistent with the results of our experiment.

The highest amount of BP occurred in response to treatments with the lowest amount of N, while the lowest BP was observed when using the highest N concentration. Applications with lower N concentrations proved optimal for preserving phenols, maintaining high levels of antioxidant and free radical scavenging activity throughout storage, albeit potentially at a reduced yield. The observed decrease in BP in fresh-cut mints with the lowest amount of N in the present experiment could be attributed to the slowing down of phenolic oxidation, thus positively influencing their accumulation [66]. The shelf life and quality of freshly harvested mint are influenced by HNS treatments and harvesting time. Additionally, the genetic background plays a role in the response of BP and So-Q content in mint. Qualitative reduction in the appearance of fresh-cut products is one of the main factors negatively affecting consumer preference and vegetable marketing [67]. Apart from yellowing, the tissue BP is one of the most important factors limiting the shelf life and marketability of processed products. Both $CaCO_3$ and K are effective elements in preventing the loss of firmness. $CaCO_3$ is one of the most essential and effective nutrients in increasing and maintaining the quality of cut plant tissues. Calcium accumulation in plant tissues strengthens polymeric bonds between the middle lamella of the cell pectocellulosic membrane while improving the network strength of cell walls and thus increasing the mechanical strength of tissues [68]. In our study, the lowest BP rate was found in the treatment with the highest K while the highest BP rate was observed in response to the highest N concentration and the lowest P and K concentration. Olivos et al. [69] reported the effective roles of P, K, and $CaCO_3$ in increasing the shelf life during the storage time. K plays a key role in the firmness of plant tissue. Thus, maintaining the balance in this element is essential [27]. Calcium together with K plays an essential role in permeability, dehydration, maintenance of inflammatory pressure, cell function and in general plant quality, storage properties, and marketing. Calcium-induced stiffness is related to its effect on cell wall components that maintain the membrane and cell wall, but K induced stiffness is related to changes in cell hydrostatic pressure [27,69,70].

Browning occurs due to the disruption of cellular integrity and the subsequent release of polyphenol oxidases, which facilitate the oxidation of phenolic compounds into quinones. The quinones mentioned above demonstrate a significant level of reactivity. They frequently endure in engaging in chemical reactions with each other, as well as with proteins. This ultimately leads to the production of brown pigments in plants. Its control has been extensively studied and reported in several plant species [70,71].

Given the utilization of mints in the food and pharmaceutical industries, it is imperative that they exhibit low microbial contamination and appear free from pathogenic germs. Additionally, they must comply with the legal requirements outlined in Commission Regulation (EC) No 2073/2005. In SCS, microbial contamination is usually lower than in TCS. In previous research on this topic, researchers reported that post-harvest microbial contamination in NFT [52] and NGS® [72] could be reduced. Also, in the present study, the contamination rate was significantly lower compared to the data on microbial contamination reported in the literature for horticultural crops cultivated in TCS. In a study by Selma et al. [72] on comparing the microbial contamination levels of *Lactuca sativa* L. cultivated in soils and NGS®, the results confirmed that contaminations in the soil system were much higher than the NGS®. In this study, it was reported that the SCS appeared as the most suitable system in terms of minimizing microbial contamination. The findings of our study also showed that the NGS® system could be very effective in reducing microbial contamination in mints. In fact, cultivation through NGS® allows the farmer to obtain a cleaner and hygienically safer product than older SCSs and traditionally soil-grown systems.

5. Conclusions

The current study systematically examined the dynamic effects of various nutrient concentrations on the growth, yield, physiological aspects, and shelf life of commercial mint at different harvesting times. It can be concluded that the yield and quality traits of mint grown in NGS® were significantly affected by the HNS and the harvesting times. NGS® is recommended for its ability to control of plant nutrition in order to manage the quality characteristics of the raw materials for fresh-cut products. The findings revealed that an increase in nitrogen concentration in HNS had a negative effect on specific quality parameters, such as higher NO_3^- content. Notably, the highest NO_3^- accumulation occurred at the third harvest, while the lowest was observed at the first harvest. Although the study indicated negligible post-harvest changes due to variations in HNS, these observations are contingent upon species differences. Optimal results, including the highest DM, were achieved using the highest concentration of HNS. N6P2K6 and N12P2K6 treatments in NGS® demonstrated potential for enhancing marketable fresh yield and fresh-cut product quality. Among the species, MP exhibited the highest DM, while MV and MR showed the highest yield. Furthermore, bacterial contamination decreased at the second and third harvests in NGS®, which is attributed to the shorter growth period and potential increased resistance of plants to bacteria. Adjusting the HNS in the growing medium according to climatic conditions and plant species is crucial for obtaining a uniform product with high yield and quality in NGS®. The study underscores the importance of tailoring hydroponic nutrient solutions to specific plant species and environmental conditions for achieving optimal yields and quality in commercial mint cultivation.

Supplementary Materials: The following supporting information can be downloaded at: https://www.mdpi.com/article/10.3390/agronomy14030610/s1, Table S1: three-way ANOVA of dry matter (DM), leaf fresh weight per plant (LFW), leaf daily growth (LDG), yield per square meter, L^*, a^* and b^* as affected by hydroponic nutrient solutions (HNS), mint species and harvest time; Table S2: three-way ANOVA of counts (CFU g^{-1}) of different groups of microorganisms and nutrient content as affected by hydroponic nutrient solutions (HNS), mint species and harvest time.

Author Contributions: Conceptualization, S.N.; methodology, G.P., M.C., S.N. and S.H.; validation, S.H. and S.N.; formal analysis, S.H. and G.P.; investigation, S.H., M.C., S.N. and G.P.; resources, S.N.; data curation, S.H., G.P., M.C., S.J.H. and S.N.; writing—original draft preparation, S.H., G.P., S.J.H. and S.N.; writing—review and editing, S.H., G.P. and S.N.; visualization, S.J.H.; supervision, S.N.; project administration, S.N. All authors have read and agreed to the published version of the manuscript.

Funding: This research was funded by the VEGMAP-DISAFA, the University of Turin.

Data Availability Statement: The data presented in this study are available upon request from the corresponding author.

Acknowledgments: The authors would like to thank the Inhortosanitas Lab of DISAFA, University of Turin, for supporting the research and the Azarbaijan Shahid Madani University for the collaboration.

Conflicts of Interest: The authors declare no conflict of interest.

References

1. Hosseini, S.J.; Tahmasebi-Sarvestani, Z.; Mokhtassi-Bidgoli, A.; Keshavarz, H.; Kazemi, S.; Khalvandi, M.; Pirdashti, H.; Hashemi-Petroudi, S.H.; Nicola, S. Functional quality, antioxidant capacity and essential oil percentage in different mint species affected by salinity stress. *Chem. Biodivers.* **2023**, *20*, e202200247. [CrossRef]
2. Hosseini, S.J.; Tahmasebi-Sarvestani, Z.; Pirdashti, H.; Modarres-Sanavy, S.A.M.; Mokhtassi-Bidgoli, A.; Hazrati, S.; Nicola, S. Investigation of yield, phytochemical composition, and photosynthetic pigments in different mint ecotypes under salinity stress. *Food Sci. Nutr.* **2021**, *9*, 2620–2643. [CrossRef]
3. Tucker, A.O.; Naczi, R.F.C. Mentha: An overview of its classification and relationships. In *Mint: The Genus Mentha;* CRC Press: Boca Raton, FL, USA, 2006; pp. 1–39. [CrossRef]
4. Djenane, D.; Aïder, M.; Yangüela, J.; Idir, L.; Gómez, D.; Roncalés, P. Antioxidant and antibacterial effects of *Lavandula* and *Mentha* essential oils in minced beef inoculated with *E. coli* O157:H7 and *S. aureus* during storage at abuse refrigeration temperature. *Meat Sci.* **2012**, *92*, 667–674. [CrossRef]
5. Fazal, H.; Akram, M.; Ahmad, N.; Qaisar, M.; Kanwal, F.; Rehman, G.; Ullah, I. Nutritionally rich biochemical profile in essential oil of various *Mentha* species and their antimicrobial activities. *Protoplasma* **2022**, *267*, 557–570. [CrossRef]
6. Nair, K.P. Japanese mint. In *Herbal and Acidulant Tree Spices;* Springer: Cham, Switzerland, 2022; pp. 75–89.
7. Tibaldi, G.; Fontana, E.; Nicola, S. Postharvest management affects spearmint and calamint essential oils. *J. Sci. Food Agric.* **2013**, *93*, 580–586. [CrossRef]
8. Pignata, G.; Casale, M.; Nicola, S. *Water and Nutrient Supply in Horticultural Crops Grown in Soilless Culture: Resource Efficiency in Dynamic and Intensive Systems;* Springer: Cham, Switzerland, 2017; pp. 183–219. [CrossRef]
9. Nicola, S.; Pignata, G.; Casale, M.; Hazrati, S.; Ertani, A. Setting up a lab-scale pilot plant to study the new growing system (NGS®) for leafy vegetable and culinary herb growth. *Horticulturae* **2021**, *7*, 90. [CrossRef]
10. Zhan, L.; Bulgari, R.; Pignata, G.; Casale, M.; Nicola, S. The mixing ratio and filling-amount affect the tissue browning and antioxidant properties of fresh-cut baby leaf lettuce (*Lactuca sativa* L.) and rocket (*Eruca sativa* Mill.) grown in floating growing systems. *Foods* **2022**, *11*, 3515. [CrossRef]
11. Luna, M.C.; Martínez-Sánchez, A.; Selma, M.V.; Tudela, J.A.; Baixauli, C.; Gil, M.I. Influence of nutrient solutions in an open-field soilless system on the quality characteristics and shelf life of fresh-cut red and green lettuces (*Lactuca sativa* L.) in different seasons. *J. Sci. Food Agric.* **2013**, *93*, 415–421. [CrossRef]
12. Yang, T.; Kim, H.J. Characterizing nutrient composition and concentration in tomato-, basil-, and lettuce-based aquaponic and hydroponic systems. *Water* **2020**, *12*, 1259. [CrossRef]
13. Fontana, E.; Nicola, S. Traditional and soilless culture systems to produce corn salad (*Valerianella olitoria* L.) and rocket (*Eruca sativa* Mill.) with low nitrate content. *J. Food Agric. Environ.* **2009**, *7*, 405–410.
14. Azizoglu, U.; Yilmaz, N.; Simsek, O.; Ibal, J.C.; Tagele, S.B.; Shin, J.H. The fate of plant growth-promoting rhizobacteria in soilless agriculture: Future perspectives. *3Biotech* **2021**, *11*, 382. [CrossRef]
15. Iglesias, M.J.; López, J.G.; Collados Luján, J.F.; Ortiz, F.L.; Pereznieto, H.B.; Toresano, F.; Camacho, F. Effect of genetic and phenotypic factors on the composition of commercial marmande type tomatoes studied through HRMAS NMR spectroscopy. *Food Chem.* **2014**, *142*, 382. [CrossRef]
16. Walters, K.J. Quantifying the Effects of Hydroponic Systems, Nutrient Solution, and Air Temperature on Growth and Development of Basil (*Ocimum* L.) Species. Master's Thesis, Iowa State University, Ames, IO, USA, 2015. [CrossRef]
17. Nicola, S.; Pignata, G.; Tibaldi, G. The floating growing system can assure a low microbial contamination of baby leaf vegetables at harvest. *Acta Hortic.* **2018**, *1209*, 57–63. [CrossRef]
18. Pignata, G.; Niñirola, D.; Casale, M.; Lo Turco, P.E.; Egea-Gilabert, C.; Fernández, J.A.; Nicola, S. Inherent quality and safety of watercress grown in a floating system using *Bacillus subtilis*. *Hortic. J.* **2016**, *85*, 148–153. [CrossRef]
19. Brown, B.; Hart, J.M.; Wescott, M.P.; Christensen, N.W.C. The critical role of nutrient management in mint production. *Better Crops* **2003**, *87*, 9–11.
20. Skálová, H.; Moravcová, L.; Dixon, A.F.G.; Kindlmann, P.; Pyšek, P. Effect of temperature and nutrients on the growth and development of seedlings of an invasive plant. *AoB Plants* **2015**, *7*, plv044. [CrossRef]
21. Loera-Muro, A.; Troyo-Diéguez, E.; Murillo-Amador, B.; Barraza, A.; Caamal-Chan, G.; Lucero-Vega, G.; Nieto-Garibay, A. Effects of vermicompost leachate versus inorganic fertilizer on morphology and microbial traits in the early development growth stage in mint (*Mentha spicata* L.) and rosemary (*Rosmarinus officinalis* L.) plants under closed hydroponic system. *Horticulturae* **2021**, *7*, 100. [CrossRef]
22. Hoque, M.M.; Ajwa, H.; Othman, M.; Smith, R.; Cahn, M. Yield and postharvest quality of lettuce in response to nitrogen, phosphorus, and potassium fertilizers. *HortScience* **2010**, *45*, 1539–1544. [CrossRef]
23. Hassan, S.M.; El-Bebany, A.F.; Salem, M.Z.; Komeil, D.A. Productivity and post-harvest fungal resistance of hot pepper as affected by potassium silicate, clove extract foliar spray and nitrogen application. *Plants* **2021**, *10*, 662. [CrossRef]
24. Poulsen, N.; Johansen, A.S.; Sørensen, J.N. Influence of growth conditions on the value of crisphead lettuce—4. Quality changes during storage. *Plant Foods Hum. Nutr.* **1995**, *47*, 157–162. [CrossRef]

25. Adel Mahmoodabad, H.; Hokmalipoor, S.; Shaban, M.; Ashrafi Parchin, R. Effect of foliar spray of urea and soil application of vermicompost on essential oil and chlorophyll content of green Mint (*Mentha spicata* L.). *Int. J. Adv. Biol. Biomed. Res.* **2014**, *2*, 2104–2108.
26. Parrey, Z.A.; Shah, S.H.; Fayaz, M.; Casini, R.; Elansary, H.O.; Mohammad, F. Nitrogen supplementation modulates morphological, biochemical, yield and quality attributes of peppermint. *Plants* **2023**, *12*, 809. [CrossRef]
27. Hasanuzzaman, M.; Bhuyan, M.H.M.B.; Nahar, K.; Hossain, M.S.; Al Mahmud, J.; Hossen, M.S.; Masud, A.A.; Moumita, C.; Fujita, M. Potassium: A vital regulator of plant responses and tolerance to abiotic stresses. *Agronomy* **2018**, *8*, 31. [CrossRef]
28. Ryu, D.H.; Cho, J.Y.; Yang, S.H.; Kim, H.Y. Effects of harvest timing on phytochemical composition in Lamiaceae plants under an environment-controlled system. *Antioxidants* **2023**, *12*, 1909. [CrossRef]
29. Marcum, D.B.; Hanson, B.R. Effect of irrigation and harvest timing on peppermint oil yield in California. *Agric. Water Manag.* **2006**, *82*, 118–128. [CrossRef]
30. Machiani, M.A.; Javanmard, A.; Morshedloo, M.R.; Maggi, F. Evaluation of yield, essential oil content and compositions of peppermint (*Mentha piperita* L.) intercropped with faba bean (*Vicia faba* L.). *J. Clean. Prod.* **2018**, *171*, 529–537. [CrossRef]
31. Luetic, S.; Knezovic, Z.; Jurcic, K.; Majic, Z.; Tripkovic, K.; Sutlovic, D. Leafy vegetable nitrite and nitrate content: Potential health effects. *Foods* **2023**, *12*, 1655. [CrossRef]
32. Bian, Z.; Wang, Y.; Zhang, X.; Li, T.; Grundy, S.; Yang, Q.; Cheng, R. A review of environment effects on nitrate accumulation in leafy vegetables grown in controlled environments. *Foods* **2020**, *9*, 732. [CrossRef]
33. Nicola, S.; Hoeberechts, J.; Fontana, E. Comparison between traditional and soilless culture systems to produce rocket (*Eruca sativa*) with low nitrate content. *Acta Hortic.* **2005**, *697*, 549–555. [CrossRef]
34. Nicola, S.; Pignata, G.; Casale, M.; Lo Turco, P.E.; Gaino, W. Overview of a lab-scale pilot plant for studying baby leaf vegetables grown in soilless culture. *Hortic. J.* **2016**, *85*, 97–104. [CrossRef]
35. Couture, R.; Cantwell, M.I.; Ke, D.; Saltveit, M.E. Physiological attributes related to quality attributes and storage life of minimally processed lettuce. *HortScience* **1993**, *28*, 723–725. [CrossRef]
36. Loaiza-Velarde, J.G.; Saltveit, M.E. Heat shocks applied either before or after wounding reduce browning of lettuce leaf tissue. *J. Am. Soc. Hortic. Sci.* **2001**, *126*, 227–234. [CrossRef]
37. Fageria, N.K. *Nitrogen Management in Crop Production*; CRC Press: Boca Raton, FL, USA, 2014.
38. Mangal, V.; Lal, M.K.; Tiwari, R.K.; Altaf, M.A.; Sood, S.; Kumar, D.; Aftab, T. Molecular insights into the role of reactive oxygen, nitrogen and sulphur species in conferring salinity stress tolerance in plants. *J. Plant Growth Regul.* **2003**, *42*, 554–574. [CrossRef]
39. Baek, J.P.; Park, K.W.; Craker, L.E.; Kang, H.M. Changes in growth and quality of three mint cultivars at different harvesting periods. *Hortic. Environ. Biotechnol.* **2016**, *57*, 207–212. [CrossRef]
40. Bindraban, P.S.; Dimkpa, C.; Nagarajan, L.; Roy, A.; Rabbinge, R. Revisiting fertilisers and fertilisation strategies for improved nutrient uptake by plants. *Biol. Fertil. Soils* **2015**, *51*, 897–911. [CrossRef]
41. Hosseini, S.J.; Tahmasebi-Sarvestani, Z.; Pirdashti, H.; Modarres-Sanavy, S.A.M.; Mokhtassi-Bidgoli, A.; Hazrati, S.; Nicola, S. Assessment of salinity indices to identify mint ecotypes using intelligent and regression models. *Int. J. Hortic. Sci. Technol.* **2020**, *7*, 119–137. [CrossRef]
42. Savvas, D.; Adamidis, K. Automated management of nutrient solutions based on target electrical conductivity, pH, and nutrient concentration ratios. *J. Plant Nutr.* **1999**, *22*, 1415–1432. [CrossRef]
43. Zeljković, S.Ć.; Aucique-Perez, C.E.; Štefelová, N.; De Diego, N. Optimizing growing conditions for hydroponic farming of selected medicinal and aromatic plants. *Food Chem.* **2022**, *375*, 131845. [CrossRef]
44. Roosta, H.R. Effects of foliar spray of K on mint, radish, parsley and coriander plants in aquaponic system. *J. Plant Nutr.* **2014**, *37*, 2236–2254. [CrossRef]
45. Kalve, S.; Sizani, B.L.; Markakis, M.N.; Helsmoortel, C.; Vandeweyer, G.; Laukens, K.; Sommen, M.; Naulaerts, S.; Vissenberg, K.; Prinsen, E.; et al. Osmotic stress inhibits leaf growth of *Arabidopsis thaliana* by enhancing ARF-mediated auxin responses. *New Phytol.* **2020**, *226*, 1766–1780. [CrossRef]
46. Atieno, J.; Li, Y.; Langridge, P.; Dowling, K.; Brien, C.; Berger, B.; Varshney, R.K.; Sutton, T. Exploring genetic variation for salinity tolerance in chickpea using image-based phenotyping. *Sci. Rep.* **2017**, *7*, 1300. [CrossRef] [PubMed]
47. Miceli, A.; Moncada, A.; Vetrano, F.; D'Anna, F. First results on yield and quality response of Basil (*Ocimum basilicum* L.) grown in a floating system. *Acta Hortic.* **2003**, *609*, 377–381. [CrossRef]
48. Petropoulos, S.A.; El-Nakhel, C.; Graziani, G.; Kyriacou, M.C.; Rouphael, Y. The effects of nutrient solution feeding regime on yield, mineral profile, and phytochemical composition of spinach microgreens. *Horticulturae* **2021**, *7*, 162. [CrossRef]
49. Hutsol, T.; Priss, O.; Kiurcheva, L.; Serdiuk, M.; Panasiewicz, K.; Jakubus, M.; Kukharets, M. Mint plants (*Mentha*) as a promising source of biologically active substances to combat hidden hunger. *Sustainability* **2023**, *15*, 11648. [CrossRef]
50. Rizzo, V.; Muratore, G. Effects of packaging on shelf life of fresh celery. *J. Food Eng.* **2009**, *90*, 124–128. [CrossRef]
51. Brindisi, L.J.; Simon, J.E. Preharvest and postharvest techniques that optimize the shelf life of fresh basil (*Ocimum basilicum* L.): A review. *Front. Plant Sci.* **2023**, *14*, 1237577. [CrossRef]
52. Scuderi, D.; Giuffrida, F.; Noto, G. Effects of nutrient solution EC on yield, quality and shelf-life of lettuce grown in floating system. *Acta Hortic.* **2009**, *807*, 221–226. [CrossRef]
53. Giménez, A.; Fernández, J.A.; Pascual, J.A.; Ros, M.; Saez-Tovar, J.; Martinez-Sabater, E.; Gruda, N.S.; Egea-Gilabert, C. Promising composts as growing media for the production of baby leaf lettuce in a floating system. *Agronomy* **2020**, *10*, 1540. [CrossRef]

54. Moorby, H.; Nye, P.H.; White, R.E. The effect of phosphate nutrition of young rape plants on nitrate reductase activity and xylem exudation, and their relation to H ion efflux from the roots. *Plant Soil* **1988**, *105*, 257–263. [CrossRef]
55. Xu, X.; Du, X.; Wang, F.; Sha, J.; Chen, Q.; Tian, G.; Zhu, Z.; Ge, S.; Jiang, Y. Effects of potassium levels on plant growth, accumulation and distribution of carbon, and nitrate metabolism in apple dwarf rootstock seedlings. *Front. Plant Sci.* **2020**, *11*, 904. [CrossRef]
56. Fallovo, C.; Rouphael, Y.; Rea, E.; Battistelli, A.; Colla, G. Nutrient solution concentration and growing season affect yield and quality of *Lactuca sativa* L. var. acephala in floating raft culture. *J. Sci. Food Agric.* **2009**, *89*, 1682–1689. [CrossRef]
57. Vanegas, D.E.; Contreras, A.; Bustamante, A.I.; Tapia, M.L.; Lizana, L.A.; Escalona, V.H. Effect of nitrogen concentration in the nutritional solution and harvest time on nitrate content in baby leaf Swiss chard crop in a hydroponic system. *Acta Hortic.* **2016**, *1141*, 191–198. [CrossRef]
58. Hachiya, T.; Sakakibara, H. Interactions between nitrate and ammonium in their uptake, allocation, assimilation, and signaling in plants. *J. Exp. Bot.* **2017**, *68*, 2501–2512. [CrossRef]
59. Rasmusson, A.G.; Escobar, M.A.; Hao, M.; Podgórska, A.; Szal, B. Mitochondrial NAD(P)H oxidation pathways and nitrate/ammonium redox balancing in plants. *Mitochondrion* **2020**, *53*, 158–165. [CrossRef]
60. Wang, W.; Hu, B.; Li, A.; Chu, C. NRT1.1s in plants: Functions beyond nitrate transport. *J. Exp. Bot.* **2020**, *71*, 4373–4379. [CrossRef]
61. Atkin, O.K.; Cummins, W.R. The effect of root temperature on the induction of nitrate reductase activities and nitrogen uptake rates in arctic plant species. *Plant Soil* **1994**, *159*, 187–197. [CrossRef]
62. Liu, Z.; Yan, H.; Wang, K.; Kuang, T.; Zhang, J.; Gui, L.; An, X.; Chang, W. Crystal structure of spinach major light-harvesting complex at 2.72 A resolution. *Nature* **2004**, *428*, 287–292. [CrossRef]
63. Conesa, E.; Fernandez, J.A.; Ninirola, D.; Egea-Gilabert, C. Nutrient solution aeration and growing cycles affect quality and yield of fresh-cut baby leaf red lettuce. *Agric. Food Sci.* **2015**, *24*, 313–322. [CrossRef]
64. Zhao, H.; Li, X.; Jiang, Y. Response of nitrogen losses to excessive nitrogen fertilizer application in intensive greenhouse vegetable production. *Sustainability* **2019**, *11*, 1513. [CrossRef]
65. Colla, G.; Kim, H.J.; Kyriacou, M.C.; Rouphael, Y. Nitrate in fruits and vegetables. *Sci. Hortic.* **2018**, *237*, 221–238. [CrossRef]
66. IS Mahlangu, R.; Maboko, M.M.; Mudau, F.N.; Amoo, S.O. Nitrogen levels, plant density and postharvest storage duration affect phytochemical and antioxidant properties of field-grown basil and rocket crops. *Int. J. Veg. Sci.* **2021**, *27*, 515–525. [CrossRef]
67. Giacalone, G.; Chiabrando, V. Effect of different treatments with calcium salts on sensory quality of fresh-cut apple. *J. Food Nutr. Res.* **2013**, *52*, 79–86.
68. Larbi, A.; Kchaou, H.; Gaaliche, B.; Gargouri, K.; Boulal, H.; Morales, F. Supplementary potassium and calcium improves salt tolerance in olive plants. *Sci. Hortic.* **2020**, *260*, 108912. [CrossRef]
69. Olivos, A.; Johnson, S.; Xiaoqiong, Q.; Crisosto, H.C. Fruit phosphorous and nitrogen deficiencies affect "grand pearl" nectarine flesh browning. *HortScience* **2012**, *47*, 391–394. [CrossRef]
70. Thor, K. Calcium—nutrient and messenger. *Front. Plant Sci.* **2019**, *10*, 440. [CrossRef]
71. Abbott, J.A.; Buta, J.G. Effect of antibrowning treatment on color and firmness of fresh-cut pears. *J. Food Qual.* **2002**, *25*, 333–341. [CrossRef]
72. Selma, M.V.; Luna, M.C.; Martínez-Sánchez, A.; Tudela, J.A.; Beltrán, D.; Baixauli, C.; Gil, M.I. Sensory quality, bioactive constituents and microbiological quality of green and red fresh-cut lettuces (*Lactuca sativa* L.) are influenced by soil and soilless agricultural production systems. *Postharvest Biol. Technol.* **2012**, *63*, 16–24. [CrossRef]

agronomy

Article

Origanum dictamnus Essential Oil in Vapour or Aqueous Solution Application for Pepper Fruit Preservation against *Botrytis cinerea*

Nikolaos Tzortzakis

Department of Agricultural Sciences, Biotechnology and Food Science, Cyprus University of Technology, Limassol 3036, Cyprus; nikolaos.tzortzakis@cut.ac.cy

Abstract: The use of synthetic sanitizers for fresh commodities preservation is of concern, with eco-friendly alternatives, including essential oils (EOs), attracting research and industry interest. Dittany (*Origanum dictamnus*—DIT) oil was applied, either through vapour or dipping, on pepper fruit or in vitro against *Botrytis cinerea*, and compared to untreated (control) or chlorine (CHL)-treated fruits stored at 11 °C. Direct DIT vapours (up to 6 d) suppressed lesion growth, spore germination, and spore production compared to the untreated fruits. The antimicrobial properties of EOs were evidenced in pre-exposed-to-DIT oil vapours (residual effect), resulting in fruit lesion suppression. However, DIT-pretreated fruits had the same spore production and spore germination as the control and CHL applications. In in vitro tests, DIT vapours decreased colony growth and spore production when fungi were grown on Potato Dextrose Agar (PDA) or PDA pre-exposed to DIT following *B. cinerea* inoculation. This evidenced that the disease suppression after DIT vapour application primarily affected the interaction of the fruit–pathogen and/or residual responses on fruit tissue and/or PDA media. Fungal biomass in Potato Dextrose Broth (PDB) was evaluated after DIT and CHL applications (10–50–100–500–5000 μL L^{-1}) and decreased with the DIT oil. Additionally, DIT or CHL sanitary dipping on pepper was ineffective against *B. cinerea* compared to vapour application. Therefore, DIT vapours revealed antimicrobial properties and could be an alternative postharvest sanitiser. DIT oil application should also be evaluated at semi-commercial scale for further optimizations, prior to commercialisation.

Keywords: grey mould; essential oil; fruit decay; *Capsicum annuum*; sanitizers

1. Introduction

Fresh produce preservation is still challenging and attracting the attention of researchers and application strategies on the postharvest community. The need for better fresh commodities' commercial storage/transit conditions has increased recently due to the financial costs associated with spoiling food, consumer worries regarding the safety of foods containing synthetic substances, and an increase in fresh produce consumption [1]. The challenge is to control the spread of postharvest diseases and preserve fruit quality and safety. The most often utilized sanitizers for postharvest washing are chlorine and peroxyacetic acid [2], because of their well-known antimicrobial properties and simplicity of use [3,4]. In fact, commonly, fresh produce is preserved with chemical treatments, such as disinfectants containing chlorine or bromine [2,5]. However, existing sanitation means have given rise to significant health and environmental issues because of the possible cancer-causing residues from pesticide inputs and their lack of effectiveness on a variety of microorganisms [6,7]. Innovative preservation solutions, both environmentally and consumer-safe, are therefore, required.

Natural components have received great attention and application in the food industry [6,8,9]. The majority of natural products are easily accessible, safe for the environment, less likely to cause pest resistance, less dangerous for non-target organisms, less detrimental to plant growth, and, best of all, sometimes, less costly than certain chemical treatments [10–12]. Essential oils (EOs), hydrosol (Hyd; enriched water after the distillation process), and extracts from aromatic plants that have antimicrobial properties have garnered significant attention lately [2,13–15]. These products are thought to be safe for humans and the environment [16]. They can be used as agrochemicals [11] by controlling pathogens and/or microorganisms that produce toxins in food [17] but also observing stimulatory effects on crops [18]. Both EOs and their constituents reveal efficiency against insects [19,20], nematodes, fungi, and bacteria [15,21–23].

One advantage of EOs is their bioactivity as a volatile mean, which could make them effective fumigants for the preservation of commodities during storage [1,15,24]. Many research studies have documented the beneficial responses of EOs on the preservation of fresh produce, including strawberry (*Fragaria vesca* L.), tomato (*Solanum lycopersicum* L.), cucumber (*Cucumis sativus* L.), apricot (*Prunus armeniaca* L.), pear (*Pyrus communis* L.), apples (*Malus domestica* L.), and avocado (*Persea americana* L.) fruit, to name a few [1,9,11,25–29]. Essential oil and/or hydrosol might trigger a defence response, by increasing defence-related enzymes and total phenolics of the fresh commodities [9,15]. The consumer preference for natural plant components, such as EOs, over synthetic fungicides may contribute to the high acceptance of these novel treatments [2].

Origanum dictamnus (L.), commonly well known as dittany (DIT), is an indigenous plant species found in Crete, Greece, and well known for its antimicrobial [30,31], antioxidant [6,32], antiproliferative [6], aphrodisiac, oxytocic, antirheumatic, and healing properties [6,33]. Dittany properties are basically attributed to the main EO components, the phenol carvacrol [6]. Carvacrol is capable of suppressing the growth of bacteria and to inhibit the flux toxin production by *Bacillus cereus* [17] and control spoilage yeast in foodstuffs [23].

Pepper (*Capsicum annuum*) crop is widely cultivated, and fruit may be consumed raw or cooked [34,35]. Since it contains ingredients, including capsaicin, phenolic molecules, and vitamins A and C, it has also been linked to a lower risk of cancer and/or other disorders. Peppers are susceptible to low storage temperatures, i.e., less than 7 °C, and to opportunistic microbes at the preharvest and postharvest stages, causing significant fruit deterioration and alterations in secondary metabolism because of physiologic modifications, which promote pathogen growth. These are the main causes of pepper postharvest loss [36]. The encapsulation of chitosan containing *Heracleum persicum* fruit EO extended sweet bell pepper fruit shelf life due to the antioxidant capacity of the investigated oil [35].

Among the most significant fruit disorders globally, grey mould (*Botrytis cinerea* Pers.: Fr; teleomorph: *Botryotinia fuckeliana*) is a widespread and prevalent rot that affects a variety of vegetables [31,37]. The scope of this study was to (1) explore the impact of dittany EO towards the rotting of pepper fruit caused by *B. cinerea*, as opposed to using chlorine, a widely used sanitizer, (2) explore any residual effect of the EO in pre-treated fruits towards fungi development, and (3) evaluate the EO application method on pepper fruit vapours *versus* dipping.

2. Materials and Methods

2.1. Plant Material and Microorganism

Pepper (*C. annuum* cv. Bounty F1) fruits were purchased from a commercial greenhouse in Limassol, Cyprus. Plants were grown in soil, under vertical scheme for a period of 6 months, under 2-branch pruning scheme, with common cultivation practices applied. Fruits were selected to have uniform size, colour, ripeness ("immature-green" stage), and checked for no defects or injuries. Essential oils of dittany (*Origanum dictamnus*) were employed, while chlorine (CHL; 5% NaOCl) was utilized as a commercial chemical sanitizer.

Essential oils were obtained from dittany plants grown in nature, harvested in a hilly area of Crete-Chania, dried in air-ventilated oven at 35 °C, and then EOs were hydrodistilled by using a Clevenger apparatus for 2.5 h [38]. The composition of EO was evaluated using

Gas Chromatography–Mass Spectrometry (GC-MS) according to Xylia et al. [14]. Essential oil components were identified through the comparison of EO's retention time indices and mass spectra with authentic standards when assessable and based on the comparison of their mass spectra with spectral data of Mass Spectra Library NIST08 and published sources [39].

B. cinerea isolated from tomato (strain number: 169558) was supplied by CABI (CABI Bioscience UK Centre, Bakeham Lane, Egham, UK) and aseptically sub-cultured in Potato Dextrose Agar (PDA); it was then purified, and cultures were stored at 4 °C for long-term use, as described previously [31].

2.2. Experimental Set up and Applications

The EOs and CHL impacts on fungal biomass development and sporulation were tested in Potato Dextrose Broth (PDB) in various levels (10–50–100–500–5000 μL L^{-1}). For the in vitro studies, two mycelia plugs were inoculated in 30 mL PDB medium containing EOs or CHL in the above-mentioned concentrations in a 250 mL glass container. The containers were closed and parafilm-sealed after the inoculation.

Regarding the fungal growth, DIT (0.35 mL L^{-1}) oil was diluted in distilled H_2O (with 5% (v/v) Tween-20) and mixed with PDA media before solidification. DIT oil application was compared with CHL (48 mL L^{-1}) in PDA as chemical control and an untreated PDA, as a negative control treatment. The DIT and CHL concentrations were based on previous research [40,41]. The complete experimental set up is presented at Table 1. For the in vitro studies, the disc diffusion method was employed for fungal colony growth on PDA medium, as described previously [31]. In brief, PDA media inoculated with a mycelial disc from 4–5-day-old culture of *B. cinerea* and aliquots (100 μL) of each volatile solution (DIT or CHL) were soaked into a filter paper (3.0 cm × 3.0 cm). The paper discs were placed on the inverted lid of each Petri dish; the lids were closed and sealed with parafilm. The number of spores produced and spores' viability were tested by adding aliquots of 30 mL of DIT or CHL solutions, in individual small tubes, which were placed inside the plastic containers just before the lids were closed. Wet filter paper was placed in each container to maintain high relative humidity (RH ~95%) during the storage period.

Table 1. The experimental set up of the volatiles (chlorine or dittany essential oil) on *Botrytis cinerea* development in vitro (PDA medium) or in vivo (pepper fruit). In curative effect (CE) fruits (or plates) inoculated with *B. cinerea* and then exposed to vapours. A second sample of fruits (or plates) pre-exposed to vapours (residual effect-RE), inoculated with fungi and transferred/stored to ambient air (control). Treatments maintained throughout at 11 °C and 95% RH.

In Vitro	Conditions	Treatment	Section
Colony growth (CE-VAP)	at 20 °C for 2 h, at 11 °C for 6 d	Inoc.-6 d in VAP	2.3
Colony growth (RE-VAP)	at 20 °C for 2 h, at 11 °C for 6 d, at 11 °C for 6 d	6 d in VAP-Inoc.-6 d in air	2.3
Spore production (CE-VAP)	at 11 °C for 10 d	Inoc.-10 d in VAP	2.6
Spore production (RE-VAP)	at 11 °C for 6 d, at 11 °C for 10 d	6 d in VAP-Inoc.-10 d in air	2.6
Spore germination (CE-VAP)	at 11 °C for 10 d, at 11 °C for 24 h	Inoc.-10 d in VAP-Inoc. in PDA-24 h in air	2.7 a
Spore germination (RE-VAP)	at 11 °C for 6 d, at 11 °C for 10 d, at 11 °C for 24 h	6 d in VAP-Inoc.-10 d in air-Inoc. in PDA-24 h in air	2.7 b
In Vivo			
Lesion growth (CE-VAP)	at 20 °C for 2 h, at 11 °C for 12 d	Inoc.-12 d in VAP	2.4 a
Lesion growth (RE-VAP)	at 20 °C for 2 h, at 11 °C for 6 d, at 11 °C for 6 d	Inoc.-6 d in VAP-6 d in air	2.4 b
Lesion growth (RE-VAP)	at 20 °C for 2 h, at 11 °C for 6 d, at 11 °C for 6 d	6 d in VAP-Inoc.-6 d in air	2.4 c
Lesion growth (CE-DIP)	at 20 °C for 24 h, at 11 °C for 12 d	Inoc.-1 d at RT-30 min in DIP-12 d in air	2.5
Lesion growth (RE-DIP)	at 20 °C for 24 h, at 11 °C for 12 d	30 min in DIP-Inoc.-1 d at RT-12 d in air	2.5
Spore production (CE-VAP)	at 11 °C for 15 d	Inoc.-15 d in VAP	2.6
Spore production (RE-VAP)	at 11 °C for 6 d, at 11 °C for 15 d	6 d in VAP-Inoc.-15 d in air	2.6
Spore germination (CE-VAP)	at 11 °C for 15 d, at 11 °C for 24 h	Inoc.-15 d in VAP-Inoc. in PDA-24 h in air	2.7 c
Spore germination (RE-VAP)	at 11 °C for 6 d, at 11 °C for 10 d, at 11 °C for 24 h	6 d in VAP-Inoc.-15 d in air-Inoc in PDA-24 h in air	2.7 c

VAP: vapours; DIP: dipping; RE: residual effect; CE: curative effect; PDA: potato dextrose agar; Inoc.: inoculation. Letters a, b, c under the "section" heading correspond with sections in the text.

Two sub-experiments were carried out for pepper fruit, as detailed in Sections 2.4 and 2.5, in order to investigate the proper applications in vivo (either volatiles or dipping). Twelve distinct fruits were subjected to each volatile compound treatment in sub-experiment I, resulting in three treatments: untreated control, chemical control treated with CHL (48 mL L^{-1}), and fruit treated with DIT (0.35 mL L^{-1}). Fruits were placed into 4.5 L polystyrene containers (2 fruit/container) with snap-on lids for each individual experiment. Filter paper (Whatman No. 1; 3.0 cm × 3.0 cm) was soaked with 5 mL of the desired concentration of each volatile solution, and filter paper was placed inside in the containers. Moreover, wet filter paper with distilled H_2O was used for 95% RH maintenance. The volatiles were allowed to vaporize inside the containers spontaneously at 20 °C for 2 h. Control samples were handled similarly apart from the volatile treatment (controls consisted of distilled H_2O with 5% (v/v) Tween-20). In vitro and in vivo experiments were repeated twice.

The sub-experiment II involved dipping 12 individual fruit per treatment into the EO or CHL solutions (same concentration as vapour treatments) for 30 min, drying the fruit for 1 h at room temperature (RT) and storing the fruit in ambient air conditions for up to 12 d.

2.3. Impact of Dittany and Chlorine Enrichment on Pathogen Growth In Vitro

Two mycelial plugs (1 cm^2), obtained from the periphery of 4–5 d old culture of *B. cinerea* at 25 °C, were placed in the centre of containers with 30 mL PDB. Following inoculation, containers were incubated for 9 d in the dark at 11 °C and 95% RH. Fungal biomass was collected, dried, and weighed, and the results were expressed as dry weight (g), while inhibition of sporulation was calculated (considering the area measured using digital imaging tools (Adobe Photoshop CS6 analysis tools)) compared to the sporulation of the control (without DIT or CHL compounds).

A mycelial plug (2.5 mm diameter), as described above, was placed in the centre of PDA plates. Following inoculation, plates were incubated for 6 d in the containers exposed to air or volatiles in the dark at 11 °C, 95% RH (curative effect). In a second experiment, PDA plates were exposed to air or volatiles for 6 d; then, PDA medium was inoculated with the fungi (as described above), lids replaced on the plates, and plates transferred to air for an additional 6 d period (residual effect). Colony diameter was measured and results monitored as colony area (cm^2) development.

2.4. Impact of Dittany and Chlorine Enrichment on Grey Mould Development in Wound-Inoculated Fruit

Two wounds (3 mm diameter and 1–2 mm deep) were made on opposite sides, and at the top and bottom, of commercial immature green pepper, using a sterilised spike. *B. cinerea* was inoculated as described previously [42], by inserting a mycelial plug (2.5 mm diameter) in each wound of the pepper fruit. The following sub-experiments implemented the following:

(a) Peppers inoculated with *B. cinerea* were placed in containers and exposed continually to air or volatiles for 6 d or 12 d (examine curative effect).
(b) Peppers inoculated with *B. cinerea* were placed in containers and exposed to air or volatiles for 6 d and then transferred to air for an additional 6 d (examine residual effect due to partial exposure to volatiles).
(c) Peppers were first exposed to air or volatiles for 6 d, then removed from volatiles and inoculated with *B. cinerea* and transferred to air for an additional 6 d (examine residual effect on fruit; no fungal exposure to volatiles).

Fumigation was conducted in the dark at 11 °C, 95% RH. Lesion development (expressed in cm^2) was measured at the end of the experiments.

2.5. Impact of Sanitary Dips on Grey Mould Development in Wound-Inoculated Fruit

In sub-experiment II, fruits were dipped into DIT or CHL solution, at the same concentrations as the vapour's treatments. In the first treatment, the fruit inoculated with *B. cinerea* were incubated for 24 h at RT and then dipped into the EOs or CHL solution for 30 min, dried for 1 h at RT and transferred to air for 12 d (curative effect). In the second treatment, fruits were dipped in the EOs or CHL solutions for 30 min, dried for 1 h at RT, inoculated with *B. cinerea*, incubated for 24 h at RT, and transferred to storage conditions in air for 12 d (preventative effect). Incubation was performed in the dark at 11 °C, 95% RH and lesion development (cm^2) measured every 2 d throughout the storage period. Both experiments were repeated twice.

2.6. Effect of Dittany and Chlorine Enrichment on Spore Production

The *B. cinerea* spore suspension was collected from a sporulated PDA medium (as described in Tzortzakis, 2010 [42]) by harvesting the spores using an L-shaped spreader. Spores were concentrated and counted using a haemocytometer slide. Plates (PDA medium) were inoculated in the centre with 20 μL spore solution (1.8×10^4 spores mL^{-1}), transferred to containers, and exposed to air or volatiles (CHL or EO) for 10 d (curative effect). In the second experiment, plates were first placed in containers, exposed to air or volatiles (EO or CHL) for 6 d, and then inoculated with freshly prepared fungal suspension (preventative effect). In both experiments, the lids were removed from the plates during incubation. After exposure, the plates were closed and transferred to air for 10 d (residual effect). Following exposure to volatiles (CHL or EO) or air, spores were collected with an L-shaped spreader with 20 mL dH_2O (with Tween 80; 0.1% v/v) for 5 min and concentrated to a final 1 mL volume, as described by Tzortzakis [42].

Wounded pepper fruit were inoculated with 15 μL spore suspension (2×10^4 spores mL^{-1}) of *B. cinerea* in each wound, placed in containers, and exposed to air or volatiles (CHL or EO) for 15 d until spores formed. In the second experiment, healthy fruits were exposed to air or volatiles (CHL or EO) for 6 d, inoculated with *B. cinerea* (as above), and transferred to air for 15 d until spores formed. Fumigation was performed in the dark at 11 °C, 95% RH. Following exposure to volatiles (CHL or EO) or air, fruit was shaken for 25 min in 80 mL dH_2O (with Tween 80 (0.1% v/v) for better spore separation) to remove the spores. Spore suspensions were concentrated to a final volume of 1 mL. A haemocytometer slide was used for microscope quantification counts.

2.7. Effect of Dittany and Chlorine Enrichment on Spore Germination

Spore germination tests were implemented in several treatments:

(d) Spores from 9 d old *B. cinerea* colonies were collected with an L-shaped spreader and inoculated on PDA medium (2–3 mm thick). Plates were placed in containers, the lids were removed, and they were exposed to air or volatiles (CHL or EO) for 24 h (curative effect).

(e) Plates were placed in containers, the lids removed, and they were exposed to air or volatiles (CHL or EO) for 24 h and then inoculated with the fungal suspension). Following exposure, lids were replaced, and plates were transferred to air for 24 h (residual effect).

(f) Spore suspensions were made from inoculated and volatile-exposed pepper fruits, inoculated on PDA medium, and incubated in air for 24 h in the dark at 11 °C, 95% RH. Measurements were recorded after 24 h incubation. Within each of the 9 replicates (4 sets of 16 squares; 44% of the haemocytometer slide was measured, in two different areas for the PDA medium or the fruit per replicate), 100 spores were examined, and the percentage of germinated spores was calculated.

2.8. Statistical Analysis

Values were presented as means ± standard errors (SE) from 9–12 independent values per treatment. Data were subjected to analysis of variance and Duncan's Multiple Range test ($p < 0.05$) using IBM SPSS v.25 for Windows.

3. Results

The composition of the main (>2.0%) EO components is presented at Table 2. Twenty-four components were identified, indicating 99.87% of the oil composition. The main components were carvacrol (69.89%), p-cymene (12.92%), and γ-Terpinene (7.12%), in agreement with previous reports [43]. The main category of the EO was oxygenated monoterpenes, averaged in 72.87%, followed by monoterpene hydrocarbons (24.31%) and then by sesquiterpene hydrocarbons (2.69%).

Table 2. The chemical and percentage composition of the major components from the essential oil of *Origanum dictamnus*.

No	Compound	R.I. [b]	Retention Time	Percentage (%) Composition
1	*a*-Thujene	930	9.507	0.67
2	*a*-Pinene [a]	939	9.746	0.57
3	Sabinene	975	11.253	0.06
4	*b*-Pinene [a]	979	11.353	0.10
5	1-Octen-3-ol	979	11.530	0.10
6	*b*-Myrcene	990	11.933	1.11
7	*a*-Phellandrene	1002	12.402	0.10
8	*a*-Terpinene	1017	12.855	1.49
9	p-cymene [a]	1024	13.163	12.92
10	Limonene [a]	1029	13.302	0.37
11	γ-Terpinene [a]	1059	14.400	7.12
12	*cis*-Sabinene hydrate	1070	14.701	0.72
13	Linalool [a]	1096	15.841	0.53
14	Terpinen-4-ol	1177	18.479	0.29
15	Thymoquinone	1252	20.816	0.19
16	Thymol [a]	1290	22.007	0.12
17	Carvacrol [a]	1299	22.382	69.89
18	*a*-Cubenene	1348	23.768	0.08
19	*a*-Copaene	1376	24.565	0.58
20	*b*-Caryophyllene	1419	25.883	1.38
21	*β*-Bisabolene	1505	27.302	0.14
22	*δ*-Cadinene	1523	29.561	0.29
23	Thymohydro quinone	1555	30.761	1.00
24	Caryophyllene oxide	1583	32.548	0.20
	Total (%)			99.87
	Monoterpene hydrocarbons			24.31
	Oxygenated monoterpenes			72.87
	Sesquiterpene hydrocarbons			2.69

([a]) Identification by comparison of retention times and co-injection with authentic compound. ([b]) R.I. (Retention Indices) from experimental using a SBP-5 column using a homologous series of n-alkanes (C9–C25).

3.1. Effects of Dittany Oil on Fungal Biomass In Vitro

The *B. cinerea* biomass decreased as the concentration of the chlorine or dittany oil increased in the PDB media (Figure 1). In detail, the application of CHL decreased fungal biomass from 28.4% up to 57.1% at 10 $\mu L\ L^{-1}$ and 500 $\mu L\ L^{-1}$ applications, respectively, compared with the control treatment. The relevant decreases in fungal biomass found after the dittany oil applications were 14.3% at DIT 10 $\mu L\ L^{-1}$ and 91.4% at DIT 500 $\mu L\ L^{-1}$. Spore production accelerated at CHL $\geq$ 100 $\mu L\ L^{-1}$ but was eliminated in the DIT applications (see Figure S1).

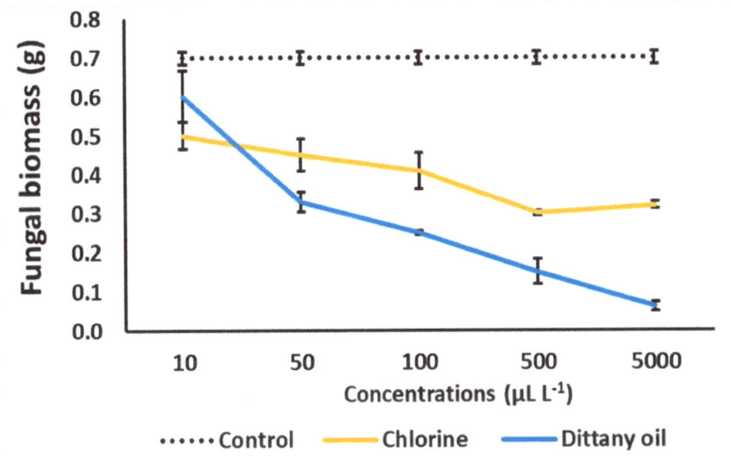

Figure 1. Impact of chlorine or dittany oil on fungal biomass (g) of *Botrytis cinerea* grown in Potato Dextrose Broth (PDB). Inoculated PDB with *B. cinerea* was exposed to vapours and incubated at 11 °C and 95% RH. Values represent means of measurements made on six independent replicates per treatment.

3.2. Effects of Vapours on Grey Mould Development In Vitro

DIT and CHL vapours decreased the *B. cinerea* colony growth (vegetative stage) in vitro (i.e., fungi inoculated and cultured on PDA prior exposure to volatiles, named "curative effect") following 4 and 6 d of incubation compared to the control treatment (Figure 2). However, when PDA media were pre-exposed to volatiles (named "residual effect") and then PDA was inoculated with the fungi, only DIT oil vapor decreased fungal colony growth, while CHL vapours did not affect the colony growth, which was at similar levels as the colony growth in the control treatment (Figure 2).

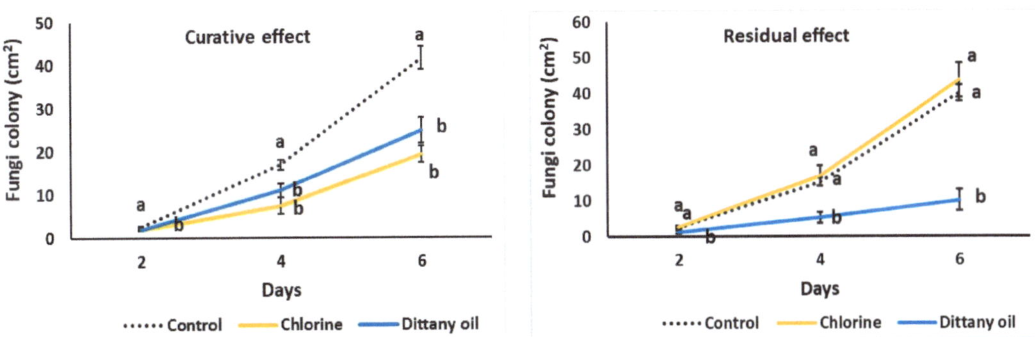

Figure 2. Impact of curative or residual vapour enrichment of chlorine or dittany oil on *Botrytis cinerea* colony growth (cm^2) on PDA media. Values represent means (±SE) of measurements made on nine independent plates per treatment. Values followed by the same letter on each day do not differ significantly at $p = 0.05$.

The effects of CHL or DIT vapours on the reproductive phases of the fungi growth, including spore production and spore germination/viability, are illustrated in Table 3. The number of spores produced decreased (up to 50.8%) in DIT vapour enrichment compared to the control and/or CHL treatments (Table 3). In pre-exposed PDA media to vapours, spore production was suppressed by up to 55.2% and 81.9% for the CHL and DIT applications, respectively, compared with the control. Residual effects were also evidenced for the spore's viability, as spore germination decreased (up to 43.9%) for vaporized media with CHL or DIT oil. However, spores treated with vapours (vapour exposed PDA medium) did not affect their viability (reproductive phase).

Table 3. Impact of curative and residual vapour enrichment of chlorine or dittany oil on *Botrytis cinerea* spore germination and spore production in vitro on PDA medium. Treatments maintained throughout at 11 °C and 95% RH. In each row, mean (±SE) values (n = 9) of plates for the individual vapour enrichment, followed by the same letter, do not differ significantly at p = 0.05.

In Vitro	Control	Chlorine	Dittany Oil
	Curative Vapour Enrichment *		
Spore production × 10^5 (mL^{-1})	95.29 ± 6.60 a	111.64 ± 10.89 a	46.87 ± 14.35 b
Spore germination (%)	68.58 ± 5.03 a	60.88 ± 6.21 a	73.11 ± 4.26 a
	Vapour-Induced Residual Effect **		
Spore production × 10^5 (mL^{-1})	141.36 ± 16.42 a	63.26 ± 6.67 b	25.63 ± 7.93 c
Spore germination (%)	88.51 ± 2.96 a	47.99 ± 4.63 b	61.54 ± 7.69 b

* Curative vapour enrichment: plates initially inoculated with *B. cinerea* and then subjected to vapours. ** Vapour-induced residual effect: plates pre-exposed to vapours, inoculated with fungi and then subjected to ambient air (control).

3.3. Effects of Vapours on Grey Mould Development on Pepper Fruit

The fungal lesion development of inoculated pepper fruits with *B. cinerea* decreased after 6 d of storage when subjected to DIT oil vapours; however, this effect did not persist after 12 d of storage at 11 °C and 95% RH (Figure 3). Wound-inoculated fruit subjected to vapours (CHL or DIT) for 6 d and then additionally stored for 6 d in ambient air (in total 12 d of storage) failed to decrease the fungal lesion development, indicating no residual effect after 12 d of storage. However, the fungal lesion area was supressed (up to 44.7%) in pre-exposed fruits to DIT vapours for 6 d followed by *B. cinerea* inoculation and storage for 6 d in ambient air, indicating a residual lasting effect for the 6 d of storage in ambient air (Figure 3). The fruit exhibited no apparent signs of injury or any other abnormalities.

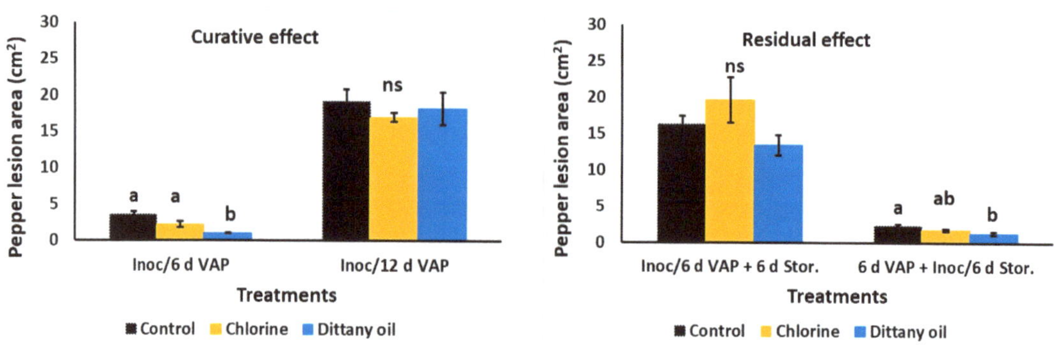

Figure 3. Impact of curative or residual vapour (VAP) enrichment of chlorine or dittany oil on *Botrytis cinerea* lesion growth (cm^2) on pepper fruit. Values represent means (±SE) of measurements made on nine pepper fruits per treatment. Values followed by the same letter in each treatment/application do not differ significantly at p = 0.05. The ns indicates no significance.

The impact of vapours on the grey mould development on pepper fruit is shown in Table 4. Notably, the production of spores significantly ($p < 0.05$) decreased during the exposure of the inoculated fruits to CHL and DIT when compared to the control (peppers exposed to ambient air throughout). However, spore germination decreased in the case of DIT-treated pepper fruits compared to control or CHL-treated fruits during the vapour applications. In peppers pre-exposed to vapours (residual effect), both spores' production and germination were unaffected and averaged as 1.87×10^5 spores mL^{-1} and 98.14%, respectively (Table 4).

Table 4. Impact of curative and residual vapour enrichment of chlorine or dittany oil on *Botrytis cinerea* spore germination and spore production on pepper fruit, stored at 11 °C and 95% RH. In each row, mean (±SE) values (n = 9) of fruits for the individual vapour enrichment followed by the same letter do not differ significantly at $p = 0.05$.

	Curative Vapour Enrichment *		
In Vivo	**Control**	**Chlorine**	**Dittany Oil**
Spore production $\times 10^5$ (mL^{-1})	34.97 ± 6.25 a	18.37 ± 3.09 b	14.11 ± 2.05 b
Spore germination (%)	99.60 ± 0.24 a	97.20 ± 1.01 a	91.25 ± 2.39 b
	Vapour-Induced Residual Effect **		
Spore production $\times 10^5$ (mL^{-1})	2.85 ± 0.82 a	1.52 ± 0.41 a	1.04 ± 0.10 a
Spore germination (%)	98.20 ± 1.56 a	99.20 ± 0.20 a	96.76 ± 1.65 a

* Curative vapour enrichment: fruits inoculated with *B. cinerea* and then subjected to vapours. ** Vapour-induced residual effect: fruits pre-exposed to vapours, then inoculated with fungi, transferred and stored in ambient air.

Indeed, the antimicrobial activity of the tested vapours was evidenced, as when pure CHL and DIT oil was assessed, fungal lesion growth decreased up to 69.4% and 53.0% for CHL and DIT applications, respectively (Table 5). The antimicrobial properties of the CHL and DIT oil were over expressed at the spore viability, as spore germination decreased up to 97.3% and 99.7% for CHL and DIT applications, respectively.

Table 5. Impact of curative vapour enrichment of pure (100%) chlorine or dittany oil on *Botrytis cinerea* lesion growth and spore germination on pepper fruit, stored at 11 °C and 95% RH. In each row, mean (±SE) values (n = 9) of fruits for the individual vapour enrichment followed by the same letter do not differ significantly at $p = 0.05$.

	Curative Vapour Enrichment *		
In Vivo	**Control**	**Chlorine**	**Dittany Oil**
Lesion growth (cm^2)	9.48 ± 0.96 a	3.33 ± 0.68 b	4.46 ± 0.56 b
Spore germination (%)	96.54 ± 1.06 a	2.63 ± 0.67 b	0.25 ± 0.25 b

* Curative vapour enrichment: fruits inoculated with *B. cinerea* and then subjected to pure (100%) vapours.

3.4. Sanitary Dips on Grey Mould Development In Vivo

Apart from the vapour application in the present study, sanitary dips of fruit in CHL or DIT solution were also assessed. Sanitary dips on fruits previously inoculated with the fungi did not affect the lesion growth compared to the control treatment, while DIT application slightly accelerated the fungal growth (Figure 4). Similarly, sanitary dips did not reveal any residual effects, as pre-treated fruits with vapours that were inoculated had similar fungi development. Pepper fruits subjected to dipping retained a negligible odour of the sanitary solution (CHL or DIT), but fruits retained their colour as no discolouration was evidenced.

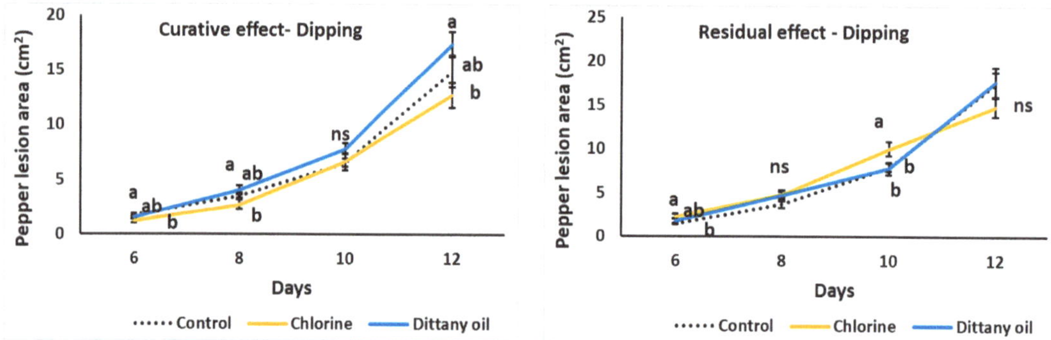

Figure 4. Impact of curative or residual dipping-enrichment of chlorine or dittany oil on *Botrytis cinerea* lesion growth (cm^2) on pepper fruits stored at 11 °C and 95% RH. Values represent means (±SE) of measurements made on nine independent fruits per treatment. Values followed by the same letter on each day do not differ significantly at $p = 0.05$. The ns indicates no significance.

4. Discussion

Fungicide treatments for postharvest decay prevention have been limited in the EU due to severe regulations regarding concerns about food safety [15,44]. Unlike their usage as food flavourings, EOs' potential use as natural antibacterial, antioxidant, and anticancer agents has received less research and approval, which has limited their use in the industry. This study underscores the potential use of EOs for controlling postharvest diseases caused by the *B. cinerea* in pepper fruit. Natural volatiles demonstrated efficacy in mitigating grey mould in pepper fruit during both the vegetative (mycelial development) and reproductive (spore germination and production) phases of *B. cinerea* in vitro. This interference with the fungal disease cycle is crucial for effective disease control. Interestingly, the antifungal efficacy of the DIT oil was evidenced in pre-exposed PDA media to the EOs, indicating a residual effect, which was evidenced for both the vegetative (Figure 2) and reproductive (Table 3) stages of the tested fungi. In previous studies, cinnamon (*Cinnamomum zeylanicum* L.) oil at $\leq$500 μL L^{-1} inhibited several *Aspergilus* species (*A. flavus*, *A. parasiticus*, *A. ochraceus*) and *Fusarium moniliforme* on PDA [45], being in line with the current observations. An *in vitro* test of *Origanum vulgare* (L.), *Thymus capitatus* (L.), *Cymbopogon citratus* (L.), and *Coriandrum sativum* (L.) vapours (0.5–1.0 mL L^{-1}) inhibited *B. cinerea* and *Alternaria arborescens* [46]. The high *C. citratus* EO levels (2800 μL L^{-1}) completely inhibited *Aspergillus flavous*, while the production of mycotoxins (aflatoxin) was inhibited at 100 μL L^{-1} of *C. citratus* treatments [47].

The antimicrobial properties of the DIT oil mirrored the pepper fruit as well, but the efficacy seemed to last for the first 6 d of application, as longer duration in both curative and residual applications was not promising (Figure 3). In contrast, the antimicrobial properties of DIT oil when applied in eggplant persisted up to 14 d, and the higher the concentration, the greater the suppression against grey mould [31]; this is probably related to the different commodity used. DIT oil has carvacrol as the main component based on the EO analysis (Table 2) and previous reports [43]. Carvacrol has proven antimicrobial and antioxidant properties [48,49], reflecting the DIT oils' biocidal activities. When an EO mixture that contained carvacrol–oregano–cinnamon at a ratio of 7:1:2 was used, the controlled release of EO vapours from the active box prolonged the storage period of tomatoes and decreased decay [50]. The combination of EOs and/or other postharvest means (i.e., chitosan) improved the efficacy of the preservation of fresh produce [51]. When chitosan and thyme oil were combined in a 3:1 ratio, they controlled *Colletotrichum gloeosporioides* in avocado fruit during storage [29] so they can be considered a suitable alternative to the currently adopted prochloraz applications. The application of the EOs as nanoemulsions could be a possible strategy for improving the efficiency of natural fungicides [52]. Several studies on tomato, strawberry, and cucumber fruit confirmed the antifungal properties of EOs derived from oregano, thyme, clove, and lemon plants, towards various important postharvest pathogens (*B. cinerea*, *Penicillium italicum*, and *Penicillium digitatum*) [41,53], being in agreement with the results of the current work. Notably, the advantages associated with volatile enrichment persisted in fruit exposed to DIT oil vapours, leading to the suppression of lesion growth and direct impacts on the fungal spore germination and spore production. The viability of spores in pepper fruit, inoculated through wounds, was comparatively less affected than spore production.

Interestingly, a 6 d DIT oil application before inoculation revealed a residual effect and decreased the grey mould development after the fungi inoculation (Figure 3). These observations suggest that vapour applications may exert a lasting effect, possibly through fruit priming in anticipation of subsequent challenges [54] and potentially fostering defence-related metabolism in the fruit [55]. This provided evidence that pre-treating fruit with DIT oil during storage in chilled conditions might prevent the establishment and growth of the fungi. Molecular-based studies would be useful for comprehending the secondary metabolisms and potential mechanisms involved. Indeed, evidence shows that the EOs' antimicrobial and antioxidant activities are related to the synthesis of phenolic components and the role of phenylalanine ammonia-lyase (PAL) as a primary enzyme that is involved in the phenolics biosynthesis [29]. Similarly, pathogen-related proteins (chitinase and b-1,3-glucanase) have a vital role in plant defence-related mechanisms against fungi by facilitating the biochemical mechanisms involved in fungi cell polymer hydrolysation [29,41]. Thyme oil decreased pathogen development and spore viability by impacting the active sites of enzymes and cellular metabolism [56]. However, DIT oil seems to pose fungistatic effects on the reproductive (spore germination and production) phase of the fungi, rather than fungicidal effects (Table 4). Nevertheless, the spore morphology, the moisture content, and the substrate are recognized as crucial factors influencing fruit susceptibility. The reduced spore germination and production emphasized in this study would impede the fungus' spread, diminishing its ability to generate spores. Consequently, natural products become highly significant for postharvest sanitation means. In previous studies, EOs' antimicrobial efficacy has been reported to prolong the storage life of various vegetables and fruits [29,52,57,58].

Essential oils offer a distinct advantage in their bioactivity when applied in the vapour phase, a characteristic that renders them valuable as potential sanitizer for safeguarding stored commodities. In many commodities, sanitary dipping is an effective and straightforward way to preserve fresh commodities, while in some commodities, i.e., grapes and strawberries, the dipping technique is less preferable at it may fail to protect the fresh commodities, accelerating pathogen development. In the present study, sanitary dipping was less successful in fruit sanitation compared to vapour application (Figure 4), while fruits dipped in DIT or CHL solution retained a faint odour in the sanitary solution. It has been reported previously that sage EO dipping application was ineffective for the preservation of pepper fruit [59], while sanitary dipping of kiwi (*Actinidia chinensis*) fruit against *B. cinerea* at higher rates (i.e., 500 $\mu L\ L^{-1}$ aqueous solution of *Origanum* spp. oil) than the ones used in the present study (i.e., 350 $\mu L\ L^{-1}$) revealed a fungal decrease in the fruit flesh but severely reduced fruit quality [60]. Moreover, *Satureja montana* and *Origanum vulgare* EO emulsion was effective to control *Escherichia coli* O157:H7 in the postharvest washing of lettuce [3]; however, the effects were varied according to the different lettuce types, romaine and butterhead *versus* crisp head lettuce, with greater antimicrobial activity for the latter. High EO levels (i.e., 500 $\mu L\ L^{-1}$) may accelerate fruit metabolism, increase fruit decay, and decrease aroma/marketability, as evidenced on vapourised tomatoes at the breaker ripening stage when exposed to sage (*Salvia trilova* L.) EO, while lower EO concentration of 50 $\mu L\ L^{-1}$ maintained fruit texture, respiration, and ethylene production [55]. In that study, it was also observed that even 2 d of sage EO application can affect the fruit metabolism and storability, while residual effects were evidenced in pre-exposed tomatoes to sage EO for 7 d, following an additional 7 days of storage at 11 °C and 90% RH [55].

Essential oil addition in foodstuffs might have limitations in terms of not only aroma and taste but also colour and texture, and the general visual perception may be altered [23]. In that sense, EO application should balance the impacts on fresh produce quality, safety, and acceptability with the antimicrobial efficacy of the EO.

The application of DIT EOs ≥ 50 μL L^{-1} led to a decrease in fungal biomass, which almost zeroed at the highest tested concentration of 5000 μL L^{-1}. The DIT presence affected the fungal development (fungal biomass (Figure 1)) and mycelial growth (Figure 2) per se but also possibly altered the culture medium composition/properties. This was confirmed by exposing the PDA medium to vapours prior the fungi inoculation, where the fungi colony decreased in pre-exposed PDA to DIT vapours, indicating residual effects. Further investigation is needed to examine the different DIT oil concentrations and the time of duration/exposure to vapours. There is no doubt that DIT oil poses strong antimicrobial activity, as pure DIT oil decreased, by almost half, the fungal vegetative growth and, most importantly, nearly prevented the fungi spread, as the spore viability was almost zeroed (Table 5). The mechanism by which volatiles affect fungal spores (germination and production) is not well understood. The EO fungitoxic properties are likely associated with the combined action of various components that the EOs are composed of [25]. This complex matrix of EO components may hinder pathogens from developing resistance, providing a valuable advantage for postharvest storage. Further research is needed to understand any synergistic action of plant molecules, both in vitro and in vivo, and application should be further explored under semi-commercial and full-commercial (i.e., packinghouses) scales and not only under laboratory premises.

Reduced spore production and spore germination, as demonstrated in this study, could effectively inhibit fungal proliferation. This positions EOs as useful postharvest sanitation tools, offering a substitute to commercial chemical-based chlorine [61]. Essential oils, approved as food additives, present a more straightforward registration process for postharvest use compared to novel synthetic pesticides. Thymol, thyme EO, and thyme (as a spice) are involved in the lists of the US Food and Drug Administration as both food and food additives, with thymol first registered as a pesticide in the US in 1964 [62]. When applying EOs, achieving an optimized concentration, duration, and exposure method is crucial to prevent product deterioration. The thickness of fresh commodity is linked to tainting, with thicker fruit skins being less susceptible to such issues [63]. Since the effectiveness of a single natural sanitizer might be low and may not be as broadly effective as synthetic pesticides, during postharvest applications, their efficacy can be enhanced by combining them with specially designed packaging. This approach avoids leaving detectable residues on the surface, addressing concerns about the available sanitation means, which are primarily chlorine-based. Additionally, applying EOs via the vapour phase is likely to be more cost-effective than dipping.

While utilizing EOs, it is important to keep in mind that their effectiveness towards fungal infections and/or other pathogens attacks relies on the EO composition, concentration, methods of applications, timing, pathogen type, and the features of the commodity [64]. Generally, the concentrations of EOs and their components are required to hinder microbial growth to be greater in food compared to culture media. Any commercial application should take this phenomenon into account, as it is linked to interactions among phenolic ingredients and the food matrix [65]. However, the effects of EO application on fresh produce quality can be contradictory. For instance, the use of savoury and thyme EOs in vapour form on peaches and nectarines benefited the fruit's qualitative features (i.e., less weight loss and significant losses of carotenoid and ascorbic acid content) [66]. However, the high *Salvia trilova* EO concentrations may have a negative impact on quality factors, including the sensory qualities (i.e., odour, taste, and colour) of the treated fresh produce [55]. Upon utilization, EOs might have direct (lysis of membrane permeability) or indirect (induce resistance) activity. Essential oils, due to their lipophilicity, can easily disrupt the permeability of the membrane and/or enter it [67,68].

5. Conclusions

EOs, as a natural product, are an eco-friendly substitute for chemical preservatives in the food sector, attracting attention as a natural green preservative. The present study demonstrates the potential use of DIT oil for postharvest disease management, both in the vegetative and reproductive phases of *Botrytis cinerea*. DIT oil decreased fungal lesion growth and prevented the spore germination and spore production to some extent, even better in most cases than the commonly used sanitizer, chlorine, for the postharvest storage of pepper fruit. However, these effects were observed for the 6 d vapour application and not for 12 d, indicating the lasting effect of the vapours on fresh produce preservation. DIT oil application reveals residual effects that could possibly protect the fruit from opportunistic pathogens during the storage of fresh commodities. The EO of DIT may be considered as curative and residual alternatives to typical fresh produce preservation procedures. In comparison to sanitary dips, vapour application was found to be more successful in pepper fruit preservation. Before commercial use, each commodity must be evaluated individually, and the volatile concentration and sanitizing procedure must be optimized.

Supplementary Materials: The following supporting information can be downloaded at: https://www.mdpi.com/article/10.3390/agronomy14020257/s1, Figure S1. Effect of chlorine or dittany essential oil on fungal (*Botrytis cinerea*) sporulation grown in Potato Dextrose Broth (PDB).

Funding: This research received no external funding.

Data Availability Statement: The authors declare that data are available only upon request.

Conflicts of Interest: The author declares no conflicts of interest.

References

1. Cao, Z.; Zhou, D.; Ge, X.; Luo, Y.; Su, J. The role of essential oils in maintaining the postharvest quality and preservation of peach and other fruits. *J. Food Biochem.* **2022**, *46*, e14513. [CrossRef]
2. Domingos, M.M.; Valiati, B.S.; Uliana, D.S.; Leal, R.M.L.; Guimarães, A.P.; Lepaus, B.M.; Geringer, S.J.; de São José, J.F.B. Essential oils against fruit spoilage fungi. In *Plant Essential Oils*; Prakash, B., Dubey, N.K., de São José, J.F.B., Eds.; Springer: Singapore, 2023; pp. 105–124, ISBN 978-981-99-4369-2.
3. Pizzo, J.S.; Pelvine, R.A.; da Silva, A.L.B.R.; Mikcha, J.M.G.; Visentainer, J.V.; Rodrigues, C. Use of essential oil emulsions to control *Escherichia coli* O157:H7 in the postharvest washing of lettuce. *Foods* **2023**, *12*, 2571. [CrossRef]
4. López-Gómez, A.; Ros-Chumillas, M.; Antolinos, V.; Buendía-Moreno, L.; Navarro-Segura, L.; Sánchez-Martínez, M.J.; Martínez-Hernández, G.B.; Soto-Jover, S. Fresh culinary herbs decontamination with essential oil vapours applied under vacuum conditions. *Postharvest Biol. Technol.* **2019**, *156*, 110942. [CrossRef]
5. Yeni, F.; Yavaş, S.; Alpas, H.; Soyer, Y. Most common foodborne pathogens and mycotoxins on fresh produce: A review of recent outbreaks. *Crit. Rev. Food Sci. Nutr.* **2016**, *56*, 1532–1544. [CrossRef]
6. Mitropoulou, G.; Fitsiou, E.; Stavropoulou, E.; Papavassilopoulou, E.; Vamvakias, M.; Pappa, A.; Oreopoulou, A.; Kourkoutas, Y. Composition, antimicrobial, antioxidant, and antiproliferative activity of *Origanum dictamnus* (dittany) essential oil. *Microb. Ecol. Health Dis.* **2015**, *26*, 26543. [CrossRef] [PubMed]
7. Coroneo, V.; Carraro, V.; Marras, B.; Marrucci, A.; Succa, S.; Meloni, B.; Pinna, A.; Angioni, A.; Sanna, A.; Schintu, M. Presence of trihalomethanes in ready-to-eat vegetables disinfected with chlorine. *Food Addit. Contam. Part A Chem. Anal. Control. Expo. Risk Assess* **2017**, *34*, 2111–2117. [CrossRef] [PubMed]
8. Martínez-Romero, D.; Castillo, S.; Guillén, F.; Díaz-mula, H.M.; Zapata, P.J.; Valero, D.; Serrano, M. *Aloe vera* gel coating maintains quality and safety of ready-to-eat pomegranate arils. *Postharvest Biol. Technol.* **2013**, *86*, 107–112. [CrossRef]
9. Tzortzakis, N.; Xylia, P.; Chrysargyris, A. Sage essential oil improves the effectiveness of *Aloe vera* gel on postharvest quality of tomato fruit. *Agronomy* **2019**, *9*, 635. [CrossRef]
10. Khursheed, A.; Rather, M.A.; Jain, V.; Wani, A.R.; Rasool, S.; Nazir, R.; Malik, N.A.; Majid, S.A. Plant based natural products as potential ecofriendly and safer biopesticides: A comprehensive overview of their advantages over conventional pesticides, limitations and regulatory aspects. *Microb. Pathog.* **2022**, *173*, 105854. [CrossRef]
11. Ibáñez, M.D.; Blázquez, M.A. Phytotoxic effects of commercial essential oils on selected vegetable crops: Cucumber and tomato. *Sustain. Chem. Pharm.* **2020**, *15*, 100209. [CrossRef]
12. Inselberg, H.; do Nascimento Nunes, M.C. Using Cannabidiol as a potential postharvest treatment to maintain quality and extend the shelf life of strawberries. *Postharvest Biol. Technol.* **2021**, *173*, 111416. [CrossRef]
13. Jakubczyk, K.; Tuchowska, A.; Janda-Milczarek, K. Plant hydrolates—Antioxidant properties, chemical composition and potential applications. *Biomed. Pharmacother.* **2021**, *142*, 112033. [CrossRef]

14. Xylia, P.; Clark, A.; Chrysargyris, A.; Romanazzi, G.; Tzortzakis, N. Quality and safety attributes on shredded carrots by using *Origanum majorana* and ascorbic acid. *Postharvest Biol. Technol.* **2019**, *155*, 120–129. [CrossRef]
15. Cindi, M.D.; Soundy, P.; Romanazzi, G.; Sivakumar, D. Different defense responses and brown rot control in two *Prunus persica* cultivars to essential oil vapours after storage. *Postharvest Biol. Technol.* **2016**, *119*, 9–17. [CrossRef]
16. Alam, A.; Tripathi, A.; Sharma, V.; Sharma, N. Essential oils: A novel consumer and eco-friendly approach to combat postharvest phytopathogens. *J. Adv. Biol. Biotechnol.* **2017**, *11*, 1–16. [CrossRef]
17. Vergis, J.; Gokulakrishnan, P.; Agarwal, R.K.; Kumar, A. Essential oils as natural food antimicrobial agents: A review. *Crit. Rev. Food Sci. Nutr.* **2015**, *55*, 1320–1323. [CrossRef] [PubMed]
18. Chrysargyris, A.; Charalambous, S.; Xylia, P.; Litskas, V.; Stavrinides, M.; Tzortzakis, N. Assessing the biostimulant effects of a novel plant-based formulation on tomato crop. *Sustainability* **2020**, *12*, 8432. [CrossRef]
19. Isman, M.B.; Wilson, J.A.; Bradbury, R. Insecticidal activities of commercial rosemary oils (*Rosmarinus officinalis*) against larvae of *Pseudaletia unipuncta* and *Trichoplusia ni* in relation to their chemical compositions. *Pharm. Biol.* **2008**, *46*, 82–87. [CrossRef]
20. Di Sotto, A.; Gullì, M.; Acquaviva, A.; Tacchini, M.; Di Simone, S.C.; Chiavaroli, A.; Recinella, L.; Leone, S.; Brunetti, L.; Orlando, G.; et al. Phytochemical and pharmacological profiles of the essential oil from the inflorescences of the *Cannabis sativa* L. *Ind. Crop. Prod.* **2022**, *183*, 114980. [CrossRef]
21. Ebadollahi, A.; Ziaee, M.; Palla, F. Essential oils extracted from different species of the Lamiaceae plant family as prospective bioagents against several detrimental pests. *Molecules* **2020**, *25*, 1556. [CrossRef]
22. Xylia, P.; Chrysargyris, A.; Miltiadous, P.; Tzortzakis, N. *Origanum dubium* (cypriot oregano) as a promising sanitizing agent against *Salmonella enterica* and *Listeria monocytogenes* on tomato and cucumber fruits. *Biology* **2022**, *11*, 1772. [CrossRef] [PubMed]
23. Osimani, A.; Garofalo, C.; Harasym, J.; Aquilanti, L. Use of essential oils against foodborne spoilage yeasts: Advantages and drawbacks. *Curr. Opin. Food Sci.* **2022**, *45*, 100821. [CrossRef]
24. Serrano, M.; Martínez-Romero, D.; Castillo, S.; Guillén, F.; Valero, D. The use of natural antifungal compounds improves the beneficial effect of MAP in sweet cherry storage. *Innov. Food Sci. Emerg. Technol.* **2005**, *6*, 115–123. [CrossRef]
25. Xylia, P.; Chrysargyris, A.; Ahmed, Z.F.R.; Tzortzakis, N. Application of rosemary and eucalyptus essential oils and their main component on the preservation of apple and pear fruits. *Horticulture* **2021**, *7*, 479. [CrossRef]
26. Hassani, A.; Fathi, Z.; Ghosta, Y.; Abdollahi, A.; Meshkatalsadat, M.H.; Marandi, R.J. Evaluation of plant essential oils for control of postharvest brown and gray mold rots on apricot. *J. Food Saf.* **2012**, *32*, 94–101. [CrossRef]
27. Mohammadi, L.; Ramezanian, A.; Tanaka, F.; Tanaka, F. Impact of *Aloe vera* gel coating enriched with basil (*Ocimum basilicum* L.) essential oil on postharvest quality of strawberry fruit. *J. Food Meas. Charact.* **2021**, *15*, 353–362. [CrossRef]
28. Rabiei, V.; Shirzadeh, E.; Rabbiangourani, H.; Sharafi, Y. Effect of thyme and lavender essential oils on the qualitative and quantitative traits and storage life of apple "Jonagold" cultivar. *J. Med. Plant Res.* **2011**, *5*, 5522–5527.
29. Bill, M.; Sivakumar, D.; Korsten, L.; Thompson, A.K. The efficacy of combined application of edible coatings and thyme oil in inducing resistance components in avocado (*Persea americana* Mill.) against anthracnose during post-harvest storage. *Crop. Prot.* **2014**, *64*, 159–167. [CrossRef]
30. Sakkas, H.; Papadopoulou, C. Antimicrobial activity of basil, oregano, and thyme essential oils. *J. Microbiol. Biotechnol.* **2017**, *27*, 429–438. [CrossRef]
31. Stavropoulou, A.; Loulakakis, K.; Magan, N.; Tzortzakis, N. *Origanum dictamnus* oil vapour suppresses the development of grey mould in eggplant fruit in vitro. *Biomed Res. Int.* **2014**, *2014*, 11. [CrossRef]
32. Skotti, E.; Anastasaki, E.; Kanellou, G.; Polissiou, M.; Tarantilis, P.A. Total phenolic content, antioxidant activity and toxicity of aqueous extracts from selected Greek medicinal and aromatic plants. *Ind. Crop. Prod.* **2014**, *53*, 46–54. [CrossRef]
33. Chorianopoulos, N.; Kalpoutzakis, E.; Aligiannis, N.; Mitaku, S.; Nychas, G.J.; Haroutounian, S.A. Essential oils of *Satureja*, *Origanum*, and *Thymus* species: Chemical composition and antibacterial activities against foodborne pathogens. *J. Agric. Food Chem.* **2004**, *52*, 8261–8267. [CrossRef] [PubMed]
34. Olatunji, T.L.; Afolayan, A.J. The suitability of chili pepper (*Capsicum annuum* L.) for alleviating human micronutrient dietary deficiencies: A review. *Food Sci. Nutr.* **2018**, *6*, 2239–2251. [CrossRef] [PubMed]
35. Taheri, A.; Behnamian, M.; Dezhsetan, S.; Karimirad, R. Shelf life extension of bell pepper by application of chitosan nanoparticles containing *Heracleum persicum* fruit essential oil. *Postharvest Biol. Technol.* **2020**, *170*, 111313. [CrossRef]
36. González-Saucedo, A.; Barrera-Necha, L.L.; Ventura-Aguilar, R.I.; Correa-Pacheco, Z.N.; Bautista-Baños, S.; Hernández-López, M. Extension of the postharvest quality of bell pepper by applying nanostructured coatings of chitosan with *Byrsonima crassifolia* extract (L.) Kunth. *Postharvest Biol. Technol.* **2019**, *149*, 74–82. [CrossRef]
37. Romanazzi, G.; Feliziani, E.; Santini, M.; Landi, L. Effectiveness of postharvest treatment with chitosan and other resistance inducers in the control of storage decay of strawberry. *Postharvest Biol. Technol.* **2013**, *75*, 24–27. [CrossRef]
38. Znini, M.; Cristofari, G.; Majidi, L.; Mazouz, H.; Tomi, P.; Paolini, J.; Costa, J. Antifungal activity of essential oil from *Asteriscus graveolens* against postharvest phytopathogenic fungi in apples. *Nat. Prod. Commun.* **2011**, *6*, 1763–1768. [CrossRef] [PubMed]
39. Adams, R.P. *Identification of Essential oil Components by Gas Chromatography/Mass Spectrometry*, 5th ed.; Allured Publishing Corporation: Carol Stream, IL, USA, 2007; ISBN 9781932633214.
40. Tzortzakis, N. Impact of cinnamon oil-enrichment on microbial spoilage of fresh produce. *Innov. Food Sci. Emerg. Technol.* **2009**, *10*, 97–102. [CrossRef]

41. Chen, C.; Cai, N.; Chen, J.; Wan, C. Clove essential oil as an alternative approach to control postharvest blue mold caused by *Penicillium italicum* in citrus fruit. *Biomolecules* **2019**, *9*, 197. [CrossRef]
42. Tzortzakis, N.G. Ethanol, vinegar and *Origanum vulgare* oil vapour suppress the development of anthracnose rot in tomato fruit. *Int. J. Food Microbiol.* **2010**, *142*, 14–18. [CrossRef]
43. Liolios, C.C.; Graikou, K.; Skaltsa, E.; Chinou, I. Dittany of Crete: A botanical and ethnopharmacological review. *J. Ethnopharmacol.* **2010**, *131*, 229–241. [CrossRef]
44. Martínez-Romero, D.; Serrano, M.; Bailén, G.; Guillén, F.; Zapata, P.J.; Valverde, J.M.; Castillo, S.; Fuentes, M.; Valero, D. The use of a natural fungicide as an alternative to preharvest synthetic fungicide treatments to control lettuce deterioration during postharvest storage. *Postharvest Biol. Technol.* **2008**, *47*, 54–60. [CrossRef]
45. Soliman, K.M.; Badeaa, R.I. Effect of oil extracted from some medicinal plants on different mycotoxigenic fungi. *Food Chem. Toxicol.* **2002**, *40*, 1669–1675. [CrossRef]
46. Plotto, A.; Roberts, D.D.; Roberts, R.G. Evaluation of plant essential oils as natural postharvest disease control of tomato (*Lycopersicon esculentum*). *Acta Hortic.* **2003**, *628*, 737–745. [CrossRef]
47. Paranagama, P.A.; Abeysekera, K.H.T.; Abeywickrama, K.; Nugaliyadde, L. Fungicidal and anti-aflatoxigenic effects of the essential oil of *Cymbopogon citratus* (DC.) Stapf. (lemongrass) against *Aspergillus flavus* Link. isolated from stored rice. *Lett. Appl. Microbiol.* **2003**, *37*, 86–90. [CrossRef] [PubMed]
48. Rodriguez-Garcia, I.; Silva-Espinoza, B.A.; Ortega-Ramirez, L.A.; Leyva, J.M.; Siddiqui, M.W.; Cruz-Valenzuela, M.R.; Gonzalez-Aguilar, G.A.; Ayala-Zavala, J.F. Oregano essential oil as an antimicrobial and antioxidant additive in food products. *Crit. Rev. Food Sci. Nutr.* **2016**, *56*, 1717–1727. [CrossRef] [PubMed]
49. de Medeiros Barbosa, I.; da Costa Medeiros, J.A.; de Oliveira, K.Á.R.; Gomes-Neto, N.J.; Tavares, J.F.; Magnani, M.; de Souza, E.L. Efficacy of the combined application of oregano and rosemary essential oils for the control of *Escherichia coli, Listeria monocytogenes* and *Salmonella Enteritidis* in leafy vegetables. *Food Control.* **2016**, *59*, 468–477. [CrossRef]
50. Buendía-Moreno, L.; Ros-Chumillas, M.; Navarro-Segura, L.; Sánchez-Martínez, M.J.; Soto-Jover, S.; Antolinos, V.; Martínez-Hernández, G.B.; López-Gómez, A. Effects of an active cardboard box using encapsulated essential oils on the tomato shelf life. *Food Bioprocess Technol.* **2019**, *12*, 1548–1558. [CrossRef]
51. Correa-Pacheco, Z.N.; Bautista-Baños, S.; Valle-Marquina, M.Á.; Hernández-López, M. The Effect of Nanostructured Chitosan and Chitosan-thyme Essential Oil Coatings on *Colletotrichum gloeosporioides* Growth *in vitro* and on cv Hass Avocado and Fruit Quality. *J. Phytopathol.* **2017**, *165*, 297–305. [CrossRef]
52. Naserzadeh, Y.; Mahmoudi, N.; Pakina, E. Antipathogenic effects of emulsion and nanoemulsion of cinnamon essential oil against Rhizopus rot and grey mold on strawberry fruits. *Foods Raw Mater.* **2019**, *7*, 210–216. [CrossRef]
53. Vitoratos, A.; Bilalis, D.; Karkanis, A.; Efthimiadou, A. Antifungal activity of plant essential oils against *Botrytis cinerea, Penicillium italicum* and *Penicillium digitatum*. *Not. Bot. Horti Agrobot. Cluj-Napoca* **2013**, *41*, 86–92. [CrossRef]
54. Conrath, U.; Pieterse, C.M.J.; Mauch-mani, B. Priming in plant–pathogen interactions. *Trends Plant Sci.* **2002**, *7*, 210–216. [CrossRef]
55. Chrysargyris, A.; Rousos, C.; Xylia, P.; Tzortzakis, N. Vapour application of sage essential oil maintain tomato fruit quality in breaker and red ripening stages. *Plants* **2021**, *10*, 2645. [CrossRef]
56. Arrebola, E.; Sivakumar, D.; Bacigalupo, R.; Korsten, L. Combined application of antagonist *Bacillus amyloliquefaciens* and essential oils for the control of peach postharvest diseases. *Crop. Prot.* **2010**, *29*, 369–377. [CrossRef]
57. Tornuk, F.; Cankurt, H.; Ozturk, I.; Sagdic, O.; Bayram, O.; Yetim, H. Efficacy of various plant hydrosols as natural food sanitizers in reducing *Escherichia coli* O157:H7 and *Salmonella* Typhimurium on fresh cut carrots and apples. *Int. J. Food Microbiol.* **2011**, *148*, 30–35. [CrossRef] [PubMed]
58. Lopez-Reyes, J.G.; Spadaro, D.; Prelle, A.; Garibaldi, A.; Gullino, M.L. Efficacy of plant essential oils on postharvest control of rots caused by fungi on different stone fruits in vivo. *J. Food Prot.* **2013**, *76*, 631–639. [CrossRef] [PubMed]
59. Tzortzakis, N.; Chrysargyris, A.; Sivakumar, D.; Loulakakis, K. Vapour or dipping applications of methyl jasmonate, vinegar and sage oil for pepper fruit sanitation towards grey mould. *Postharvest Biol. Technol.* **2016**, *118*, 120–127. [CrossRef]
60. Thanassoulopoulos, C.C.; Yanna, L. On the biological control of *Botrytis cinerea* on kiwifruit cv "Hayward" during storage. *Acta Hortic.* **1997**, *444*, 757–764. [CrossRef]
61. Pablos, C.; Fernández, A.; Thackeray, A.; Marugán, J. Effects of natural antimicrobials on prevention and reduction of bacterial cross-contamination during the washing of ready-to-eat fresh-cut lettuce. *Food Sci. Technol. Int.* **2017**, *23*, 403–414. [CrossRef]
62. Tripathi, P.; Dubey, N.K. Exploitation of natural products as an alternative strategy to control postharvest fungal rotting of fruit and vegetables. *Postharvest Biol. Technol.* **2004**, *32*, 235–245. [CrossRef]
63. Burt, S. Essential oils: Their antibacterial properties and potential applications in foods—A review. *Int. J. Food Microbiol.* **2004**, *94*, 223–253. [CrossRef] [PubMed]
64. Vieira, A.M.F.D.; Steffens, C.A.; Argenta, L.C.; do Amarante, C.V.T.; Oster, A.H.; Casa, R.T.; Amarante, A.G.M.; Espíndola, B.P. Essential oils for the postharvest control of blue mold and quality of "Fuji" apples. *Pesqui. Agropecu. Bras.* **2018**, *53*, 547–556. [CrossRef]
65. Karampoula, F.; Giaouris, E.; Deschamps, J.; Doulgeraki, A.I.; Nychas, G.J.E.; Dubois-Brissonnet, F. Hydrosol of *Thymbra capitata* is a highly efficient biocide against *Salmonella enterica* serovar Typhimurium biofilms. *Appl. Environ. Microbiol.* **2016**, *82*, 5309–5319. [CrossRef] [PubMed]

66. Santoro, K.; Maghenzani, M.; Chiabrando, V.; Bosio, P.; Gullino, M.L.; Spadaro, D.; Giacalone, G. Thyme and savory essential oil vapor treatments control brown rot and improve the storage quality of peaches and nectarines, but could favor gray mold. *Foods* **2018**, *7*, 7. [CrossRef]
67. Perricone, M.; Arace, E.; Corbo, M.R.; Sinigaglia, M.; Bevilacqua, A. Bioactivity of essential oils: A review on their interaction with food components. *Front. Microbiol.* **2015**, *6*, 76. [CrossRef]
68. Snoussi, M.; Noumi, E.; Trabelsi, N.; Flamini, G.; Papetti, A.; De Feo, V. *Mentha spicata* Essential oil: Chemical composition, antioxidant and antibacterial activities against planktonic and biofilm cultures of *vibrio* spp. strains. *Molecules* **2015**, *20*, 14402–14424. [CrossRef]

agronomy

MDPI

Article

Metabolic Characterization of Four Members of the Genus *Stachys* L. (Lamiaceae)

Ekaterina-Michaela Tomou [1,*], Anastasia Karioti [2], Giorgos Tsirogiannidis [1], Nikos Krigas [3] and Helen Skaltsa [1]

1 Department of Pharmacognosy and Chemistry of Natural Products, School of Health Sciences, Faculty of Pharmacy, National and Kapodistrian University of Athens, Panepistimiopolis Zografou, 15771 Athens, Greece; giorgostsirogiannidis@gmail.com (G.T.); skaltsa@pharm.uoa.gr (H.S.)

2 Laboratory of Pharmacognosy, School of Pharmacy, Aristotle University of Thessaloniki, University Campus, 54124 Thessaloniki, Greece; akarioti@pharm.auth.gr

3 Institute of Plant Breeding and Genetic Resources, Agricultural Organization Demeter, P.O. Box 60458, 57001 Thessaloniki, Greece; nikoskrigas@gmail.com

* Correspondence: ktomou@pharm.uoa.gr

Abstract: Several members of *Stachys* L. (among the largest Lamiaceae genera) have been traditionally used as medicinal plants. With 54 *Stachys* taxa (species and subspecies) occurring in mainland and/or insular Greece, the present study aimed to investigate the metabolic profiling of four range-restricted local *Stachys* members: *Stachys candida* and *S. chrysantha* (protected and endangered local Greek endemics), *S. leucoglossa* subsp. *leucoglossa* (local Balkan endemic), and *S. spinulosa* (local Balkan subendemic). In this investigation, the infusions of their above-ground parts were characterized using NMR and HPLC-PDA-MS techniques. Thus, 1D- and 2D-NMR spectra were obtained to compare the chemical fingerprints of these plants. Furthermore, previously isolated compounds from *Stachys* spp. were used to identify specific constituents. NMR screening revealed the presence of: (i) phenylethanoid glycosides, mainly acteoside in *S. candida* and *S. chrysantha* (section *Candida*, *Swainsoniana* phyloclade), and (ii) flavone 7-*O*-allosylglucoside (isoscutellarein 7-*O*-[6‴-*O*-acetyl-β-D-allopyranosyl]-(1→2)-β-D-glucopyranoside) and iridoids (monomelittoside or/and melittoside) in *S. leucoglossa* subsp. *leucoglossa* (section *Olisia*, *Swainsoniana/Olisia* phyloclade, *Swainsoniana* phyloclade) and caffeoylquinic acid (chlorogenic acid) in *S. spinulosa* (section *Campanistrum*, *Stachys* phyloclade). In total, 26 compounds were detected by HPLC-PDA-MS belonging to flavonoids, phenylethanoid glycosides, and phenolic acids. Among them, chlorogenic acid was identified in all samples as one of their main metabolites. The present study complements previous studies with first reports of constituents detected in the studied taxa, reports for the first time on the metabolic characterization of *S. spinulosa*, and discusses the chemotaxonomic significance of such findings.

Keywords: *Stachys spinulosa*; infusions; metabolic profiling; NMR; HPLC-PDA-MS; Greece

1. Introduction

The genus *Stachys* L. (Lamiaceae family) comprises about 365 species, which are mainly distributed in the northern hemisphere [1–3] and are arranged in 19 sections [4] or different phyloclades [2]. In Greece, 54 species and subspecies of *Stachys* are distributed in parts of the mainland and/or insular country [5], and 41% of them are local endemics confined to Greece, while 28% are Balkan endemics or subendemics extending to Turkey and/or Italy.

Several *Stachys* taxa (species and subspecies) are used as infusions and decoctions in folk medicine, and their intended uses mainly concern the treatment of infections, the common cold, gastrointestinal disorders, inflammation, skin diseases, and wounds or as a remedy for asthma and anxiety implications [6]. Many of these folk uses (and related ancient or old recipes) date back to Dioscorides times in ancient Greece and Rome, traditional

Chinese medicine, or Middle Eastern (Iranian and Turkish) folk phytotherapy [6]. Previous phytochemical investigations of *Stachys* spp. have reported several constituents belonging to different chemical categories, including terpenoids, iridoids, phytosterols, polyphenols (e.g., flavonoids, phenylethanoid glycosides, and phenolic acids), polysaccharides, and others [6,7]. Many *Stachys* taxa, as well as their isolated compounds, are known to exhibit diverse pharmacological properties, such as antioxidant, anti-inflammatory, anti-diabetic, anti-microbial, anti-proliferative, and cytotoxic activities, among others [6,7].

In our continuing endeavor to explore and document in phytochemical terms the different *Stachys* taxa growing in Greece [6], we have performed a chemical investigation herein in four members of the genus *Stachys* without bracteoles (or minute bracteoles), namely *S. candida* Bory & Chaub., *S. chrysantha* Boiss. & Heldr., *S. leucoglossa* Griseb. subsp. *leucoglossa*, and *S. spinulosa* Sm. More specifically, we focused on: (i) the endangered *S. candida* and *S. chrysantha* [8], which are local Greek (Peloponnese) endemics protected by the Greek Presidential Decree 67/1981 that belong to the section *Candida* [3,5] or the phyloclade *Swainsoniana* [2], with the first being an unarmed perennial with orbicular to ovate–orbicular leaves, calyx teeth longer than wide, and white corolla with purple spots, with the other being a non–spiny perennial with elliptic–ovate to suborbicular leaves and yellow tomentose corolla; (ii) the range-restricted local Balkan endemic *S. leucoglossa* subsp. *leucoglossa* belonging to the section *Olisia* or the *Swainsoniana/Olisia* phyloclade [2], an unarmed perennial with white or pale pink corolla in remote 2–6-flowered verticillasters with an almost glabrous calyx and very small floral leaves; and (iii) the local Balkan subendemic *S. spinulosa* which belongs to the section *Campanistrum* [3,5] or the *Stachys* phyloclade [2], a hispid annual with $\pm$ scabrid stem on angles, spiny bracts and crowded verticillasters in dense spikes, and coarsely spined cauline leaves, and lower leaves cordate at the base.

S. candida and *S. chrysantha* have been previously studied regarding their phytochemical composition, as well as the anti-inflammatory activity of their methanol extracts and their flavonoids [6]. Specifically, xanthomicrol, chrysoeriol, calycopterin, chrysoeriol-7-*O*-β-D-(3″-E-p-coumaroyl)-glucopyranoside, chrysoeriol-7-*O*-β-D-glucopyranoside, and isoscutellarein-7-*O*-[6‴-*O*-acetyl-β-D-allopyranosyl]-(1→2)-6″-*O*-acetyl-β-D-glucopyra noside were isolated from both plants, while luteolin-7-*O*-β-D-glucopyranoside was found only in *S. chrysantha* methanol extract. Additionally reported in *S. candida* were: four flavonoids (apigenin-7-*O*-β-D-glucopyranoside, isoscutellarein-7-*O*-[6‴-*O*-acetyl-β-D-allopyranosyl-(1→2)]-β-D-glucopyranoside, 4′-methyl-isoscutellarein-7-*O*-[6‴-*O*-acetyl-β-D-allopyranosyl-(1→2)]-β-D-glucopyranoside, and 4′-methyl-hypolaetin-7-*O*-[6‴-*O*-acetyl-β-D-allopyranosyl-(1→2)]-β-D-glucopyranoside), one phenylethanoid glycoside (acteoside), and one phenolic acid (chlorogenic acid). However, *S. leucoglossa* (not determined by subspecies) has been chemically characterized only once regarding its content in flavonoids (namely, isoscutellarein-7-*O*-[6‴-*O*-acetyl-β-D-allopyranosyl-(1→2)]-β-D-glucopyranoside, 4′-methyl-isoscutellarein-7-*O*-[6‴-*O*-acetyl-β-D-allopyranosyl-(1→2)]-β-D-glucopyranoside, and 4′-*O*-methyl-hypolaetin-7-*O*-[6‴-*O*-acetyl-β-D-allopyranosyl-(1→2)]-β-D-glucopyranoside) and iridoids (i.e., melittoside, harpagide, acetyl-harpagide, and ajugol) [9]. Nonetheless, *S. spinulosa* has not been subjected to any phytochemical study to date. Therefore, the present study aimed to investigate the metabolic profiles of the infusions of these four Greek *Stachys* taxa through NMR and HPLC-PDA-MS analyses.

2. Materials and Methods

2.1. Plant Material

The flowering above-ground parts of the studied plant species and subspecies were collected from wild-growing populations in different locations in Greece (Figure S1), and the living material was maintained ex situ at the premises of the Institute of Plant Breeding and Genetic Resources, Agricultural Organization Demeter (Table 1). Voucher specimens were identified by Dr. Nikos Krigas and deposited in the Herbarium of the Balkan Botanic

Garden of Kroussia (BBGK), Institute of Plant Breeding and Genetic Resources, Agricultural Organization Demeter.

Table 1. Investigated *Stachys* taxa (species and subspecies) with abbreviations used, sample origin, date of collections, and living or voucher specimen.

Species	Abbreviation	Sample Origin	Date	Living and/or Voucher Specimen
S. candida	SCA	Mt. Taygetos	29/5/2020	GR-1-BBGK-20,164-A
S. chrysantha	SCH	Mt. Taygetos	28/2/2020	GR-1-BBGK-20,96
S. leucoglossa subsp. *leucoglossa*	SLE	Mt. Karpouzi	13/6/2020	GR-1-BBGK-20,97
S. spinulosa	SSP	Sparta	29/5/2020	GR-1-BBGK-20,164-B

2.2. Preparation of the Infusions

Precisely 4.0 g of dried comminuted aerial parts of each plant sample was added to 200 mL of boiled water for 5 min, separately. Due to the fact that *Stachys* infusions in Greece have been traditionally prepared and consumed as 'mountain tea', referring to different *Sideritis* spp., the preparation of the infusions was based on the monograph of the European Medicines Agency (EMA) concerning different species and subspecies of the genus *Sideritis* [10]. Then, the samples were filtered and concentrated to dryness using a rota evaporator under reduced pressure to yield residues of 2.3 g for *S. candida*, 2.8 g for *S. chrysantha*, 2.0 g for *S. leucoglossa* subsp. *leucoglossa*, and 1.8 g for *S. spinulosa*. Distilled water was used as a solvent for the infusions, avoiding any additive/impurity in the samples.

2.3. NMR Analysis

In the NMR experiments, a part of each sample (5.0 mg) was dissolved in 600 μL of CD_3OD. The 1D- and 2D-NMR spectra of the samples were recorded on a Bruker 400 DRX instrument at 300 K. Chemical shifts are given in ppm (δ) and are referenced to the solvent signal at 3.31/ 49.0 ppm (CD_3OD) for ^1H- and ^{13}C-NMR, respectively. COSY (COrrelation SpectroscopY) and HSQC (Heteronuclear Single Quantum Correlation) experiments were performed using standard Bruker microprograms.

2.4. HPLC-PDA-MS Analysis

HPLC-PDA-MS analysis was performed on a Thermo Finnigan system (Palo Alto, CA, USA), which consisted of an LC Pump Plus, Autosampler, and Surveyor PDA Plus Detector. The HPLC system was interfaced with an ESI MSQ Plus (Thermo Finnigan, San Jose, CA, USA) operating with Xcalibur software (version 2.1). The mass spectrometer operated in negative and positive ionization modes; the scan spectra were from m/z 100 to 1000, the gas temperature was at 350 °C, the nitrogen flow rate was at 10 L/min, and the capillary voltage was 3000 V. The cone voltage was in the range of 60–110 V. The column was an SB-Aq Zorbax (Agilent, Santa Clara, CA, USA) RP-C18 column (150 mm × 3.5 mm) with a particle size of 5 μm maintained at 30 °C. The eluents were H_2O at pH 2.8 by formic acid (0.05% v/v) (A) and acetonitrile (B) and with a flow rate of 0.4 mL/min. The gradient program was as follows: 0–7 min, 90–85%A; 7–12 min, 85–82%A; 12–25 82%A; 25–27 min, 82–75%A; 27–32 min, 75%A; 32–42 min, 75–60%A; 42–49 min, 60%A; 49–53 min, 60–90%A; 53–60 min, 90%. The injected volume of the samples was 5 μL of solution. The UV-vis spectra were recorded between 220 and 600 nm, and the chromatographic profiles were registered at 280, 330, and 350 nm.

3. Results and Discussion

The present study reports on the chemical fingerprints of the infusions of four *Stachys* taxa (*S. candida, S. chrysantha, S. leucoglossa* subsp. *leucoglossa*, and *S. spinulosa*) growing wild in Greece by means of NMR and HPLC-PDA-MS techniques. In general, LC–MS (Liquid Chromatography-Mass Spectrometry) and NMR (Nuclear Magnetic Resonance) are commonly used techniques for metabolic characterization in plants. By employing both techniques, the qualitative and quantitative strategy can be considerably improved, rendering the identification of plant extracts' constituents feasible [11].

3.1. NMR Analysis

A preliminary screening of the *Stachys* spp. infusions was first obtained by ^{1}H-NMR spectra, and the chemical categories of their constituents were identified based on peaks in specific regions. The comparative 1D-NMR fingerprints are presented in Figure 1. In addition, 2D-NMR spectra (COSY and HSQC) were acquired to provide a better overview (Figures S1–S5). In all four ^{1}H-NMR spectra, signals of polysaccharides (region: 5.40–3.10 ppm) were noticed. However, some differences in the chemical fingerprints were observed among the samples.

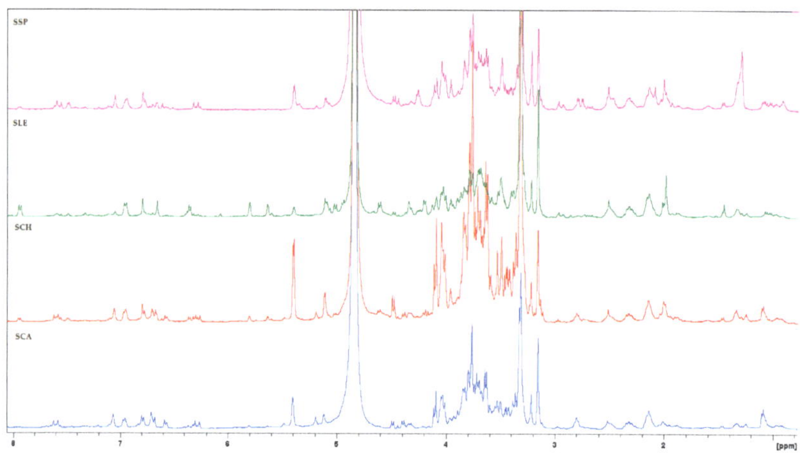

Figure 1. Overlaid ^{1}H-NMR spectra of the four investigated *Stachys* infusions in the range of δ_H 8.00–0.90: *Stachys candida* (SCA, blue color), *Stachys chrysantha* (SCH, red color), *Stachys leucoglossa* subsp. *leucoglossa* (SLE, green color), and *Stachys spinulosa* (SSP, purple color).

In the ^{1}H-NMR spectra of *S. candida* (SCA) and *S. chrysantha* (SCH) belonging to the section *Candida* or the *Swainsoniana* phyloclade, mainly signals from phenylethanoid glycosides were detected (Figure 2A,B). Specifically, at δ_H 7.60 (d, J = 16.0 Hz) and 6.28 (d, J = 16.0 Hz), we found *trans*-coupling olefinic signals ascribable to double bonds (HSQC: δc 146.8 and 113.5, respectively), at the δ_H range of 7.07–6.57, we observed signals which could belong to protons of aromatic tri-substituted rings, and at δ_H around 2.80, we spotted signals of benzylic methylene protons (HSQC: δc 35.0). In the ^{1}H-^{1}H-COSY spectra of SCA and SCH, the principal correlation peaks among protons corresponding to phenylethanoid glycosides were also detected (Figures S2a and S3a). Their HSQC spectra are presented in Figures S2b and S3b. As an effort to interpret the observed different signals in the 1D-/2D-NMR spectra of SCA and SCH infusions, we compared them with the NMR spectra of previously isolated compounds sourced from our own previous works in *Stachys* spp. Through the careful screening of the ^{1}H-NMR spectra of different compounds, we noticed that the main proton signals in the spectra of *S. candida* and *S. chrysantha* infusions could be attributed to the phenylethanoid glycoside, namely acteoside. The overlaid ^{1}H-NMR spectra of both infusions and acteoside are illustrated in Figure 2. Acteoside has been previously found in several *Stachys* taxa [6,7], including *S. candida* [6]. Furthermore, the presence of flavones has been previously reported from members of the genus *Stachys* [6,7]. In *Stachys* species of the section *Candida*, chrysoeriol and isoscutellarein derivatives have been found [6]. It should be mentioned that minor signals of isoscutellarein derivatives were also detected in the SCH sample. However, it was not feasible to identify these constituents with previously isolated compounds in the NMR spectra due to signal overlapping and their low concentration.

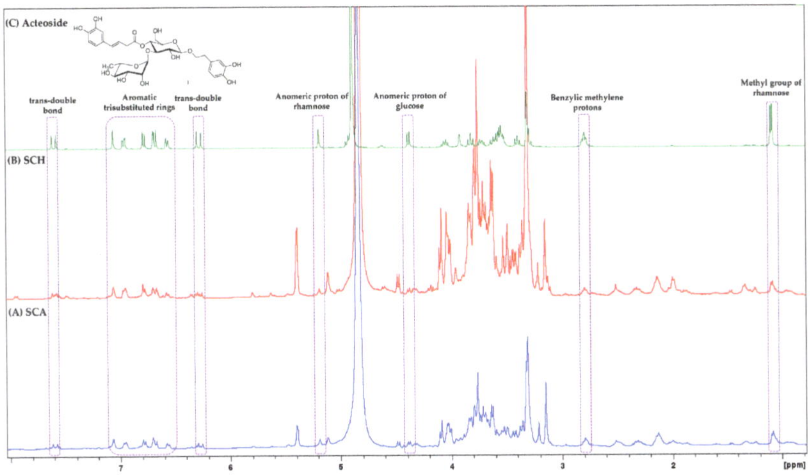

Figure 2. Overlaid ^{1}H-NMR spectra of (**A**) *Stachys candida* (SCA, blue color), (**B**) *Stachys chrysantha* (SCH, red color) infusions, and (**C**) acteoside (green color). Specific signals are indicated by purple boxes.

In the ^{1}H-NMR spectrum of *S. leucoglossa* subsp. *leucoglossa* (SLE) belonging to the section *Olisia* or the *Swainsoniana/Olisia* phyloclade, mainly signals from flavonoids and iridoids were detected (Figures 3 and 4). In the downfield region (δ_H 7.95–6.34), signals of flavones [δ_H: 7.94 (d, J = 8.7 Hz), 6.96 (d, J = 8.7 Hz), 6.79 (s), 6.66 (s)] were observed (Figure 3). In addition, a double peak (J = 6.5 Hz) appeared at δ_H 6.35. This assignment, along with the specific signals at δ_H 5.80 (s), 5.63 (d), and 5.10 (d), could belong to iridoids (Figure 4). In the ^{1}H-^{1}H-COSY spectrum, the principal correlation peaks among vicinal protons corresponding to flavones and iridoids were also detected (Figure S4a). The HSQC spectrum of SLE is presented in Figure S4b. By carefully screening the ^{1}H-NMR spectra of the SLE infusion with those of previously isolated compounds of *Stachys* spp., we noticed that the main proton signals in the spectrum of this infusion could be assigned to isoscutellarein derivatives, namely isoscutellarein-7-*O*-[6‴-*O*-acetyl-β-D-allopyranosyl]-(1→2)-β-D-glucopyranoside, as well as to monomelittoside or/and its 5-glucoside derivative, namely melittoside (Figures 3 and 4). The presence of flavone 7-*O*-allosylglucosides has been previously reported from members of the genus *Stachys* [6,7]. In the *Stachys* species of the section *Olisia*, monoacetyl and diacetyl derivatives of isoscutellarein have been found, including reports of *S. leucoglossa* [6,9]. Furthermore, iridoids are known to be among the main metabolites in *Stachys* spp. [6,7], and previous studies have reported their presence in the members of the section *Olisia*, such as *S. recta* L. and *S. spinosa* L. [12,13], while melittoside has been found in *S. leucoglossa* [9].

In the ^{1}H-NMR spectrum of *S. spinulosa* (SSP) belonging to the section *Campanistrum* or the *Stachys* phyloclade, mainly signals from caffeoylquinic acid derivatives were detected (Figure 5). Specifically, at δ_H 7.58 (d, J = 16.2 Hz) and 6.28 (d, J = 16.2 Hz), we found *trans*-coupling olefinic signals ascribable to double bonds (HSQC: δc 147.0 and 115.7, respectively), and at the δ_H range of 7.05–6.77, the observed signals could belong to protons of aromatic tri-substituted rings. Although the signal overlapped in the middle area of the ^{1}H-NMR spectrum, signals of the oxymethine protons of the quinic acid moiety were spotted at δ_H 5.36 (HSQC: 72.3), 4.10 (HSQC: 73.0), and 3.67 (HSQC: 74.5), with its methylene protons also appearing at the δ_H range of around 2.15–1.95 (HSQC: δc 40.5). In the ^{1}H-^{1}H-COSY spectrum, the principal correlation peaks among protons of the caffeoyl and quinic acid moieties were detected (Figure S5a). The HSQC spectrum of SSP is presented in Figure S5b. By carefully screening the ^{1}H-NMR spectra of the SLE infusion with those of previously isolated compounds of *Stachys* spp., we noticed that the main proton signals in the spectrum of this infusion could be assigned to chlorogenic acid. The overlaid ^{1}H-NMR spectra of the SSP infusion and this compound are illustrated in Figure 5. Caffeoylquinic acids, especially chlorogenic acid, have been found in many *Stachys* species [6]. Moreover, flavones have been previously reported from two *Stachys* species belonging to the section *Campanistrum* [14,15]. Even though minor signals of flavones and other derivatives were also detected in the SSP sample, we were not able to identify specific constituents with previously isolated compounds in the NMR spectra due to signal overlapping and their low concentration. It should be mentioned that this is the first study addressing the phytochemical characterization of *S. spinulosa*.

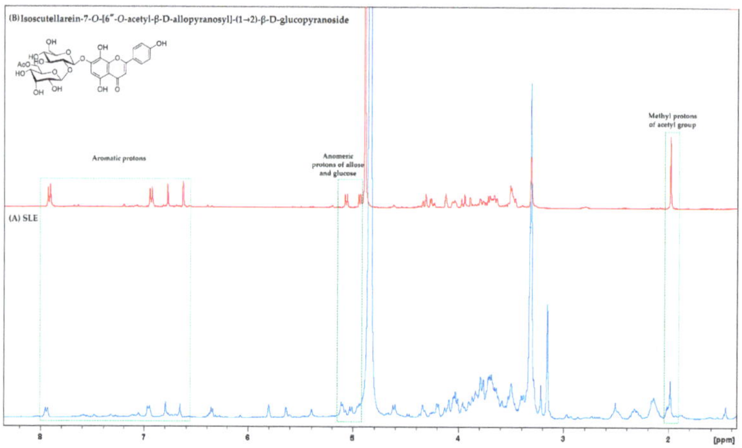

Figure 3. Overlaid ^{1}H-NMR spectra of (**A**) *Stachys leucoglossa* subsp. *leucoglossa* infusion (SLE, blue color) and (**B**) isoscutellarein-7-*O*-[6‴-*O*-acetyl-β-D-allopyranosyl]-(1→2)-β-D-glucopyranoside (red color). Specific signals are indicated by green boxes.

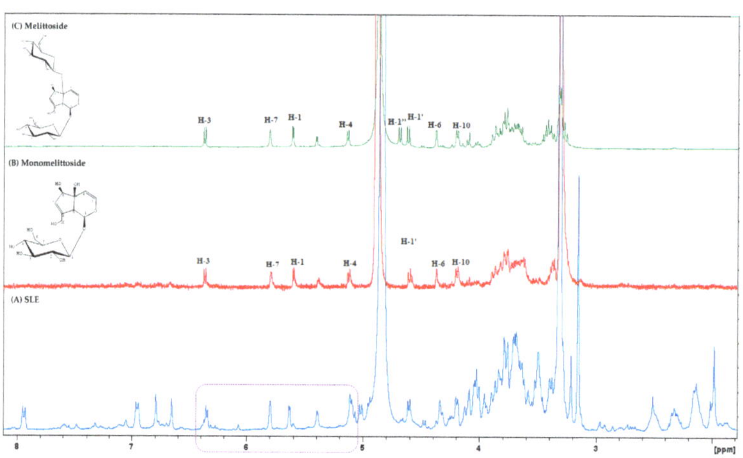

Figure 4. Overlaid ^{1}H-NMR spectra of (**A**) *Stachys leucoglossa* subsp. *leucoglossa* infusion (SLE, blue color), (**B**) monomelittoside (red color), and (**C**) melittoside (green color). Specific signals are indicated by a purple box.

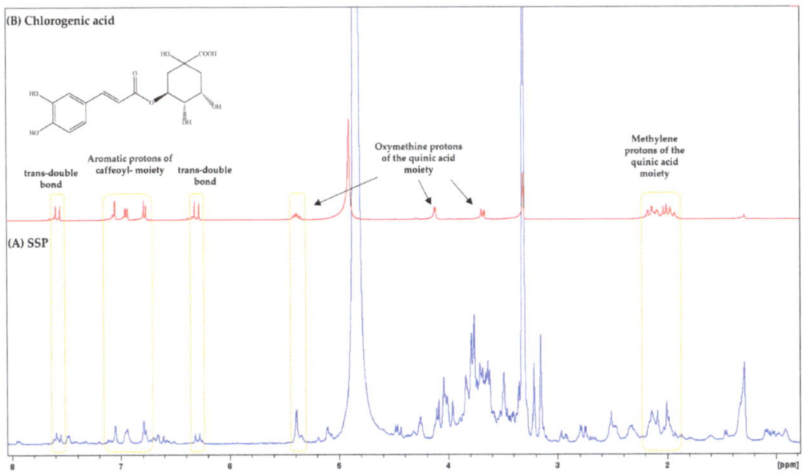

Figure 5. Overlaid ^{1}H-NMR spectra of (**A**) *Stachys spinulosa* infusion (SSP, blue color) and (**B**) chlorogenic acid (red color). Specific signals are indicated by orange boxes.

3.2. HPLC-PDA-MS Analysis

In total, 26 compounds were detected by HPLC-PDA-MS (Table 2), with them belonging to three main classes, namely flavonoids, phenylethanoid glycosides, and phenolic acids. The HPLC-PDA chromatograms of the *Stachys* infusions are illustrated in Figure S6.

Chlorogenic acid (**2**) was identified in all samples as one of the main metabolites in the *Stachys* infusions under study. The identification of the latter was based on a reference standard. An isobaric compound at an earlier retention time was assigned as an isomer, probably with a 4-substitution of the quinic acid group, as suggested by a fragment at *m/z* 173 [16]. Chlorogenic acid and isomers have previously been identified in the *Stachys* taxa of several sections, such as the sections *Candida* (*S. candida* and *S. horvaticii* Micevski, previously known as *S. iva* Griseb.), *Eriostomum*, and *Olisia* (*S. atherocalyx* K. Koch and *S. recta* L.), and members of the section *Stachys* [6]. However, this is the first report on the presence of chlorogenic acid and isomers in *S. chrysantha*, *S. leucoglossa* subsp. *leucoglossa*, and *S. spinulosa*.

Phenylethanoid glycosides are the main constituents of the genus *Stachys* [6,7]. In this study, nine phenylethanoid glycosides were identified in the four investigated infusions. Three detected phenylethanoid glycosides, namely lavandulifolioside (**4**), acteoside (**5**), and leucosceptoside A (**12**), were unambiguously identified by co-chromatography using compounds isolated in previous works by our group [6]. Although these compounds have previously been found in various *Stachys* taxa [6], they were detected for the first time in *S. chrysantha*, *S. leucoglossa* subsp. *leucoglossa*, and *S. spinulosa*. Acteoside has been previously isolated from *S. candida* [6], while lavandulifolioside and leucosceptoside A have not been identified in this species before. Peaks **6–8** were assigned tentatively as isomers of lavandulifolioside and acteoside as they had similar fragmentation patterns. Different substitution patterns of the acyl groups on the sugars might account for the differences in the retention times. Peaks **10** and **16** were tentatively assigned to stachysoside B and/or isomers, as suggested by the difference in the molecular weight (by 14 amu) when compared to lavandulifolioside. Furthermore, they were eluted at considerably longer retention times compared to lavandulifolioside. These compounds have been reported in *S. affinis* Bunge (synonym: *S. sieboldii* Miq.) [17]. Likewise, peak **19** was assigned tentatively to stachysoside C due to its pseudomolecular ion at *m/z* 783.0, suggesting the presence of two extra methyl groups in comparison to lavandulifolioside. The small amount of the latter in the infusion did not permit supporting further fragmentations. However, its retention time of almost

36 min agrees with this hypothesis. Such compounds are common in other *Stachys* species, such as *S. affinis* (synonym: *S. sieboldii*) [17] and *S. plumosa* Griseb. from Serbia [18].

The flavone glycosides detected in the infusions belong to the group of isoscutellarein/hypolaetin derivatives with some differentiations. Peaks **20**, **22**, and **24–26**, being the main metabolites, were identified based on co-chromatography with previously isolated compounds from both laboratories [6]. Accordingly, peaks **9**, **13**, **14**, **18**, and **23** were identified either as their deacetylated counterparts or isomers resulting from different acylation sites or, ultimately, as diacetylated derivatives. Regarding the section *Campanistrum*, 8-hydroxyflavone-allosylglucosides have been found in *S. arvensis* (L.) L. and *S. ocymastrum* (L.) Briq. (=*S. hirta* L.) [15]. However, there is no previous report on such compounds in *S. spinulosa*. Regarding the section *Candida*, isoscutellarein/hypolaetin derivatives have been previously reported in *S. candida*, *S. chrysantha*, and *S. horvaticii* Micevski (previously known as *S. iva*) [6]. Furthermore, monoacetyl and diacetyl derivatives of isoscutellarein have been found in *Stachys* species of the section *Olisia*, including *S. leucoglossa* [6,9]. *S. candida* and *S. spinulosa* also contained apigenin/luteolin derivatives in small amounts. In a previous study, chrysoeriol and its derivative, namely chrysoeriol 7-(3″-E-p-coumaroyl)-D-glucopyranoside, were isolated from *S. candida* [6]. Vicenin-2 was identified by the diagnostic successive losses of 120 amu [19], while peak **15** was identified as apigenin-7-O-glucoside previously isolated from *S. candida* [6]. It should be mentioned that vicenin-2 and apigenin-7-O-glucoside were detected only in *S. candida.* In addition, this study reports the presence of vicenin-2 in this species for the first time. Peaks **11** and **21** were detected only in *S. spinulosa* and were identified as acetylated luteolin and chrysoeriol dihexosides based on UV and MS fragmentations as well as on previously isolated compounds from *S. spinosa* and *S. aegyptiaca* Pers. [13,19].

Table 2. UV-VIS absorption and MS fragmentation data (positive and negative mode) of the compounds detected in the Greek *Stachys* infusions examined (SCA: *Stachys candida;* SCH: *Stachys chrysantha;* SLE: *Stachys leucoglossa* subsp. *leucoglossa;* SSP: *Stachys spinulosa*). The first reports of specific compounds for SCA, SCH, and SLE are indicated with an asterisk (*).

No.	Rt (min)	UV (nm)	Negative Mode, *m*/*z*	Positive Mode, *m*/*z*	Identification	SCA	SCH	SLE	SSP
1	10.90	298, 327	136.9, 172.9, 179.1, 191.1 [quinic acid-H]$^-$, **353.1** [M-H]$^-$, 707.1 [2M-H]$^-$	**354.9** [M+H]$^+$	Chlorogenic acid isomer	+ *	+ *	+ *	+
2	11.20	298, 324	191.1 [quinic acid-H]$^-$, **353.1** [M-H]$^-$, 707.1 [2M-H]$^-$	**354.9** [M+H]$^+$, **376.9** [M+Na]$^+$	Chlorogenic acid	+	+ *	+ *	+
3	15.36	271, 334	353.1, 473.0, **592.9** [M-H]$^-$	**595.0** [M+H]$^+$	Vicenin-2	+ *	-	-	-
4	22.28	291, 328	160.9, 593.2 [M-caffeoyl-H]$^-$, **755.1** [M-H]$^-$	479.1, **757.1** [M+H]$^+$, **779.1** [M+Na]$^+$	Lavandulifolioside (syn. Stachysoside A)	+ *	+ *	tr *	tr
5	23.73	291, 330	160.9 [caffeoyl group-H]$^-$, 461.1 [M-caffeoyl-H]$^-$, **623.1** [M-H]$^-$	**625.1** [M+H]$^+$, **647.1** [M+Na]$^+$	Acteoside	+	+ *	tr *	+
6	24.56	288, 329	160.9, 593.1, 623.2 [M-pentosyl-H]$^-$, **755.2** [M-H]$^-$	439.2, 625.1 [M-pentosyl+H]$^+$, **757.0** [M+H]$^+$, **779.1** [M+Na]$^+$	Isomer of lavandulifolioside	+ *	+ *	-	-

Table 2. *Cont.*

No.	Rt (min)	UV (nm)	Negative Mode, *m*/*z*	Positive Mode, *m*/*z*	Identification	SCA	SCH	SLE	SSP
7	27.26	289, 327	461.1 [M-caffeoyl-H]$^-$, **623.0** [M-H]$^-$	**647.1** [M+Na]$^+$	Acteoside isomer	+ *	+ *	-	-
8	28.57	289, 327	**623.1** [M-H]$^-$	**647.1** [M+Na]$^+$	Acteoside isomer	+ *	-	-	-
9	29.65	277, 307, 324	285.0 [A-H]$^-$, **608.9** [M-H]$^-$	**611.0** [M+H]$^+$	Isoscutellarein-7-*O*-allopyranosyl-(1→2)-glucopyranoside	-	-	+ *	-
10	32.03	288,329	137.0 [dihydroxytyrosol-H]$^-$, 593.1 [M-feruloyl-H]$^-$, 607.1 [M-rhamnose-H]$^-$, **769.4** [M-H]$^-$	**771.1** [M+H]$^+$, **793.2** [M+Na]$^+$	Phenylethyl glycoside, isomer I (stachysoside B, syn. Leonoside A) tentatively	+ *	-	-	-
11	32.40	255, 268, 348	285.0 [A-H]$^-$, **651.1** [M-H]$^-$	**653.1** [M+H]$^+$, **661.1** [M+Na]$^+$	Luteolin-acetyl-dihexoside	-	-	-	+
12	32.65–32.81	290, 329	136.9 [dihydroxytyrosol-H]$^-$, 446.9, **637.1** [M-H]$^-$	**639.2** [M+H]$^+$, **661.1** [M+Na]$^+$	Leucosceptoside A	+ *	+ *	-	+
13	32.70	277, 307, 327	**651** [M-H]$^-$	**653.1** [M+H]$^+$	Isoscutellarein-7-*O*-[6‴-acetyl-allopyranosyl-(1→2)]-glucopyranoside Isomer I	-	-	+	-
14	33.06–33.20	253, 287, 296, 334	301.2 [A-H]$^-$, **667.2** [M-H]$^-$	479.1, **669.1** [M+H]$^+$	Hypolaetin-acetylated derivative	+	+ *	+ *	+
15	33.44	268, 330	**430.9** [M-H]$^-$	**433.1** [M+H]$^+$	Apigenin-glucoside	+	-	-	-
16	33.66	282, 328	137.0 [dihydroxytyrosol-H]$^-$, 476.9, 637.1, **769.1** [M-H]$^-$	**771.1** [M+H]$^+$	Phenylethyl glycoside, isomer II (stachysoside B, syn. Leonoside A)	+*	+ *	-	-
17	34.05	267, 340 and 272, 287, 332			Mixture	+	-	-	-
18	34.62	277, 307, 327	**651** [M-H]$^-$	**653.1** [M+H]$^+$	Isoscutellarein-7-*O*-[6‴-acetyl-allopyranosyl-(1→2)]-glucopyranoside Isomer II	-	-	+	-
19	35.88	287, 329	**783** [M-H]$^-$	**785.1** [M+H]$^+$	Phenylethyl glycoside (stachysoside C) tentatively	+ *	-	-	-

Table 2. *Cont.*

No.	Rt (min)	UV (nm)	Negative Mode, *m/z*	Positive Mode, *m/z*	Identification	SCA	SCH	SLE	SSP
20	35.99–36.21	276, 306, 328	285.0, [A-H]$^-$, 429.1, **651** [M-H]$^-$	**653.1** [M+H]$^+$	Isoscutellarein-7-*O*-[6‴-acetyl-allopyranosyl-(1→2)]-glucopyranoside	+	+	+	+
21	36.49	269, 342	298.9 [A-H]$^-$, 623.1 [M-acetyl-H]$^-$, **665.0** [M-H]$^-$	**667.1** [M+H]$^+$	Chrysoeriol-7-*O*-acetyl-dihexoside (= stachyspinoside) tentatively	-	-	-	+
22	37.34–37.45	278, 296, 335	314.9 [A-H]$^-$, 625.2 [M-acetyl-H]$^-$, **681.1** [M-H]$^-$	**683.1** [M+H]$^+$	3′-hydroxy-4′-*O*-methylisoscutellarein-7-*O*-[6‴-acetyl-allopyranosyl-(1→2)]-glucopyranoside	+	+ *	+	+
23	41.55–41.70	278, 297 338	300.9 [A-H]$^-$, **709.0** [M-H]$^-$	**711.1** [M+H]$^+$	hypolaetin-7-*O*-di-acetyl-dihexoside	-	+ *	+ *	+
24	42.24	277, 307, 322	298.9 [A-H]$^-$, **665.0** [M-H]$^-$	**667.1** [M+H]$^+$	4′-*O*-methylisoscutellarein-7-*O*-[6‴-acetyl-allopyranosyl-(1→2)]-glucopyranoside	-	-	+	tr
25	43.66	276, 307, 324	**693.1** [M-H]$^-$	**695.1** [M+H]$^+$	Isoscutellarein-7-*O*-diacetyl-dihexoside	-	+ *	+ *	tr
26	44.16–44.26	276, 297, 338	314.9 [A-H]$^-$, 501.0 [M-hexosyl-acetyl-H_2O-H]$^-$, **723.1** [M-H]$^-$	**725.1** [M+H]$^+$	3′-hydroxy-4′-*O*-methylisoscutellarein-7-*O*-diacetyl-dihexoside	-	+ *	+ *	+

A: aglycon; tr: traces.

4. Conclusions

In the present study, the metabolic characterization of two protected and endangered local Greek endemic members of the genus *Stachys* (*S. candida* and *S. chrysantha*) and two local Balkan (sub-) endemic *Stachys* members (*S. leucoglossa* subsp. *leucoglossa*, and *S. spinulosa*) were investigated using NMR and HPLC-PDA-MS techniques. For the detection of specific constituents in the studied *Stachys* members, 1D- and 2D-NMR experiments were employed to compare their chemical fingerprints in combination with reference compounds isolated and identified in previous studies. In total, 26 compounds were detected by HPLC-PDA-MS as belonging to flavonoids (mainly isoscutellarein and hypolaetin derivatives), phenylethanoid glycosides, and phenolic acids (chlorogenic acid and its isomer). The chemical composition of *S. spinulosa* was investigated herein for the first time in detail, while knowledge of the metabolic profiling concerning the rest of the studied taxa was substantially supplemented herein with new reports of specific compounds not previously reported for one or more of the studied taxa. From a chemotaxonomical viewpoint, some compounds were found in all four *Stachys* members, evidencing a shared phytochemical relatedness. On the other hand, vicenin-2 and apigenin-glucoside were detected only in *S. candida*, while isoscutellarein-7-O-allopyranosyl-(1→2)-glucopyranoside was only detected in *S. leucoglossa* subsp. *leucoglossa*. Furthermore, *S. candida* and *S. chrysantha* had some compounds in common, evidencing (in phytochemical terms) their close affinity within the section *Candida* or the phyloclade *Swainsoniana*. Therefore, the data furnished herein may further contribute to ongoing phytochemical and chemotaxonomical investigations regarding the distribution of different compounds and categories thereof across different members and sections of the genus *Stachys*.

Supplementary Materials: The following supporting information can be downloaded at: https://www.mdpi.com/article/10.3390/agronomy13102624/s1. Figure S1: Representative photographs of the studied *Stachys* taxa in their original habitats in Greece: (A) *Stachys chrysantha*, (B) *Stachys candida*, (C) *Stachys leucoglossa* subsp. *leucoglossa*, and (D) *Stachys spinulosa*. Photos: Eleftherios Dariotis (reproduced with permission); Figure S2: ^{1}H-^{1}H-COSY and HSQC spectra of *Stachys candida* (SCA); Figure S3: ^{1}H-^{1}H-COSY and HSQC spectra of *Stachys chrysantha* (SCH); Figure S4: ^{1}H-^{1}H-COSY and HSQC spectra of *Stachys leucoglossa* subsp. *leucoglossa* (SLE); Figure S5: ^{1}H-^{1}H-COSY and HSQC spectra of *Stachys spinulosa* (SSP); Figure S6. HPLC-PDA chromatograms of the *Stachys* infusions at 330 nm.

Author Contributions: Conceptualization, E.-M.T. and H.S.; methodology, E.-M.T. and A.K.; formal analysis, E.-M.T. and A.K.; collections and taxonomic identification: N.K.; investigation, E.-M.T., A.K. and G.T.; data curation, E.-M.T. and A.K.; writing—original draft preparation, E.-M.T. and A.K.; writing—review and editing, E.-M.T., A.K., N.K. and H.S.; visualization, E.-M.T., A.K. and N.K.; supervision, E.-M.T. and H.S. All authors have read and agreed to the published version of the manuscript.

Funding: This research received no external funding.

Data Availability Statement: The datasets generated and/or analyzed in the current study are available from the corresponding author upon request or are included in the article and its supplementary materials.

Conflicts of Interest: The authors declare no conflict of interest.

References

1. Harley, R.M.; Atkins, S.; Budantsev, A.L.; Cantino, P.D.; Conn, B.J.; Grayer, R.; Harley, M.M.; de Kok, R.; Krestovskaya, T.; Morales, R.; et al. Labiatae. In *The Families and Genera of Vascular Plants*; Kubitzki, K., Ed.; Springer: Berlin, Germany, 2004; Volume II, pp. 167–275.
2. Salmaki, Y.; Heubl, G.; Weigend, M. Towards a new classification of tribe Stachydeae (Lamiaceae): Naming clades using molecular evidence. *Bot. J. Linn. Soc.* **2019**, *190*, 345–358. [CrossRef]
3. Plants of the World Online (POWO). Available online: https://powo.science.kew.org/ (accessed on 1 August 2023).
4. Bhattacharjee, R. Taxonomic studies in *Stachys* II. A new infragenic classification of *Stachys* L. Notes R. *Bot. Gard. Edinb.* **1980**, *38*, 65–96.
5. Vascular Flora of Greece Web (Vascular Plants of Greece-An Annotated Checklist). Available online: https://portal.cybertaxonomy.org/flora-greece/intro (accessed on 1 August 2023).
6. Tomou, E.-M.; Barda, C.; Skaltsa, H. Genus *Stachys*: A review of traditional uses, phytochemistry and bioactivity. *Medicines* **2020**, *7*, 63. [CrossRef]
7. Tundis, R.; Peruzzi, L.; Menichini, F. Phytochemical and biological studies of *Stachys* species in relation to chemotaxonomy: A review. *Phytochemistry* **2014**, *102*, 7–39. [CrossRef] [PubMed]
8. Kougioumoutzis, K.; Kokkoris, I.P.; Panitsa, M.; Strid, A.; Dimopoulos, P. Extinction risk assessment of the Greek endemic flora. *Biology* **2021**, *10*, 195. [CrossRef]
9. Lenherr, A.; Meier, B.; Sticher, O. Modern HPLC as a tool for chemotaxonomical investigations: Iridoid glucosides and acetylated flavonoids in the group of *Stachys recta*. *Planta Med.* **1984**, *50*, 403–409. [CrossRef] [PubMed]
10. Committee on Herbal Medicinal Products-HMPC (2016). EMA/HMPC/39453/2015. European Union Herbal Monograph on *Sideritis scardica* Griseb.; *Sideritis clandestina* (Bory&Chaub.) Hayek; *Sideritis raeseri* Boiss.&Heldr.; *Sideritis syriaca* L., herba. Available online: https://www.ema.europa.eu/en/documents/herbal-monograph/final-european-union-herbal-monograph-sideritis-scardica-griseb-sideritis-clandestina-bory-chaub_en.pdf (accessed on 1 August 2023).
11. Safer, S.; Cicek, S.S.; Pieri, V.; Schwaiger, S.; Schneider, P.; Wissemann, V.; Stuppner, H. Metabolic fingerprinting of *Leontopodium* species (Asteraceae) by means of ^{1}H NMR and HPLC-ESI-MS. *Phytochemistry* **2011**, *72*, 1379–1389. [CrossRef] [PubMed]
12. Háznagy-Radnai, E.; Czigle, S.; Janicsák, G.; Máthé, I. Iridoids of *Stachys* species growing in Hungary. *JPC J. Planar Chromatogr.* **2006**, *19*, 187–190. [CrossRef]
13. Kotsos, M.; Aligiannis, N.; Mitaku, S.; Skaltsounis, A.L.; Charvala, C. Chemistry of plants from Crete: Stachyspinoside, a new flavonoid glycoside and iridoids from *Stachys spinosa*. *Nat. Prod. Lett.* **2001**, *15*, 377–386. [CrossRef] [PubMed]
14. Tomás-Barberán, F.A.; Gil, M.I.; Ferreres, F.; Tomás-Lorente, F. Flavonoid p-coumaroylglucosides and 8-hydroxyflavone allosylglucosides in some Labiatae. *Phytochemistry* **1992**, *31*, 3097–3102. [CrossRef]

15. Lakhal, H.; Boudiar, T.; Kabouche, A.; Laggoune, S.; Kabouche, Z.; Topçu, G. Antioxidant activity and flavonoids of *Stachys ocymastrum*. *Chem. Nat. Compd.* **2011**, *46*, 964–965. [CrossRef]
16. Clifford, M.N.; Johnston, K.L.; Knight, S.; Kuhnert, N. Hierarchical scheme for LC-MSn identification of chlorogenic acids. *J Agric. Food Chem.* **2003**, *51*, 2900–2911. [CrossRef] [PubMed]
17. Nishimura, H.; Sasaki, H.; Inagaki, N.; Masao, C.; Chen, Z.; Mitsuhashi, H. Nine phenethyl alcohol glycosides from *Stachys sieboldii*. *Phytochemistry* **1991**, *30*, 965–969. [CrossRef] [PubMed]
18. Karioti, A.; Kukić-Marković, J.; Bilia, A.R.; Niketić, M.; Petrović, S. Chemical profiling of six *Stachys* taxa from Balkan Peninsula. *Biochem. Syst. Ecol.* **2022**, *104*, 104482. [CrossRef]
19. El-Ansari, M.A.; Abdalla, M.F.; Saleh, N.A.M.; Barron, D.; Le Quéré, J.L. Flavonoid constituents of *Stachys aegyptiaca*. *Phytochemistry* **1991**, *30*, 1169–1173. [CrossRef]

agronomy

Article

Application of Lavender and Rosemary Essential Oils (EOs), Their Mixture and Eucalyptol (EOs Main Compound) on Cucumber Fruit Quality Attributes and Microbial Load

Panayiota Xylia, Christos Goumenos, Nikolaos Tzortzakis * and Antonios Chrysargyris *

Department of Agricultural Sciences, Biotechnology and Food Science, Cyprus University of Technology, Limassol 3036, Cyprus; pa.xylia@edu.cut.ac.cy (P.X.); cd.goumenos@edu.cut.ac.cy (C.G.)
* Correspondence: nikolaos.tzortzakis@cut.ac.cy (N.T.); a.chrysargyris@cut.ac.cy (A.C.)

Abstract: Cucumber (*Cucumis sativus* L.), one of the most widely consumed vegetables, presents high perishability during storage and marketing if it is not handled and stored properly. Currently, there is an increased interest of the food industry to reduce waste (due to quality losses) and to utilize natural products for the preservation of fresh commodities. This study's goal was to evaluate the effects of lavender (Lav) and rosemary (Ros) essential oils (EOs), their mixture (Lav + Ros, 1:1 v/v) and their main compound (eucalyptol) via vapor phase on cucumber's postharvest quality. The outcomes of this study demonstrated that 200 μL/L of Lav and Ros EOs increased the respiration rate of cucumbers after 10 days of storage at 11 °C, while 100 μL/L of the EOs mixture and Eucalyptol (100 and 200 μL/L) had no effect on respiration, on the same day. The application of Eucalyptol (100 and 200 μL/L) resulted in less acceptable fruits (less pleasant aroma and unpleasant taste). A decrease in fruit firmness was found in cucumbers exposed to Lav 200 μL/L and Ros 100 μL/L. Interestingly, Eucalyptol was found to accelerate the fruit ripening index after five days of storage, and to decrease organoleptic properties of the fruit (i.e., aroma, taste) on the fifth day of storage. The fruit revealed increased oxidative stress (i.e., increased lipid peroxidation), especially at a high concentration (200 μL/L) of Eucalyptol after 10 days. This has resulted in the activation of other non-enzymatic antioxidant mechanisms such as the increase in fruit ascorbic acid content. Notably, no effects on fruit weight loss, total soluble solids and color were observed with the examined treatments. Overall, this study suggests that the investigated products (EOs and their main compound) have a putative role in postharvest storage for the preservation of cucumbers. However, further investigation is needed for the determination of the optimum application conditions (i.e., concentration, time and method of application) on cucumbers and other fresh produce.

Keywords: cucumber; *Lavandula angustifolia*; *Rosmarinus officinalis*; eucalyptol; postharvest quality

1. Introduction

The cucumber (*Cucumis sativus* L.) belongs to the *Cucurbitaceae* family, and it is widely consumed around the world. Cucumber fruit is an excellent source of nutrients, i.e., antioxidants, ascorbic acid, magnesium and dietary fibers [1]. Daily consumption of fresh fruits and vegetables has been linked to many health benefits and lower rates of chronic illnesses since vegetables such as cucumbers contain components with anti-inflammatory and antioxidant properties [2,3]. Same as other fresh produce, cucumbers present short shelf life, mainly due to the visible signs of water loss (weight loss), pigment degradation (discoloration) and postharvest diseases [4,5].

The shelf life of fresh commodities (including cucumbers) can be impacted by a series of pre- and postharvest factors. During preharvest, growing conditions such as temperature, light, irrigation and harvesting handling can influence the fruit's quality after harvesting, processing and storage [6]. Mishandling, storage temperature and relative humidity and inefficient cleaning techniques are postharvest factors that can affect the quality of fresh produce [7,8]. Washing fresh produce with chlorinated water is a crucial step in preserving fruits and vegetables because it minimizes the microbial load. However, the use of chlorine and chlorine-based agents has been linked with adverse health effects due to the formation of harmful compounds [9]. Thus, alternative and natural means and compounds have been investigated for the preservation of fresh produce (in liquid and vapor form).

Natural products such as essential oils (EOs) and other natural compounds have been exploited as alternative means for the postharvest preservation of fresh commodities due to their significant properties (i.e., antioxidant and antimicrobial (among others)) [10–13]. EOs are mixtures of secondary plant metabolites and a series of them have been characterized as "Generally Recognized as Safe" (GRAS) and can be used in the food industry [14]. Previous studies have shown the beneficial effects of lavender (*Lavandula angustifolia* Mill.) EO which include antioxidant, antimicrobial and anti-inflammatory properties [14–18]. Similarly, rosemary (*Rosmarinus officinalis* L.) EO possesses antioxidant, antimicrobial, anti-inflammatory and insecticidal activities [19–21]. The Lamiaceae family (Labiatae) is one of the most notable families of medicinal plants, having a wide range of species (over 7000 species in 236 genera) and a global distribution. As they can be purchased as fresh food, dry products, essential oils, and as a wide series of other byproducts, species from this family are economically important on a global scale [22].

A series of earlier studies have examined the effects of the application of EOs on fresh commodities and the reported results are encouraging [11–13,23,24]. Lavender and rosemary EOs have been previously applied on fresh produce (i.e., lemons, endives, apples, pears) with various application methods [18,25,26]. For instance, vapor application of mint, basil and lavender EO on lemon fruits showed significant antifungal effects against *Penicillium digitatum* while at the same time maintained fruit quality attributes (i.e., weight, firmness, pH) [18]. Application of rosemary EO (range: 200–1000 μL/L) on fresh tomatoes resulted in decreased fruit weight loss, while it enhanced fruit flavor [13]. Furthermore, combining different EOs obtained from diverse plants and applying them on fresh produce could preserve sensory attributes (i.e., aroma and flavor) while lowering the possibility of fruit degradation [12].

The postharvest application of EOs alone and/or in combination with other compounds can directly affect the organoleptic characteristics of the produce, resulting in a desirable/acceptable or undesirable product [24]. During the application of EOs, considerations about the concentration, type and time of application, and EOs composition should be made to avoid negative effects on the quality features of the produce [27]. Fruit quality and safety maintenance or improvement is mandatory when vaporized products as the EOs are applied on fresh produce; on top of that, the organoleptic test in such postharvest applications is fundamental. This study aimed to determine the effectiveness of (i) lavender or rosemary EOs vapor application, (ii) their 1:1 (*v*/*v*) mixture for evaluating potential synergistic effects and (iii) the application of synthetic eucalyptol (as their common major compound) on the preservation of cucumber's postharvest quality and spoilage during storage at 11 °C for 10 days. The innovative approach of this study is the combination of two EOs with the same main component, the direct comparison of the role of chemical synthetic eucalyptol to equal levels (as the levels present in the EOs mixtures), with possible synergistic or not effects derived by the main component of an EO. Moreover, the importance of using EOs as a preservation means for fresh produce, rather than the main component (eucalyptol), is important because biocidal properties of an EO are mirrored to the mixture of the components present in the oil profile and not only by one of them.

2. Materials and Methods

2.1. Plant Material and Essential Oil Extraction

Cucumber fruits (*Cucumis sativus* cv. PS-64; winter parthenocarpic cucumber with cold tolerance, at size of 18–19 cm long) were collected from a local greenhouse farm (coordinates: 34°46′10.61″ N; 33°11′22.08″ E, 217 m; Limassol, Cyprus). The crop was cultivated using common cultivation practices in a clay sandy loam soil, and was drip-irrigated and fertigated according to the crop's needs. Cucumber plants were trained on a string based on vertical single pruning system (the main stem grew vertically, and lateral shoots were removed) and were grown for approximately 3.5 months. The crop cultivation took place during autumn/winter months in an unheated plastic greenhouse and the temperature ranged between 17 °C and 28 °C. Cucumbers (190 fruits) were randomly selected according to size (uniform), appearance and the absence of any physical defects. After their transfer to the laboratory, fruits were washed with chlorinated water (0.05% NaOCl), rinsed with distilled water and were uniformly divided into each treatment.

Lavender (*Lavandula angustifolia* Mill.) and rosemary (*Rosmarinus officinalis* L.) plants were collected from the experimental farm of Cyprus University of Technology (coordinates: 34°42′0.50″ N; 32°59′3.44″ E, 98 m; Limassol, Cyprus) in early October with temperature averaged at 27.6 °C and relevant humidity of 48.4%. Plant materials (chosen from 10 plants from each species; ~20 kg of fresh biomass) were transferred to the laboratory and were air dried at 42 °C in an air-ventilated oven. The EOs were obtained with hydrodistilation (Clevenger apparatus for 3 h) and their composition was determined with Gas Chromatography/Mass Spectrometry (GCMS; GC/MSQP-2010 Plus, Shimadzu, Tokyo, Japan) as previously described [28]. The major components found at lavender leaves EO were eucalyptol (59.40%), borneol (8.58%), camphor (7.98%) and β pinene (3.54%), whereas the major components found at rosemary EO were eucalyptol (31.09%), camphor (20.61%), α pinene (12.21%) and camphene (8.49%). The mixture of the two EOs was prepared in the ratio of 1:1 (v/v) from each EO. The major common compound of the two EOs, i.e., eucalyptol (Sigma–Aldrich, Darmstadt, Germany) was used at 45% as the mean value of the eucalyptol of the two EOs mixtures (1:1 v/v).

2.2. Procedure

Fruits were placed in a 5 L polypropylene (PP) plastic container (five fruits/container, two containers per treatment for each of the sampling time point (Day 5 and 10)). An appropriate volume of EOs or eucalyptol (absorbed by a paper strip for slow release of the volatiles) was placed inside the container to reach 100 and 200 µL/L. The EOs levels were selected based on preliminary tests and previous studies [26,29] and targeted to maintain the quality-related attributes during storage conditions. Distilled water was used for the control treatment (0% EO). The containers were sealed with their corresponding lid and were left at room temperature for 1 h to assist the evaporation of the volatile components of the oils. The containers were then stored in an experimental refrigerator at 11 °C and relative humidity of 90% (RH) (achieved by displacing a moist paper inside each container) for the examined storage period of 5 and 10 days [29–31]. To avoid any unwanted increase in CO_2 and decrease in O_2 throughout the fruit respiration process, the containers were opened to received aeration every other day. The EOs were applied in a vapor phase as described previously [32].

2.3. Impact on Fruits' Quality Attributes

A series of quality attributes were evaluated in this study. Fruit weight loss was determined by recording the weight of each single fruit on the tested days (days 0, 3, 5, 7 and 10) and the percentage of total weight loss was computed for each day. The effect of the applications on the respiration rate of the fruit was determined as it has been previously described by Xylia et al. [33] and the results were given in mL of CO_2 produced per kg per h (mL CO_2/kg/h).

In order to evaluate the fruits' aroma, taste, appearance and marketability on the test days (days 0, 5, and 10), at least eight panelists were used [33]. The evaluation of the aroma and taste were conducted using a 10-point scale (1 interval) as follow: 1: not cucumber-like and quite unpleasant aroma/taste, 3: not cucumber-like and lightly unpleasant aroma/taste, 5: not cucumber-like but pleasant aroma/taste, 8: less cucumber-like aroma/taste and 10: intense cucumber-like aroma/taste. Appearance (visual quality and color) was evaluated using a 10-point scale (1 interval) as follows: 1: yellow color of 50%; 3: yellow–green; 5: light green; 8: green; 10: deep green. To assess the marketability of the fruits (indicating overall quality) a scale of 1–10 (1 interval) was employed, where 1: not marketable quality (i.e., malformation, wounds, infection); 3: low marketability with malformation; 5: marketable with few defects, i.e., small size, decolorization (medium quality); 8: marketable (good quality); 10: marketable with no defects (extra quality).

The color on the surface of the fruits was determined with a colorimeter (Chroma meter CR400 Konica Minolta, Tokyo, Japan) by measuring the L*, a* and b* values (CIELAB uniform color space). Hue (h) value was determined in degrees (°): $h = 180 + \tan^{-1}(b^*/a^*)$ [34,35]. The chroma value (C) was computed as $C = (a^{*2} + b^{*2})^{1/2}$, whereas color index (CI) was computed as $CI = (a^* \times 1000)/(L^* \times b^*)$ [35]. The browning index (BI) was computed as $BI = 100 \times (X - 0.31)/0.17$ where: $X = (a^* + 1.75 \times L^*)/(5.645 \times L^* + a^* - 3.012 \times b^*)$, while the yellowing index (YI) was calculated as $YI = (142.86 \times b^*)/L^*$ [36].

Fruit firmness was determined at two distinct sites on each fruit with a texture analyzer (TA.XT plus, Stable Micro Systems, Surrey, UK) equipped with a 3 mm diameter probe (travelling speed: 2 mm/s and penetration depth: 12 mm) [33]. The force needed to crack the pericarp of the fruit was measured in Newtons (N). Drops of the extracted fruit juice were used to assess the total soluble solids content (TSS) with the use of a digital pocket refractometer (Atago, Tokyo, Japan). Results of TSS were expressed as °Brix. Fruit juice was also used to calculate titratable acidity (TA) via titration with 0.1 N NaOH using the aforementioned procedure [37], and the results were expressed as g of malic acid per L of juice. The sweetness of fruits (i.e., ripening index) was determined as the ratio of TSS over TA (TSS/TA). Cucumbers' ascorbic acid (AA) content was assayed using the 2,6-dichlorophenolindophenol titration method [37] and results were expressed as mg of AA per 100 g of fresh weight (mg/100 g).

2.4. Impact on Fruits' Polyphenols and Antioxidant Activity

The method for extraction of total polyphenols and antioxidants was assayed as described by Chrysargyris et al. [38]. The determination of total polyphenol content was created using the Folin–Ciocalteu method according to Chrysargyris et al. [28] and results were expressed as μg of gallic acid equivalents per g of fresh weight (μg GAE/g). The antioxidant activity of the methanolic fruit extracts was estimated using three different assays as follows: (i) the 2,2-diphenyl-1-picrylhydrazyl (DPPH), (ii) the ferric reducing antioxidant power (FRAP) and (iii) the 2,2′-azinobis-(ethylbenzothiazoline-6-sulfonic acid) (ABTS) method. The DPPH and FRAP were assayed based on Chrysargyris et al. [28]. The ABTS assay was performed according to Wojdylo et al. [39]. For all three tested methods, results were expressed as μg of trolox per g of fresh weight (μg trolox/g).

2.5. Determination of Fruit Damage Index and Enzymatic Antioxidant Activity

The oxidative degradation of membrane lipids (lipid peroxidation) in cucumbers was assessed in terms of malondialdehyde content using the thiobarbituric acid reaction [40], while hydrogen peroxide content was measured according to the method described by Loreto and Velikova [41]. Both indicators reflected the fruit damage indices in this study. Results were expressed in nmol MDA and μmol H_2O_2 per g of fresh weight, respectively.

The antioxidant enzymes activities of superoxide dismutase (SOD) (EC 1.15.1.1) and catalase (CAT) (EC 1.11.1.6) were determined, as described previously [42], and the absorbance was determined at 560 nm for SOD and at 240 nm for CAT. Peroxidase activity (POD) (EC 1.11.1.6) was determined following the increase in absorbance at 430 nm [42]. Results were expressed as enzyme units per mg of protein. The protein content was determined using the Bradford method and bovine serum albumin was used as reference.

2.6. Impact on Microbial Load

Cucumbers' microbial load (i.e., total viable count-TVC, yeast and mold) was determined using Plate count agar (PCA, Merck, Darmstadt, Germany) and Dichloran-rose bengal chloramphenicol Agar (DRBC agar, Merck, Darmstadt, Germany), respectively, as previously described by Xylia et al. [43]. After analysis, results were expressed as log CFU per g of fresh weight (log CFU/g).

2.7. Statistical Analysis

The data were subjected to one-way analysis of variance (one-way ANOVA) and a comparison of the treatment on each day was completed using Tukey's multiple range test ($p = 0.05$) with IBM SPSS version 25.0. An independent samples t-test was also used for comparing control data on the initial (Day 0) and final (Day 10) days.

3. Results and Discussion

Currently, fresh produce (fruits, vegetables, herbs) are highly consumed due to their highly valued and well-appreciated nutritional properties. Consumers are requesting fresh produce of high quality and safety to be available throughout the year, with the use of less or no synthetic chemicals during plant growth and storage and, of course, at a reasonable cost. It is challenging to effectively satisfy these numerous requirements; however, significant efforts are made in the direction to cover the utmost of them. Preserving the quality of fresh produce is of great importance since it influences the purchasing choices of the consumers. Postharvest treatment of fresh produce with natural products such as EOs, plant extracts and other natural components can maintain the quality-related attributes of fruits and vegetables [44,45]. Previous studies have shown the positive effects of EOs on fresh commodities by enhancing their sensory and organoleptic characteristics and igniting physicochemical processes (i.e., increase in antioxidants) that result in higher nutritional value products [24,38].

3.1. Impacts on Fruits' Quality Attributes

One of the main causes leading to weight loss of fresh commodities during storage is moisture loss (water loss) due to the processes of respiration and transpiration of the produce. Based on their respiratory profiles, and on how they produce and react to ethylene, fruits are classified into the following two broadly defined categories in relation to their ripening process: climacteric and non-climacteric. The cucumber is classified to the latter one [46]. Non-climacteric fruits, during ripening and senescence, continue their respiration and emit ethylene at basal levels. The impacts of EOs application, their mixture, and their main compound (eucalyptol) on moisture loss (weight loss) and rates of respiration of cucumber fruit are presented in Figure 1. This study shows that the applied treatments did not affect fruit weight throughout storage (Figure 1). Based on previous reports, the maximum weight loss was 7% during storage for cucumbers to maintain a marketable appearance [6], whereas in this study, the maximum weight loss was 2.56% on average. One hypothesis suggests that the hydrophobic nature of EOs decrease in water vapor permeability of the produce surface occurs to prevent water transfer and weight losses. A previous study has also proved that there was no significant weight loss in cucumber fruits during exposure to the vapors of a green-based product made with rosemary and eucalyptus EOs when it was applied at different concentrations [47]. Another study revealed that the vapor application of lavender EO at a concentration of 43 μL/L resulted in significant weight loss of lemon fruits compared to a concentration of 86 μL/L [18]. The same study showed that EOs from mint and basil applied in vapor form also resulted in a decrease in fruit weight [18]. This might be due to the different duration of the storage period for lemons versus cucumbers, as well as due to the different concentrations of the applied EOs; higher EO concentrations could evidently retard fruits' respiration rates, and accelerate fruits' senesce [26]. The fruits' respiration rate was increased with Lav 200 μL/L on the fifth day, while at the end of the storage period (day 10), at all treatments (especially

Lav 200 μL/L, Ros 200 μL/L and Lav + Ros 200 μL/L), cucumbers revealed an increased respiration rate. In previous studies, application of rosemary EO mixed with eucalyptus EO (300 μL/L) increased the respiration rates in apple and pear fruits [26], whereas lavender and thyme EOs were found to decrease apples' respiration and ethylene production [48]. As it appears, different EOs have distinct effects on the respiration process of various commodities which could be related with alterations on the cell wall integrity and gas exchange [49] as well as with the different respiration rates that the non-climacteric fruits have, such as the cucumber, and the climacteric, such as apple and pear. In another study on tomatoes, chlorine application and high levels of EOs vapors (mixture of rosemary and eucalyptus EOs) significantly boosted respiration rates (although these changes were not observed at the dipping application) after 7 days and 14 days. This indicates that the effect of dipping was shorter than that of the vapor application [33]. Therefore, dipping application of EOs is an alternative way of using EOs as a preservation means in postharvest management of fresh produce, and relevant studies should be implemented on cucumber fruits as well.

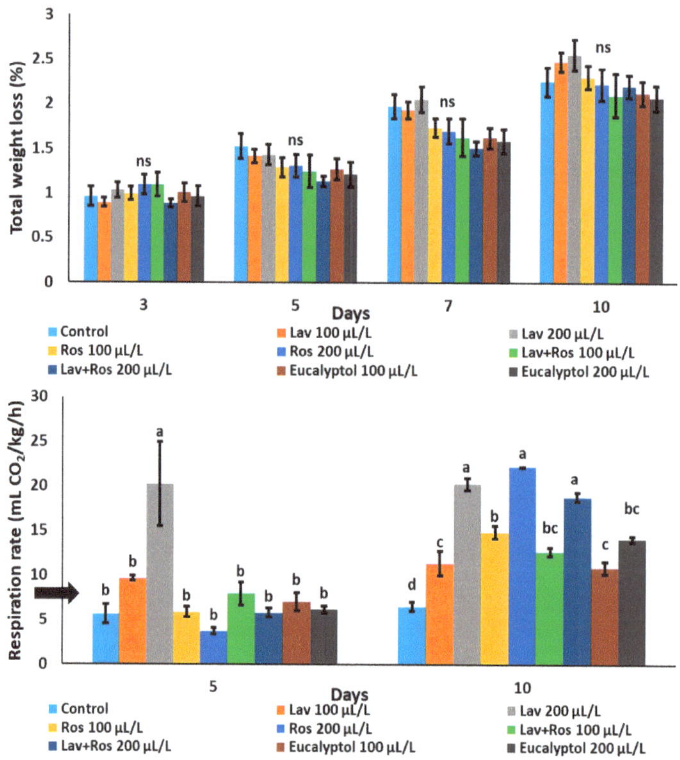

Figure 1. Impacts of lavender (Lav) and rosemary (Ros) EOs, their mixture (Lav + Ros, 1:1 v/v) and eucalyptol vapors on cucumber's weight loss and respiration rate during storage at 11 °C for 10 days. The presented values are the means (±standard errors) of six biological replicates (per treatment). Values for day 0 refer to the control (non-treated). Different Latin letters indicate statistically significant differences among treatments on each day (for each column). ns indicates non-significant. The arrow shows the initial value for the control (non-treated).

Quality and sensory characteristics such as dark green color and increased firmness are essential for consumers when purchasing cucumbers. Previous studies showed that EO application during storage of fresh commodities can affect organoleptic characteristics of produce [12,18,24–26,50]. The effects of exposure to volatiles on the cucumber's sensory attributes (aroma, appearance, taste) and marketability are presented in Figure 2 and Figure S1. Exposure to Eucalyptol 100 and 200 μL/L revealed 5.25 (not cucumber and pleasant aroma) and 3.44 (not cucumber and lightly unpleasant aroma) on the fifth day of storage, respectively. Moreover, at the end of the storage period, all treatments resulted in a not cucumber but pleasant aroma. No significant differences were observed regarding fruit appearance after 5 and 10 days of storage at 11 °C (Figure 2). All applied treatments revealed lower scores (especially Eucalyptol 100 and 200 μL/L—not cucumber and very unpleasant) than the non-treated (control) on the fifth day, while even lower scores were observed at the end of the storage period (day 10) by all the applied treatments. In a study on broccoli, it has been showed that the application of rosemary EO (1.82 mg/mL) resulted in less acceptable broccoli florets (less pleasant odor and flavor) as did the application of oregano EO (0.48 mg/mL) and their combination (0.91 and 0.24 mg/mL, respectively) [12]. In our study, after all the applied treatments, especially with Eucalyptol 100 and 200 μL/L (not marketable) on the fifth day, the panelists evaluated the produce as less marketable, while on the last day (day 10) the applied treatments resulted in a low marketable product (scores ranging between 3.50 and 5.25) (Figure 2 and Figure S1). Time of application and concentration of the applied EOs were found to significantly affect the quality-related features of a fresh produce during storage, such as aroma, color and taste, while some studies have reported negative effects on organoleptic characteristics as the concentration of the applied EOs increases [25,51]. Moreover, the impacts on fruit quality characteristics such as aroma and taste can be attributed to the possible interactions of the EOs and the volatile compounds with the food matrix components, which in some cases result in undesirable and unpleasant end products [52].

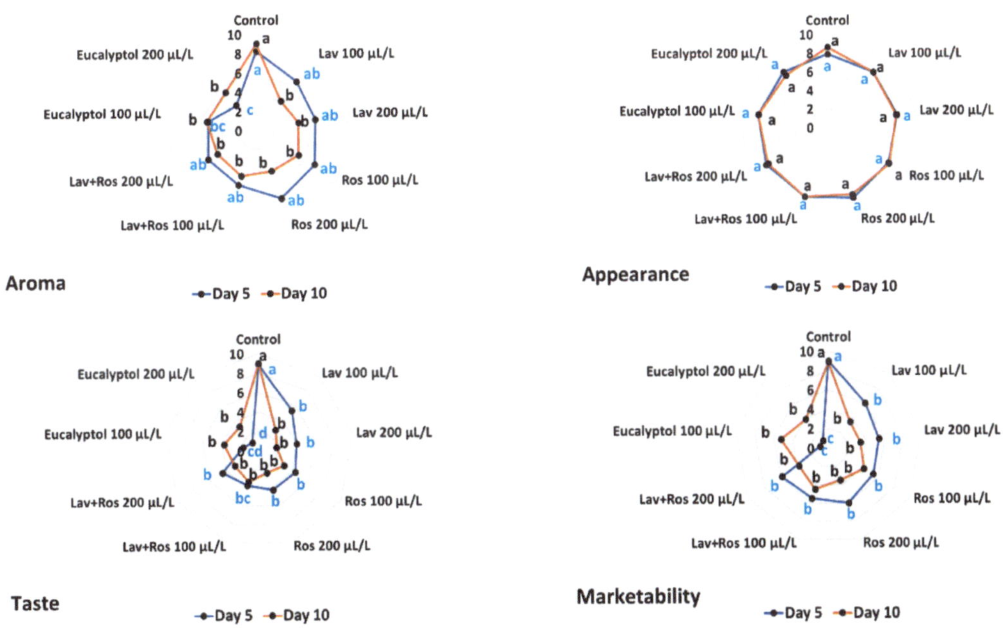

Figure 2. Impacts of lavender (Lav) and rosemary (Ros) EOs, their mixture (Lav + Ros, 1:1 *v*/*v*) and eucalyptol vapors on cucumber's sensory attributes (aroma, appearance, taste) and marketability during storage at 11 °C for 10 days. The presented values are the means of eight panelists evaluation on each treatment. Different Latin letters indicate statistically significant differences among treatments on each day (for each column).

Consumers are attracted by the dark green color of cucumbers during purchase. However, during long storage and under adverse conditions (high temperature and low relative humidity) the cucumber's green color turns to yellow due to the degradation of chlorophyll [8]. Table 1 illustrates the effects on fruit color of the exposure to volatiles of cucumber fruits after storage for 10 days. A higher a* value was found with Eucalyptol 200 μL/L in comparison to Lav 200 μL/L during the fifth day, while no significant differences were found at the end of storage. Interestingly, other color parameters that were investigated in this study (i.e., L*, b*, hue, chroma value and color index) were not affected by the applied treatments throughout the storage period, indicating that EO-treated produce were able to maintain good color attributes during the 10 days of storage. Similar observations were found in lettuce treated with thyme EO [53]. Moreover, it appears that Ros EO preserved the dark color of cucumber fruits more efficiently than the other treatments. These findings could be explained by the antioxidant activities of rosemary EO and its composition that protected chlorophylls from degradation [54]. The small changes in a* and b* values are good indicators of the absence of oxidative browning of cucumbers [55].

Table 1. Impacts of lavender (Lav) and rosemary (Ros) EOs, their mixture (Lav + Ros, 1:1 *v*/*v*) and eucalyptol vapors on cucumber's color parameters: L*, a*, b*, hue (h), chroma value (C) and color index (CI) during storage at 11 °C for 10 days.

	Treatment	L*	a*	b*	h (°)	C	CI
Day 0	Control	39.04 ± 0.87	−11.99 ± 0.71	16.30 ± 1.22	126.51 ± 0.53	20.24 ± 1.40	−19.06 ± 0.79
Day 5	Control	36.29 ± 0.48	−11.61 ± 0.16 ab	15.73 ± 0.27	126.44 ± 0.29	19.56 ± 0.29	−20.37 ± 0.38
	Lav 100 μL/L	36.40 ± 1.09	−11.28 ± 0.75 ab	15.15 ± 1.35	126.92 ± 0.66	18.90 ± 1.53	−20.82 ± 1.09
	Lav 200 μL/L	38.12 ± 1.23	−13.41 ± 0.46 b	17.32 ± 1.29	128.13 ± 1.48	21.93 ± 1.26	−21.00 ± 1.88
	Ros 100 μL/L	38.92 ± 1.17	−12.79 ± 0.60 ab	17.79 ± 1.14	125.84 ± 0.52	21.91 ± 1.28	−18.69 ± 0.81
	Ros 200 μL/L	37.31 ± 1.51	−12.48 ± 0.82 ab	17.40 ± 1.59	125.97 ± 0.80	21.42 ± 1.77	−19.75 ± 1.38
	Lav + Ros 100 μL/L	36.05 ± 1.04	−11.40 ± 0.45 ab	14.90 ± 0.76	127.49 ± 0.46	18.76 ± 0.88	−21.43 ± 0.96
	Lav + Ros 200 μL/L	37.41 ± 1.39	−12.85 ± 0.68 ab	17.61 ± 1.28	126.35 ± 0.75	21.80 ± 1.43	−19.89 ± 1.17
	Eucalyptol 100 μL/L	35.68 ± 1.04	−11.59 ± 0.57 ab	15.34 ± 1.00	127.22 ± 0.68	19.23 ± 1.12	−21.41 ± 0.90
	Eucalyptol 200 μL/L	35.24 ± 1.15	−10.28 ± 0.60 a	13.09 ± 0.92	128.24 ± 0.38	16.65 ± 1.09	−22.54 ± 1.07
Day 10	Control	36.66 ± 1.14	−11.97 ± 0.98	16.34 ± 1.74	126.55 ± 0.69	20.26 ± 1.98	−20.41 ± 1.07
	Lav 100 μL/L	39.25 ± 0.94	−13.03 ± 0.55	18.28 ± 1.02	125.59 ± 0.43	22.45 ± 1.15	−18.33 ± 0.73
	Lav 200 μL/L	37.09 ± 1.51	−13.21 ± 0.55	18.10 ± 1.22	126.30 ± 0.66	22.41 ± 1.31	−20.04 ± 1.07
	Ros 100 μL/L	39.55 ± 0.69	−13.09 ± 0.47	18.90 ± 0.81	124.75 ± 0.29	22.99 ± 0.93	−17.58 ± 0.40
	Ros 200 μL/L	38.31 ± 1.13	−12.39 ± 0.77	16.96 ± 1.46	126.43 ± 0.78	21.01 ± 1.62	−19.44 ± 1.03
	Lav + Ros 100 μL/L	39.95 ± 1.23	−12.94 ± 0.80	18.41 ± 1.55	125.32 ± 0.63	22.51 ± 1.73	−17.89 ± 0.94
	Lav + Ros 200 μL/L	38.54 ± 0.48	−11.48 ± 0.05	16.03 ± 0.29	125.63 ± 0.43	19.72 ± 0.25	−18.63 ± 0.48
	Eucalyptol 100 μL/L	36.48 ± 2.15	−11.53 ± 1.39	15.76 ± 2.40	126.90 ± 1.00	19.54 ± 2.76	−21.20 ± 1.98
	Eucalyptol 200 μL/L	38.87 ± 1.00	−13.76 ± 0.69	19.64 ± 1.40	125.23 ± 0.73	27.15 ± 2.72	−19.09 ± 0.36

The presented values are the means (±standard errors) of six biological replicates (per treatment). Values for day 0 refer to the control (non-treated). Different small Latin letters indicate statistically significant differences among treatments on each day (for each column).

When purchasing fresh produce, consumers frequently select firmer fruits since they preserve better in the refrigerator and/or have longer shelf life. During storage and fruit ripening, fruits' firmness decreases due to the weakening of the structure of their cell wall and the activation of enzymes such as wall hydrolases and pectinases, resulting in soft fruits that are not accepted by the consumers [56]. On the fifth day of storage, fruits exposed to Ros 200 μL/L revealed higher firmness than the relevant fruits exposed to Lav 200 μL/L and control fruits (Table 2), and this reflected the fruit crispness and juiciness [57]. Interestingly, a decrease in fruit firmness was found with Lav 200 μL/L and Ros 100 μL/L in comparison to the control at the end of storage.

In light of these results, the effects of the applied EOs on fruit firmness may be attributed to the secondary components of the EOs rather than the major one (i.e., eucalyptol). This can be justified if other chemical compounds of the examined EOs could be tested individually or in combination, in order to identify the corresponding effects on the preservation of fruit firmness.

Table 2. Impacts of lavender (Lav) and rosemary (Ros) EOs, their mixture (Lav + Ros, 1:1 v/v) and eucalyptol vapors on cucumber's firmness, total soluble solids (TSS), titratable acidity (TA), ripening index and ascorbic acid (AA) content during storage at 11 °C for 10 days.

	Concentration	Firmness (N)	TSS (°Brix)	TA (g Malic Acid/L)	Ripening Index (TSS/TA)	AA (mg AA/100 g)
Day 0	Control	15.24 ± 0.86	2.90 ± 0.15	1.06 ± 0.19	29.23 ± 5.27	2.61 ± 0.23 B
Day 5	Control	14.66 ± 0.47 b	3.10 ± 0.06	1.38 ± 0.24 a	24.10 ± 4.51 c	2.32 ± 0.06 e
	Lav 100 μL/L	14.90 ± 1.00 ab	3.20 ± 0.00	0.78 ± 0.01 b	41.09 ± 0.49 c	2.63 ± 0.16 de
	Lav 200 μL/L	14.31 ± 0.50 b	3.00 ± 0.17	0.53 ± 0.02 bc	56.55 ± 4.77 bc	2.23 ± 0.04 e
	Ros 100 μL/L	15.34 ± 0.32 ab	3.10 ± 0.15	0.60 ± 0.03 bc	51.71 ± 4.36 c	2.27 ± 0.13 e
	Ros 200 μL/L	18.45 ± 1.30 a	3.03 ± 0.07	0.53 ± 0.03 bc	57.84 ± 2.18 bc	3.10 ± 0.06 cd
	Lav + Ros 100 μL/L	15.38 ± 0.53 ab	3.30 ± 0.15	0.31 ± 0.10 c	81.99 ± 5.31 abc	3.65 ± 0.21 bc
	Lav + Ros 200 μL/L	16.14 ± 0.54 ab	2.87 ± 0.13	0.45 ± 0.01 cb	63.65 ± 4.68 bc	4.16 ± 0.16 ab
	Eucalyptol 100 μL/L	16.19 ± 1.19 ab	2.93 ± 0.07	0.29 ± 0.06 c	113.66 ± 32.17 ab	3.74 ± 0.14 abc
	Eucalyptol 200 μL/L	15.25 ± 0.76 ab	2.90 ± 0.20	0.21 ± 0.02 c	139.88 ± 12.52 a	4.40 ± 0.19 a
Day 10	Control	17.27 ± 0.74 a	3.07 ± 0.23	0.72 ± 0.02 a	42.99 ± 4.37	5.58 ± 0.18 abA
	Lav 100 μL/L	16.68 ± 0.66 ab	2.67 ± 0.03	0.31 ± 0.11 b	119.47 ± 51.66	4.88 ± 0.09 bcd
	Lav 200 μL/L	13.91 ± 0.81 b	2.70 ± 0.15	0.21 ± 0.07 b	149.57 ± 36.81	6.37 ± 0.10 a
	Ros 100 μL/L	14.29 ± 0.87 b	2.60 ± 0.10	0.23 ± 0.01 b	112.28 ± 9.43	5.96 ± 0.08 a
	Ros 200 μL/L	15.37 ± 0.66 ab	2.87 ± 0.12	0.33 ± 0.11 b	122.71 ± 58.93	4.45 ± 0.38 cd
	Lav + Ros 100 μL/L	14.74 ± 0.59 ab	2.73 ± 0.15	0.19 ± 0.06 b	112.75 ± 12.31	6.07 ± 0.32 a
	Lav + Ros 200 μL/L	15.88 ± 0.23 ab	2.63 ± 0.09	0.37 ± 0.03 b	73.48 ± 6.17	4.04 ± 0.15 d
	Eucalyptol 100 μL/L	15.90 ± 1.18 ab	3.03 ± 0.12	0.19 ± 0.05 b	178.93 ± 39.01	5.89 ± 0.15 ab
	Eucalyptol 200 μL/L	16.30 ± 0.50 ab	3.00 ± 0.06	0.19 ± 0.04 b	170.81 ± 40.59	5.39 ± 0.20 abc

The presented values are the means (±standard errors) of six biological replicates (per treatment). Values for day 0 refer to the control (non-treated). Different small Latin letters indicate statistically significant differences among treatments on each day (for each column). Different Latin capital letters show significant difference between control on the initial day (day 0) and the last day of storage (day 10).

A decrease in TA and an increase in TSS are generally observed during the ripening of fruits and it is expected from respiring fruits as the organic acids such as malic or citric acid are the primary substrates for the respiration process [56]. In this study, no significant changes in TSS content were found throughout storage. Given that cucumber fruit is harvested and eaten while it is still at the breaker stage (immature fruit), this fact could help to explain why we observed unchanged TSS values during the EO treatment in the short period between 5 and 10 days under storage conditions. On the other hand, all applied treatments resulted in decreased TA on both occasions (Table 2). Higher values of ripening indices were reported with Eucalyptol (100 and 200 μL/L) on the fifth day, while no changes were found at the end of storage. A previous study presented a decrease in TA in cucumber fruits exposed to the vapors of a green-based product of eucalyptus and rosemary EOs, while TSS was not affected [47]. Exposure of apples to rosemary and clove EOs resulted in decreased TA and slightly higher TSS compared to cinnamon and citronella grass EOs, even two days after their application [27]. This antithesis could be attributable to the different EOs and their applied concentrations, or to the different tested produce (i.e., climacteric and non-climacteric fruits). Since EOs could interfere with the metabolism of the fruit, they could possibly alter the ratio of TSS and TA and subsequently affect the fruit ripening process [47,56]. Interestingly, a previous study has also mentioned that eucalyptol disrupted metabolism in tomatoes [58]. The AA content of cucumbers was increased after the application of the tested treatments (except Lav 100 and 200 μL/L, and Ros 100 μL/L) on the fifth day. However, a decrease in AA content was found with Ros 200 μL/L and Lav + Ros 200 μL/L on the last day (day 10) compared to the non-treated fruits (control) (Table 2). Interestingly, AA levels were increased at the end of storage for the non-treated fruits compared to the initial day (day 0). Higher AA content was also observed in lemons after their exposure to lavender EO [18]. The increase in AA content might be connected to the antioxidant properties of the applied EOs and their main compound (eucalyptol), which protect the fruit's components from oxidation while at the same time ignite the fruit's antioxidant mechanisms, as a response to the applied abiotic stress [14,17,20]. This increase in AA is desirable since it improves the nutritional value of cucumber fruits, which is highly preferred by the consumers.

3.2. Impact on Fruits' Polyphenols and Antioxidants

As shown in Table 3, all treatments (apart from Lav 100 μL/L) resulted in lower phenolic content in cucumber fruits than the non-treated (control) fruits on day 5. After 10 days, the applications of Lav (100 and 200 μL/L), Eucalyptol (100 and 200 μL/L) and Ros 100 μL/L declined the fruits' phenols content compared to the control. Other studies on vapor application of rosemary and eucalyptus EO-based products resulted in no significant changes in phenolic content in cucumber fruits [47]. The findings of this study differ from the aforementioned work due to the different EO concentrations and composition of the applied treatments, as well as the duration of the application. On the other hand, during this study, increased antioxidants were observed with Lav 100 μL/L and Lav 200 μL/L on day 5 and 10, respectively (DPPH assay). In contrast, a decrease in antioxidants was found with exposure to Lav 200 μL/L, Ros (100 and 200 μL/L) and Eucalyptol (100 and 200 μL/L) on the fifth day (FRAP assay) (Table 3). In addition, Ros 100 μL/L, Lav + Ros 200 μL/L and Eucalyptol 100 μL/L further decreased antioxidants on the last day (FRAP assay). After five days of exposure to Ros (100 and 200 μL/L), Lav + Ros (100 and 200 μL/L) and Eucalyptol (100 and 200 μL/L), a decrease in antioxidant levels (ABTS assay) was observed. Similarly, those treatments also caused a decrease in cucumbers' antioxidant levels on the tenth day, whereas Lav 100 μL/L resulted in an increase in antioxidants (ABTS assay) (Table 3), indicating increased nutritive value produce. The decrease in cucumbers' antioxidant levels might be attributed to the protective effect of the applied EOs that shield plant tissue from biotic and abiotic stress, suppressing defense mechanisms of the plant tissue due to their antioxidant activity [14,59]. The antioxidant capacity of the fruit is not unlimited and might be exhausted after a certain point. In this sense, the cucumber fruit exposed to EO treatment might reveal higher antioxidant status even before the 5 days (where the first sampling took place). Interestingly, AA levels were increased at the end of the storage period for the non-treated fruits, and were compared to the initial day (day 0). The differences in phenolic content and antioxidant capacity of cucumbers during storage in this study might be linked to the different antioxidant activity of the applied EOs. Previous findings showed that a rosemary EO presented greater antioxidant activity than a lavender EO [60]. This could explain the decrease in antioxidants and polyphenols in rosemary-treated cucumber fruits in our case, due to reduced oxidative stress.

Table 3. Impacts of lavender (Lav) and rosemary (Ros) EOs, their mixture (Lav + Ros, 1:1 *v*/*v*) and eucalyptol vapors on cucumber's phenols content and antioxidants (examined by DPPH, FRAP and ABTS assays) during storage at 11 °C for 10 days.

	Concentration	Phenols (μg GAE/g)	DPPH (μg trolox/g)	FRAP (μg trolox/g)	ABTS (μg trolox/g)
Day 0	**Control**	52.48 ± 0.59 B	17.02 ± 1.40 A	36.83 ± 3.22	93.45 ± 0.73 A
Day 5	**Control**	63.71 ± 1.30 a	13.47 ± 0.99 b	35.21 ± 0.86 a	78.40 ± 1.02 ab
	Lav 100 μL/L	59.65 ± 0.67 ab	17.88 ± 0.53 a	33.60 ± 1.87 ab	82.93 ± 2.01 ab
	Lav 200 μL/L	50.57 ± 1.22 c	11.64 ± 0.65 b	26.34 ± 1.62 cd	83.52 ± 2.62 a
	Ros 100 μL/L	53.91 ± 2.67 bc	15.67 ± 1.32 ab	28.62 ± 1.40 bc	61.77 ± 1.37 cd
	Ros 200 μL/L	48.80 ± 2.10 c	14.67 ± 0.54 ab	27.61 ± 1.05 bcd	67.27 ± 0.82 c
	Lav + Ros 100 μL/L	53.76 ± 2.32 bc	14.41 ± 1.10 ab	27.66 ± 1.14 bcd	75.38 ± 1.46 b
	Lav + Ros 200 μL/L	46.08 ± 0.43 c	12.37 ± 0.75 b	22.44 ± 1.13 d	60.41 ± 1.44 cd
	Eucalyptol 100 μL/L	49.14 ± 0.96 c	13.18 ± 0.87 b	23.18 ± 0.77 cd	58.32 ± 1.14 d
	Eucalyptol 200 μL/L	53.24 ± 1.92 bc	13.55 ± 0.64 b	25.97 ± 0.71 cd	66.27 ± 1.61 c
Day 10	**Control**	62.73 ± 0.85 aA	8.45 ± 0.84 bcdB	36.64 ± 1.74 ab	81.37 ± 1.42 bB
	Lav 100 μL/L	51.62 ± 0.21 ab	11.40 ± 0.36 abc	32.55 ± 1.37 bc	89.92 ± 2.45 a
	Lav 200 μL/L	55.08 ± 1.15 ab	12.25 ± 0.62 a	39.07 ± 1.29 a	83.61 ± 1.13 ab
	Ros 100 μL/L	48.79 ± 1.56 c	11.61 ± 0.95 ab	31.53 ± 0.73 cd	69.29 ± 2.20 cd
	Ros 200 μL/L	64.20 ± 1.24 a	9.24 ± 0.53 abcd	36.33 ± 0.62 ab	59.86 ± 1.55 ef
	Lav + Ros 100 μL/L	66.85 ± 1.84 a	7.28 ± 0.93 d	32.76 ± 0.41 bc	56.39 ± 0.72 f
	Lav + Ros 200 μL/L	63.57 ± 1.44 a	8.12 ± 0.48 cd	29.61 ± 0.09 cd	65.23 ± 0.36 de
	Eucalyptol 100 μL/L	50.82 ± 1.60 ab	10.86 ± 0.59 abc	26.97 ± 0.47 d	72.58 ± 0.66 c
	Eucalyptol 200 μL/L	55.39 ± 1.30 b	8.57 ± 0.47 bcd	32.59 ± 0.61 bc	64.62 ± 1.40 de

The presented values are the means (±standard errors) of six biological replicates (per treatment). Values for day 0 refer to the control (non-treated). Different small Latin letters indicate statistically significant differences among treatments on each day (for each column). Different Latin capital letters show significant difference between control on the initial day (day 0) and the last day of storage (day 10).

3.3. *Fruit Damage Index and Enzymatic Antioxidant Activity*

Essential oils at high concentrations can trigger phytotoxicity and oxidative stress on fresh commodities, and this has been demonstrated by the EO tomato fruit vaporized with sage [24]. During storage, the metabolic rate of fruit increases, and abiotic and biotic stress factors usually occur, resulting in the production and accumulation of reactive oxygen species (ROS) such as hydrogen peroxide (H_2O_2). Fruit alleviates oxidative stress by the activation of both non-enzymatic (phenols, antioxidants, ascorbic acid, proline, etc.) and endogenous enzymatic (SOD, CAT, POD, ascorbate peroxidase-APX, glutathione peroxidase-GPX, etc.) metabolites. However, changes in non-enzymatic or enzymatic metabolites can take place independently or in parallel in order to detoxify fresh produce. The main endogenous (SOD) antioxidant enzymes convert the superoxide anion to H_2O_2, which is a substrate for CAT and GPX. Catalase metabolizes H_2O_2 to water and oxygen, and GPX decreases both H_2O_2 and organic hydroperoxides when reacting with glutathione (GSH) [61]. Thus, molecules such as H_2O_2 and malondialdehyde (MDA) are used for the investigation of plant tissue damage [62]. The impacts of the volatile treatments on cucumbers' damage indices and the activities of the antioxidant enzymes are presented in Table 4. Exposure of cucumbers for five days to Eucalyptol 200 μL/L resulted in decreased H_2O_2 levels compared to the control, Ros 100 μL/L, Lav + Ros 200 μL/L and Eucalyptol 100 μL/L. On the other hand, no significant differences in H_2O_2 levels were evidenced on the last day (Table 4). The production of MDA decreased with Lav (100 and 200 μL/L),

Ros 200 μL/L and Eucalyptol 200 μL/L on the fifth day, whereas Eucalyptol 200 μL/L resulted in increased MDA levels compared to the control, Lav (100 and 200 μL/L) and Ros 200 μL/L (Table 4). The lower MDA levels obtained from fruits treated with EO can be related to the enhanced activation of defense-related enzymes in response to the oxidative stress condition [63] but also to the non-enzymatic metabolites involvement, such as phenols and AA.

Table 4. Impacts of lavender (Lav) and rosemary (Ros) EOs, their mixture (Lav + Ros, 1:1 *v*/*v*) and eucalyptol vapors on cucumber's hydrogen peroxide (H_2O_2), lipid peroxidation (MDA) and the antioxidant enzyme activity of superoxide dismutase (SOD), catalase (CAT) and peroxidase (POD) during storage at 11 °C for 10 days.

	Concentration	H_2O_2 (mmol/g)	MDA (nmol/g)	SOD (units/mg of Protein)	CAT (units/mg of Protein)	POD (units/mg of Protein)
Day 0	**Control**	0.14 ± 0.00 B	9.63 ± 0.10 A	3.12 ± 0.35	4.27 ± 0.51	10.82 ± 1.72
Day 5	**Control**	0.15 ± 0.00 a	9.46 ± 0.09 abc	2.23 ± 0.16	3.26 ± 0.22 ab	15.42 ± 2.01 bc
	Lav 100 μL/L	0.14 ± 0.01 ab	8.96 ± 0.15 c	2.46 ± 0.43	3.05 ± 0.21 ab	26.88 ± 4.25 a
	Lav 200 μL/L	0.14 ± 0.00 ab	9.37 ± 0.14 bc	2.74 ± 0.04	5.69 ± 1.01 a	30.32 ± 1.76 a
	Ros 100 μL/L	0.16 ± 0.01 a	9.67 ± 0.33 abc	2.24 ± 0.16	2.81 ± 0.47 ab	14.73 ± 0.23 bc
	Ros 200 μL/L	0.14 ± 0.00 ab	8.54 ± 0.33 c	1.79 ± 0.10	4.29 ± 1.14 ab	29.73 ± 3.81 a
	Lav + Ros 100 μL/L	0.14 ± 0.01 ab	10.09 ± 0.32 abc	2.10 ± 0.09	1.69 ± 0.28 b	9.94 ± 0.24 c
	Lav + Ros 200 μL/L	0.15 ± 0.00 a	10.32 ± 0.29 abc	2.30 ± 0.15	2.50 ± 0.24 b	14.75 ± 1.14 bc
	Eucalyptol 100 μL/L	0.15 ± 0.00 a	11.60 ± 1.17 ab	1.79 ± 0.34	3.45 ± 0.43 ab	12.18 ± 1.44 c
	Eucalyptol 200 μL/L	0.12 ± 0.01 b	11.64 ± 0.07 a	1.51 ± 0.32	2.98 ± 0.62 ab	24.59 ± 3.50 ab
Day 10	**Control**	0.18 ± 0.01 A	6.98 ± 0.14 dB	3.01 ± 0.71 a	4.45 ± 0.17 de	16.54 ± 2.52 c
	Lav 100 μL/L	0.17 ± 0.01	7.70 ± 0.09 cd	1.90 ± 0.24 ab	9.58 ± 1.12 c	13.69 ± 1.88 c
	Lav 200 μL/L	0.17 ± 0.00	7.82 ± 0.14 cd	1.65 ± 0.18 ab	19.64 ± 0.59 a	37.01 ± 2.26 a
	Ros 100 μL/L	0.17 ± 0.01	8.93 ± 0.28 ab	1.41 ± 0.29 ab	4.99 ± 0.95 de	34.65 ± 7.24 ab
	Ros 200 μL/L	0.18 ± 0.00	8.08 ± 0.31 bc	1.60 ± 0.30 ab	7.70 ± 0.19 cd	26.95 ± 0.91 abc
	Lav + Ros 100 μL/L	0.18 ± 0.01	8.46 ± 0.27 abc	0.95 ± 0.05 b	3.49 ± 0.69 de	25.19 ± 3.81 abc
	Lav + Ros 200 μL/L	0.19 ± 0.01	8.52 ± 0.12 abc	2.83 ± 0.46 a	14.87 ± 1.79 b	30.18 ± 3.94 abc
	Eucalyptol 100 μL/L	0.20 ± 0.01	8.26 ± 0.13 abc	1.80 ± 0.08 ab	3.31 ± 0.20 e	19.57 ± 0.37 bc
	Eucalyptol 200 μL/L	0.17 ± 0.00	9.19 ± 0.20 a	1.52 ± 0.11 ab	5.93 ± 0.46 cde	34.81 ± 1.68 ab

The presented values are the means (±standard errors) of three biological replicates (per treatment). Values for day 0 refer to the control (non-treated). Different small Latin letters indicate statistically significant differences among treatments on each day (for each column). Different Latin capital letters show significant difference between control on the initial day (day 0) and the last day of storage (day 10).

After 5 days of storage, SOD activity remained at similar levels for all tested fruits, indicating that the antioxidant capacity of the enzyme has been exhausted since there were changes in the activities of other antioxidant enzymes, as indicated by CAT and POD activities. Therefore, the increased CAT and POD levels were associated with decreased MDA levels, reflecting the antioxidant capacity of CAT and POD to protect fruit from oxidative damage. It has been reported that when the clove EO (0.4%) was applied to citrus fruits, it increased the activity of the enzymes involved in plant defense, including POD, phenylalanine ammonia lyase (PAL), polyphenol oxidase (PPO) and lipoxygenase (LOX) [63]. Similar to this, over a 3-day storage period at 20 °C, strawberries treated with vapors of tea tree EO exhibited increased SOD, PAL and POD activity [64].

Increased H_2O_2 levels were found in untreated cucumbers (control) on the last day, in comparison to the initial day (Day 0). Moreover, decreased MDA levels were found in untreated cucumbers (control) on the last day when compared to Day 0, whereas the activity of the examined antioxidant enzymes did not differ between Day 0 and Day 10, when taking the levels of SOD, CAT and POD into consideration. As rosemary EO has significant antioxidant activity, higher than other essential oils (compared to lavender EO), applying it alone and/or in combination appeared to reduce the oxidative stress of cucumbers while they were being stored [60]. The reduction in the oxidative stress indicators by the vapor of the applied EOs is mainly attributed to their composition (i.e., phenols, flavonoids, etc.), which exhibit free radical scavenging activities while at the same time increasing the antioxidant capacity of the produce they are applied on [59].

3.4. Impact on Microbial Load

Increased microbial load could result in higher decay incidence and increased fresh produce losses [7]. Microbial control, both at preharvest and postharvest level, is of great importance for the preservation of fresh produce with high quality and safety features. As shown in Table 5, a decrease in TVC numbers was found with Ros 200 μL/L at the end of the storage period of 10 days. On the other hand, exposure to Lav 200 μL/L, Ros 100 μL/L, Lav + Ros 200 μL/L and Eucalyptol (100 and 200 μL/L) presented higher numbers on the tenth day. As reported previously, if 5 log cfu/mL (or 5.95 log cfu/g, considering the dilution rates in this study) of aerobic plate counts or yeast and mold counts is considered as the critical limit [57], only the Lav 100–200 μL/L, Ros 200 μL/L and Lav + Ros 100 μL/L were within these limits for TVC and decreased the yeast and mold numbers and extended the microbial shelf life of cucumbers after 10 days of storage. The increase in microbial numbers might be attributed to the volatile nature of the EOs components that may evaporate after a long period, and due to their antimicrobial activity weakening. Moreover, this increase might be caused by the possible leaking of nutrients, i.e., calcium, which is responsible for maintaining fruit firmness from potentially damaged fruit cell walls [65]. Yeast and mold were decreased after the exposure to Eucalyptol 200 μL/L, opposite to Lav + Ros 100 μL/L, on the fifth day of storage (Table 5). Moreover, Lav (100 and 200 μL/L) and Ros 200 μL/L also decreased yeast and mold on day 10. According to this study's results, lavender and rosemary EOs were the main factors responsible for the reduction in cucumbers' microbial load. This effect does not appear to be solely related to the eucalyptol (their major component), rather, it may also be due to the synergistic effect of other components which are present in lower concentrations in the EOs [26]. The antimicrobial activity (antibacterial and antifungal) of EOs is attributed to the hydrophobic nature of their constitutes that enables them to penetrate the microbial cell wall membrane and interfere with the cell wall permeability and structure, among other vital processes (i.e., quorum sensing) [59,66].

Table 5. Impacts of lavender (Lav) and rosemary (Ros) EOs, their mixture (Lav + Ros, 1:1 *v*/*v*) and eucalyptol vapors on cucumber's microbial load (total viable count-TVC, yeast and mold) during storage at 11 °C for 10 days.

	Concentration	TVC (log cfu/g)	Yeast and Mold (log cfu/g)
Day 0	Control	5.81 ± 0.10	4.12 ± 0.17
Day 5	Control	4.67 ± 0.03	4.59 ± 0.07 ab
	Lav 100 μL/L	4.71 ± 0.02	4.67 ± 0.18 ab
	Lav 200 μL/L	4.45 ± 0.38	4.47 ± 0.01 ab
	Ros 100 μL/L	4.70 ± 0.07	4.43 ± 0.22 ab
	Ros 200 μL/L	4.43 ± 0.10	4.65 ± 0.08 ab
	Lav + Ros 100 μL/L	4.86 ± 0.05	4.85 ± 0.03 a
	Lav + Ros 200 μL/L	4.58 ± 0.05	4.43 ± 0.04 ab
	Eucalyptol 100 μL/L	4.28 ± 0.12	4.32 ± 0.04 ab
	Eucalyptol 200 μL/L	4.21 ± 0.05	4.08 ± 0.13 b

Table 5. *Cont.*

	Concentration	TVC (log cfu/g)	Yeast and Mold (log cfu/g)
Day 10	**Control**	5.38 ± 0.15 d	4.73 ± 0.03 ab
	Lav 100 µL/L	5.69 ± 0.08 cd	3.73 ± 0.00 cd
	Lav 200 µL/L	5.88 ± 0.02 abc	3.30 ± 0.35 d
	Ros 100 µL/L	6.00 ± 0.00 abc	4.88 ± 0.01 ab
	Ros 200 µL/L	4.91 ± 0.05 e	3.11 ± 0.16 d
	Lav + Ros 100 µL/L	5.80 ± 0.05 bcd	4.41 ± 0.09 bc
	Lav + Ros 200 µL/L	6.20 ± 0.03 ab	5.18 ± 0.01 a
	Eucalyptol 100 µL/L	6.19 ± 0.08 ab	4.89 ± 0.07 ab
	Eucalyptol 200 µL/L	6.29 ± 0.15 a	4.56 ± 0.01 ab

The presented values are the means (±standard errors) of three biological replicates (per treatment). Values for day 0 refer to the control (non-treated). Different small Latin letters indicate statistically significant differences among treatments on each day (for each column).

4. Conclusions

The vapor application of lavender and rosemary EOs, their 1:1 (v/v) mixture and their major compound (eucalyptol) on cucumber fruits showed no significant differences in features such as the fruits' weight loss and color. Notably, fruits' firmness was maintained throughout storage (exceptions occurred when fruits' were exposed to Lav 200 µL/L and Ros 100 µL/L). Increased oxidative stress was found after the exposure to Eucalyptol 200 µL/L. On the other hand, the applied treatments resulted in significant increases in antioxidants and AA content (i.e., vitamin C), increasing the nutritional value of the cucumber fruit. The findings of this study indicate that the examined means (the tested EOs, their mixture and the main common compound) could be considered as alternative means for the postharvest preservation of cucumber fruits. As the Ros 200 µL/L treatment revealed increased fruit firmness and AA content after 5 days of application, and decreased TVC and yeast and mold after the 10 days of application, combining high nutritive value and low antimicrobial activity appears to be an effective application when applied during storage. However, the application of such compounds should be performed with caution as it could, in some cases, negatively affect the organoleptic characteristics (i.e., aroma, taste, appearance) of the tested fresh produce and could potentially result in phytotoxicity when used in high concentrations. In addition, the EOs mixture as a preservation means can be explored further, as the combination of the different EO components and the possible synergistic effect on fruit preservation is of high interest. Another factor to consider when using essential oils is their perceived high cost. However, if used in the recommended dosage (concentration x time of application), this could be a cost-effective method. Additional considerations for using EOs include the possibility of phytotoxicity in the applied produce and allergy issues upon consumption.

Supplementary Materials: The following supporting information can be downloaded at: https://www.mdpi.com/article/10.3390/agronomy13102493/s1, Figure S1: Impacts of lavender (Lav) and rosemary (Ros) EOs, their mixture (1:1 v/v) and eucalyptol vapors on cucumber fruits during storage at 11 °C for 10 days.

Author Contributions: Conceptualization, N.T.; methodology, P.X.; software, P.X.; validation, A.C.; formal analysis, P.X., C.G. and A.C.; investigation, P.X., C.G. and A.C.; resources, N.T.; data curation, P.X., C.G. and A.C.; writing—original draft preparation, P.X., and A.C.; writing—review and editing, A.C. and N.T.; visualization, A.C.; supervision, P.X. and N.T.; project administration, N.T.; funding acquisition, N.T. All authors have read and agreed to the published version of the manuscript.

Funding: This research was funded by the PRIMA StopMedWaste project, which is funded by PRIMA, a program supported by the European Union with co-funding by the Funding Agencies of the Research and Innovation Foundation (RIF), Cyprus.

Data Availability Statement: Not applicable.

Conflicts of Interest: The authors declare no conflict of interest.

References

1. Shi, J.; Wang, J.; Li, R.; Li, D.; Xu, F.; Sun, Q.; Zhao, B.; Mao, A.J.; Guo, Y.D. Expression patterns of genes encoding plasma membrane aquaporins during fruit development in cucumber (*Cucumis sativus* L.). *Plant Physiol. Biochem.* **2015**, *96*, 329–336. [CrossRef] [PubMed]
2. Brookie, K.L.; Best, G.I.; Conner, T.S. Intake of raw fruits and vegetables is associated with better mental health than intake of processed fruits and vegetables. *Front. Psychol.* **2018**, *9*, 487. [CrossRef] [PubMed]
3. Naureen, Z.; Dhuli, K.; Donato, K.; Aquilanti, B.; Velluti, V.; Matera, G.; Iaconelli, A.; Bertelli, M. Foods of the Mediterranean diet: Citrus, cucumber and grape. *J. Prev. Med. Hyg.* **2022**, *63*, E21–E27. [PubMed]
4. Kahramanoğlu, I.; Usanmaz, S. Improving postharvest storage quality of cucumber fruit by modified atmosphere packaging and biomaterials. *HortScience* **2019**, *54*, 2005–2014. [CrossRef]
5. Bahnasawy, A.H.; Khater, E.-S.G. Effect of Wax Coating on the Quality of Cucumber Fruits during Storage. *J. Food Process. Technol.* **2014**, *5*, 1. [CrossRef]
6. Kang, H.M.; Park, K.W.; Saltveit, M.E. Elevated growing temperatures during the day improve the postharvest chilling tolerance of greenhouse-grown cucumber (*Cucumis sativus*) fruit. *Postharvest Biol. Technol.* **2002**, *24*, 49–57. [CrossRef]
7. Xylia, P.; Botsaris, G.; Chrysargyris, A.; Skandamis, P.; Tzortzakis, N. Variation of microbial load and biochemical activity of ready-to-eat salads in Cyprus as affected by vegetable type, season, and producer. *Food Microbiol.* **2019**, *83*, 200–210. [CrossRef]
8. Jahan, S.E.; Hassan, M.K.; Roy, S.; Ahmed, Q.M.; Hasan, G.N.; Muna, A.Y.; Sarkar, M.N. Effects of different postharvest treatments on nutritional quality and shelf life of cucumber. *Asian J. Crop. Soil Sci. Plant Nutr.* **2020**, *2*, 51–61. [CrossRef]
9. Coroneo, V.; Carraro, V.; Marras, B.; Marrucci, A.; Succa, S.; Meloni, B.; Pinna, A.; Angioni, A.; Sanna, A.; Schintu, M. Presence of Trihalomethanes in ready-to-eat vegetables disinfected with chlorine. *Food Addit. Contam.—Part A Chem. Anal. Control. Expo. Risk Assess.* **2017**, *34*, 2111–2117. [CrossRef]
10. Xylia, P.; Clark, A.; Chrysargyris, A.; Romanazzi, G.; Tzortzakis, N. Quality and safety attributes on shredded carrots by using *Origanum majorana* and ascorbic acid. *Postharvest Biol. Technol.* **2019**, *155*, 120–129. [CrossRef]
11. Stavropoulou, A.; Loulakakis, K.; Magan, N.; Tzortzakis, N. *Origanum dictamnus* Oil Vapour Suppresses the Development of Grey Mould in Eggplant Fruit in Vitro. *Biomed Res. Int.* **2014**, *2014*, 562679. [CrossRef] [PubMed]
12. Alvarez, M.V.; Ortega-Ramirez, L.A.; Silva-Espinoza, B.A.; Gonzalez-Aguilar, G.A.; Ayala-Zavala, J.F. Antimicrobial, antioxidant, and sensorial impacts of oregano and rosemary essential oils over broccoli florets. *J. Food Process. Preserv.* **2019**, *43*, e13889. [CrossRef]
13. Caroline, S.I.; Sylvain Leroy, S.K.; Nachigera Gustave, M.; Essia, N.J.-J. Influence of chlorine and rosemary essential oil postharvest pre-treatments on quality parameters of fresh tomatoes during storage. *J. Postharvest Technol.* **2018**, *6*, 57–68.
14. Yuan, G.; Chen, X.; Li, D. Chitosan films and coatings containing essential oils: The antioxidant and antimicrobial activity, and application in food systems. *Food Res. Int.* **2016**, *89*, 117–128. [CrossRef]
15. De Rapper, S.; Kamatou, G.; Viljoen, A.; Vuuren, S. Van The In Vitro Antimicrobial Activity of *Lavandula angustifolia* Essential Oil in Combination with Other Aroma-Therapeutic Oils. *Evid.-Based Complement. Altern. Med.* **2013**, *2013*, 852049. [CrossRef] [PubMed]
16. Silva, G.L.D.A.; Luft, C.; Lunardelli, A.; Amaral, R.H. Antioxidant, analgesic and anti-inflammatory effects of lavender essential oil. *An. Acad. Bras. Cienc.* **2015**, *87*, 1397–1408. [CrossRef] [PubMed]
17. Cardia, F.G.E.; Silva-filho, S.E.; Silva, E.L.; Uchida, N.S.; Augusto, H.; Cavalcante, O.; Cassarotti, L.L.; Eduardo, V.; Salvadego, C.; Spironello, R.A.; et al. Effect of Lavender (*Lavandula angustifolia*) Essential Oil on Acute Inflammatory Response. *Evidence-Based Complement. Altern. Med.* **2018**, *2018*, 1413940. [CrossRef]
18. Sumalan, R.M.; Kuganov, R.; Obistioiu, D.; Popescu, I.; Sumalan, R.L.; Cocan, I. Assessment of Mint, Basil, and Lavender Essential Oil Vapor-Phase in Antifungal Protection and Lemon Fruit Quality. *Molecules* **2020**, *25*, 1831. [CrossRef]
19. Abo Ghanima, M.M.; Elsadek, M.F.; Taha, A.E.; Abd El-Hack, M.E.; Alagawany, M.; Ahmed, B.M.; Elshafie, M.M.; El-Sabrout, K. Effect of housing system and rosemary and cinnamon essential oils on layers performance, egg quality, haematological traits, blood chemistry, immunity, and antioxidant. *Animals* **2020**, *10*, 245. [CrossRef]
20. da Silva Bomfim, N.; Kohiyama, C.Y.; Nakasugi, L.P.; Nerilo, S.B.; Mossini, S.A.G.; Romoli, J.C.Z.; Graton Mikcha, J.M.; de Abreu Filho, B.A.; Machinski, M. Antifungal and antiaflatoxigenic activity of rosemary essential oil (*Rosmarinus officinalis* L.) against *Aspergillus flavus*. *Food Addit. Contam.—Part A Chem. Anal. Control. Expo. Risk Assess.* **2020**, *37*, 153–161. [CrossRef]
21. Gahruie, H.H.; Hosseini, S.M.H.; Taghavifard, M.H.; Eskandari, M.H.; Golmakani, M.T.; Shad, E. Lipid oxidation, color changes, and microbiological quality of frozen beef burgers incorporated with shirazi thyme, cinnamon, and rosemary extracts. *J. Food Qual.* **2017**, *2017*, 6350156.

22. Halmschlag, C.B.; Carneiro de Melo Moura, C.; Brambach, F.; Siregar, I.Z.; Gailing, O. Molecular and morphological survey of Lamiaceae species in converted landscapes in Sumatra. *PLoS ONE* **2022**, *17*, e0277749. [CrossRef] [PubMed]
23. Tzortzakis, N.; Xylia, P.; Chrysargyris, A. Sage essential oil improves the effectiveness of *Aloe vera* gel on postharvest quality of tomato fruit. *Agronomy* **2019**, *9*, 635. [CrossRef]
24. Chrysargyris, A.; Rousos, C.; Xylia, P.; Tzortzakis, N. Vapour application of sage essential oil maintain tomato fruit quality in breaker and red ripening stages. *Plants* **2021**, *10*, 2645. [CrossRef]
25. Xylia, P.; Chrysargyris, A.; Botsaris, G.; Tzortzakis, N. Potential application of spearmint and lavender essential oils for assuring endive quality and safety. *Crop Prot.* **2017**, *102*, 94–103. [CrossRef]
26. Xylia, P.; Chrysargyris, A.; Ahmed, Z.F.R.F.F.R.; Tzortzakis, N.; Tzortzakis, N. Application of rosemary and eucalyptus essential oils and their main component on the preservation of apple and pear fruits. *Horticulture* **2021**, *7*, 479. [CrossRef]
27. Vieira, A.M.F.D.; Steffens, C.A.; Argenta, L.C.; do Amarante, C.V.T.; Oster, A.H.; Casa, R.T.; Amarante, A.G.M.; Espíndola, B.P. Essential oils for the postharvest control of blue mold and quality of "Fuji" apples. *Pesqui. Agropecu. Bras.* **2018**, *53*, 547–556. [CrossRef]
28. Chrysargyris, A.; Panayiotou, C.; Tzortzakis, N. Nitrogen and phosphorus levels affected plant growth, essential oil composition and antioxidant status of lavender plant (*Lavandula angustifolia* Mill.). *Ind. Crops Prod.* **2016**, *83*, 577–586. [CrossRef]
29. Xylia, P.; Chrysargyris, A.; Miltiadous, P.; Tzortzakis, N. *Origanum dubium* (Cypriot Oregano) as a Promising Sanitizing Agent against *Salmonella enterica* and *Listeria monocytogenes* on Tomato and Cucumber Fruits. *Biology* **2022**, *11*, 1772. [CrossRef]
30. Tzortzakis, N.G. Maintaining postharvest quality of fresh produce with volatile compounds. *Innov. Food Sci. Emerg. Technol.* **2007**, *8*, 111–116. [CrossRef]
31. Tzortzakis, N.; Chrysargyris, A.; Sivakumar, D.; Loulakakis, K. Vapour or dipping applications of methyl jasmonate, vinegar and sage oil for pepper fruit sanitation towards grey mould. *Postharvest Biol. Technol.* **2016**, *118*, 120–127. [CrossRef]
32. Xylia, P.; Botsaris, G.; Skandamis, P.; Tzortzakis, N. Expiration Date of Ready-to-Eat Salads: Effects on Microbial Load and Biochemical Attributes. *Foods* **2021**, *10*, 941. [CrossRef] [PubMed]
33. Xylia, P.; Ioannou, I.; Chrysargyris, A.; Stavrinides, M.C.; Tzortzakis, N. Quality attributes and storage of tomato fruits as affected by an eco-friendly, essential oil-based product. *Plants* **2021**, *10*, 1125. [CrossRef]
34. Bolin, H.R.; Huxsoll, C.C. Effect of preparation procedures and storage parameters on quality retension of salad-cut lettuce. *J. Food Sci.* **1991**, *56*, 60–62. [CrossRef]
35. Goyeneche, R.; Agüero, M.V.; Roura, S.; Di Scala, K. Application of citric acid and mild heat shock to minimally processed sliced radish: Color evaluation. *Postharvest Biol. Technol.* **2014**, *93*, 106–113. [CrossRef]
36. Pathare, P.B.; Opara, U.L.; Al-Said, F.A.J. Colour Measurement and Analysis in Fresh and Processed Foods: A Review. *Food Bioprocess Technol.* **2013**, *6*, 36–60. [CrossRef]
37. AOAC International. *Official Methods of Analysis*, 18th ed.; AOAC International: Gaithersburg, MD, USA, 2007.
38. Chrysargyris, A.; Nikou, A.; Tzortzakis, N. Effectiveness of *Aloe vera* gel coating for maintaining tomato fruit quality. *N. Z. J. Crop Hortic. Sci.* **2016**, *44*, 203–217. [CrossRef]
39. Wojdyło, A.; Oszmiański, J.; Czemerys, R. Antioxidant activity and phenolic compounds in 32 selected herbs. *Food Chem.* **2007**, *105*, 940–949. [CrossRef]
40. De Azevedo Neto, A.D.; Prisco, J.T.; Enéas-Filho, J.; De Abreu, C.E.B.; Gomes-Filho, E. Effect of salt stress on antioxidative enzymes and lipid peroxidation in leaves and roots of salt-tolerant and salt-sensitive maize genotypes. *Environ. Exp. Bot.* **2006**, *56*, 87–94. [CrossRef]
41. Loreto, F.; Velikova, V. Isoprene produced by leaves protects the photosynthetic apparatus against ozone damage, quenches ozone products, and reduces lipid peroxidation of cellular membranes. *Plant Physiol.* **2001**, *127*, 1781–1787. [CrossRef]
42. Chrysargyris, A.; Hajisolomou, E.; Xylia, P.; Tzortzakis, N. Olive-mill and grape-mill waste as a substitute growing media component for unexploded vegetables production. *Sustain. Chem. Pharm.* **2023**, *31*, 100940. [CrossRef]
43. Xylia, P.; Fasko, K.G.; Chrysargyris, A.; Tzortzakis, N. Heat treatment, sodium carbonate, ascorbic acid and rosemary essential oil application for the preservation of fresh *Rosmarinus officinalis* quality. *Postharvest Biol. Technol.* **2022**, *187*, 111868. [CrossRef]
44. Cao, Z.; Zhou, D.; Ge, X.; Luo, Y.; Su, J. The role of essential oils in maintaining the postharvest quality and preservation of peach and other fruits. *J. Food Biochem.* **2022**, *46*, e14513. [CrossRef]
45. Pandey, V.K.; Islam, R.U.; Shams, R.; Dar, A.H. A comprehensive review on the application of essential oils as bioactive compounds in Nano-emulsion based edible coatings of fruits and vegetables. *Appl. Food Res.* **2022**, *2*, 100042. [CrossRef]
46. Silva, C.J.; Adaskaveg, J.A.; Mesquida-Pesci, S.D.; Ortega-Salazar, I.B.; Pattathil, S.; Zhang, L.; Hahn, M.G.; van Kan, J.A.L.; Cantu, D.; Powell, A.L.T.; et al. *Botrytis cinerea* infection accelerates ripening and cell wall disassembly to promote disease in tomato fruit. *Plant Physiol.* **2023**, *191*, 575–590. [CrossRef] [PubMed]
47. Xylia, P.; Chrysargyris, A.; Ahmed, Z.F.R.; Shahwar, D.; Tzortzakis, N. Application of Rosemary and Eucalyptus Essential Oils on the Preservation of Cucumber Fruit. *Horticulturae* **2022**, *8*, 774. [CrossRef]
48. Rabiei, V.; Shirzadeh, E.; Rabbiangourani, H.; Sharafi, Y. Effect of thyme and lavender essential oils on the qualitative and quantitative traits and storage life of apple "Jonagold" cultivar. *J. Med. Plant Res.* **2011**, *5*, 5522–5527.
49. Adams, A.-R.; Adama, A.-R.; Kuunuori Thadius, T.; Barau, B. Ginger Essential Oil for Postharvest Quality of Datterino Tomato: Effect of Immersion Duration and Storage Temperature. *J. Postharvest Technol.* **2018**, *06*, 109–121.

50. Santoro, K.; Maghenzani, M.; Chiabrando, V.; Bosio, P.; Gullino, M.L.; Spadaro, D.; Giacalone, G. Thyme and savory essential oil vapor treatments control brown rot and improve the storage quality of peaches and nectarines, but could favor gray mold. *Foods* **2018**, *7*, 7. [CrossRef]
51. Frankova, A.; Smid, J.; Bernardos, A.; Finkousova, A.; Marsik, P.; Novotny, D.; Legarová, V.; Pulkrabek, J.; Kloucek, P. The antifungal activity of essential oils in combination with warm air flow against postharvest phytopathogenic fungi in apples. *Food Control* **2016**, *68*, 62–68. [CrossRef]
52. Hyldgaard, M.; Mygind, T.; Meyer, R.L.; Debabov, D. Essential oils in food preservation: Mode of action, synergies, and interactions with food matrix components. *Front. Microbiol.* **2012**, *3*, 12. [CrossRef] [PubMed]
53. Viacava, G.E.; Ayala-Zavala, J.F.; González-Aguilar, G.A.; Ansorena, M.R. Effect of free and microencapsulated thyme essential oil on quality attributes of minimally processed lettuce. *Postharvest Biol. Technol.* **2018**, *145*, 125–133. [CrossRef]
54. Micić, D.; Đurović, S.; Riabov, P.; Tomić, A.; Šovljanski, O.; Filip, S.; Tosti, T.; Dojčinović, B.; Božović, R.; Jovanović, D.; et al. Rosemary Essential Oils as a Promising Source of Bioactive Compounds: Chemical Composition, Thermal Properties, Biological Activity, and Gastronomical Perspectives. *Foods* **2021**, *10*, 2734. [CrossRef]
55. Mohammadi, A.; Hashemi, M.; Hosseini, S.M. Postharvest treatment of nanochitosan-based coating loaded with *Zataria multiflora* essential oil improves antioxidant activity and extends shelf-life of cucumber. *Innov. Food Sci. Emerg. Technol.* **2016**, *33*, 580–588. [CrossRef]
56. Omoba, O.S.; Onyekwere, U. Postharvest physicochemical properties of cucumber fruits (*Cucumber sativus* L) treated with chitosan-lemon grass extracts under different storage durations. *Afr. J. Biotechnol.* **2016**, *15*, 2758–2766.
57. Mohammadi, A.; Hashemi, M.; Hosseini, S.M. Chitosan nanoparticles loaded with *Cinnamomum zeylanicum* essential oil enhance the shelf life of cucumber during cold storage. *Postharvest Biol. Technol.* **2015**, *110*, 203–213. [CrossRef]
58. Oms-Oliu, G.; Hertog, M.L.A.T.M.; Van de Poel, B.; Ampofo-Asiama, J.; Geeraerd, A.H.; Nicolai, B.M. Metabolic characterization of tomato fruit during preharvest development, ripening, and postharvest shelf-life. *Postharvest Biol. Technol.* **2011**, *62*, 7–16. [CrossRef]
59. Namiota, M.; Bonikowski, R. The current state of knowledge about essential oil fumigation for quality of crops during postharvest. *Int. J. Mol. Sci.* **2021**, *22*, 13351. [CrossRef]
60. Yang, S.A.; Jeon, S.K.; Lee, E.J.; Shim, C.H.; Lee, I.S. Comparative study of the chemical composition and antioxidant activity of six essential oils and their components. *Nat. Prod. Res.* **2010**, *24*, 140–151. [CrossRef]
61. Munteanu, I.G.; Apetrei, C. Analytical methods used in determining antioxidant activity: A review. *Int. J. Mol. Sci.* **2021**, *22*, 3380. [CrossRef]
62. Lu, X.; Ma, Y.; Liu, X. Effects of hydrogen peroxide accumulation, lipid peroxidation, and polyphenol oxidation during superficial scald development in 'Fuji' apples. *Hortic. Environ. Biotechnol.* **2014**, *55*, 299–307. [CrossRef]
63. Chen, C.; Cai, N.; Chen, J.; Wan, C. Clove essential oil as an alternative approach to control postharvest blue mold caused by *Penicillium italicum* in citrus fruit. *Biomolecules* **2019**, *9*, 197. [CrossRef] [PubMed]
64. Shao, X.; Wang, H.; Xu, F.; Cheng, S. Effects and possible mechanisms of tea tree oil vapor treatment on the main disease in postharvest strawberry fruit. *Postharvest Biol. Technol.* **2013**, *77*, 94–101. [CrossRef]
65. López-Gómez, A.; Ros-Chumillas, M.; Antolinos, V.; Buendía-Moreno, L.; Navarro-Segura, L.; Sánchez-Martínez, M.J.; Martínez-Hernández, G.B.; Soto-Jover, S. Fresh culinary herbs decontamination with essential oil vapours applied under vacuum conditions. *Postharvest Biol. Technol.* **2019**, *156*, 110942. [CrossRef]
66. Yap, P.S.X.; Krishnan, T.; Yiap, B.C.; Hu, C.P.; Chan, K.G.; Lim, S.H.E. Membrane disruption and anti-quorum sensing effects of synergistic interaction between *Lavandula angustifolia* (lavender oil) in combination with antibiotic against plasmid-conferred multi-drug-resistant *Escherichia coli*. *J. Appl. Microbiol.* **2014**, *116*, 1119–1128. [CrossRef]

agronomy

MDPI

Article

Enhancing the Extraction Process Efficiency of Thyme Essential Oil by Combined Ultrasound and Microwave Techniques

Adina I. Gavrila , Ciprian G. Chisega-Negrila, Laura Maholea, Mircea L. Gavrila, Oana C. Parvulescu and Ioana Popa *

Faculty of Chemistry Engineering and Biotechnologies, National University of Science and Technology Politehnica Bucharest, 011061 Bucharest, Romania; adinagav@yahoo.com (A.I.G.); cipchisega@yahoo.com (C.G.C.-N.); maholealaura@yahoo.com (L.M.); mircea.gavrila@live.com (M.L.G.); oana.parvulescu@yahoo.com (O.C.P.)

* Correspondence: asofiei.ioana@yahoo.com

Abstract: In this paper, the essential oil (EO) extraction from thyme by the consecutive use of ultrasound and microwave treatments is presented. The aim of this study was to apply an ultrasound pre-treatment of thyme leaves to enhance the thymol content and the extraction yield of the EO obtained by microwave-assisted hydro-distillation (MWHD). Compared with conventional hydro-distillation (CHD), the consecutive use of ultrasound pre-treatment and microwave extraction resulted in a 72% lower extraction time. When the ultrasound pre-treatment (using the ultrasonic processor with an amplitude of 70%) was applied, the EO content was 23% higher compared to the extraction without pre-treatment (2.67 ± 0.06 g EO/100 g DM for the MWHD with ultrasound pre-treatment compared with 2.18 ± 0.07 g EO/100 g DM for the MWHD without pre-treatment). The EO samples were analyzed by GC/MS. The results showed that the major component, thymol, varied from 43.54% (by CHD) to 65.94% (by the consecutive use of ultrasound and microwave treatments).

Keywords: ultrasound pre-treatment; microwave hydro-distillation; thyme; essential oil; thymol

1. Introduction

Thyme (*Thymus officinalis* L.) is a plant that belongs to the Lamiaceae family and is used for food flavoring and preservation. This plant can be found in Europe, North Africa, and Asia [1]. Thyme is rich in essential oil (EO), which is part of the secondary metabolites of the plants (up to 2.5% EO for the dried herb [2]). It is involved in the defense mechanism of the herb (repelling the phytophagous organisms) and in attracting pollinators. Among the constituents of thyme EO (carvacrol, *p*-cymene, α-pinene, γ-terpinene, etc.), thymol is the major one [3]. This compound has been shown to have antioxidant and anti-inflammatory [4], antibacterial [5], antimicrobial [6], antifungal [7], antiparasitic [8], immunological [9], and anticancer [10] properties. Thyme also contains polyphenolic compounds (rosmarinic, caffeic, p-hydroxybenzoic, and procatechuic acids) and carotenoids (β-carotene, lutein, and zeaxanthin) [11]. These bioactive compounds enhance the valuable properties of thymol, providing thyme extracts with a beneficial effect on human health. Also, if added to foodstuffs, thyme EO can improve their quality. For example, Medina et al. proved that thyme EO decreased oxidation in supplemented minced beef [3].

The extraction of EOs from plants can be performed by conventional methods, such as cold pressing [12], hydro-distillation, and steam distillation [13]. These methods present several drawbacks, such as long extraction times, high energy and solvent consumption, etc. [14]. Thus, in recent years, in order to overcome these shortcomings, alternative methods have been developed. The new approaches include microwave-assisted hydro-distillation [15], ultrasound [16] and enzymatic [17,18] pre-treatments before the extraction of EOs, supercritical [19] and subcritical [20] fluid extractions, solvent-free extraction [21], and microwave hydro-diffusion and gravity extraction [22,23].

Microwave-assisted hydro-distillation (MWHD) combined with the ultrasound pre-treatment of aromatic herbs to extract EOs was successfully employed [24,25]. This strategy combines the advantages of ultrasound and microwave extraction. The latter can provide volumetric and selective heating. During microwave irradiation, the heterogeneous extraction mixture is heated as a whole volume and the vegetal material can be heated selectively. Although the microwave heating of extraction mixture occurs rapidly, in the overall rate of the process, the mass transfer is limited [26,27]. On the other hand, the cavitation phenomena can promote the breakage of the cellular tissue, which will increase the mass transfer rate [28].

Our work describes the extraction of EO from thyme leaves by the consecutive use of ultrasound and microwave treatments. Thus, the aim of this study was to apply an ultrasound pre-treatment of the extraction mixture in order to increase the thymol content and the extraction yield of the EO obtained using MWHD. The influence of the ultrasound pre-treatment on the extraction process of EO was studied using both an ultrasound bath and an ultrasonic processor. To our knowledge, MWHD combined with ultrasound pre-treatment to extract EO was not used for thyme leaves. There are only a few studies regarding the ultrasound pre-treatment of thyme, and neither of them reported the comparison between different sonication equipment. Kowalski et al. applied an ultrasonic pre-treatment of the thyme leaves using an ultrasound bath, but the EO was further extracted by conventional steam distillation. In this study, a maximum yield of 2.4% (v/w) was achieved, meaning approximatively 2.15 g EO/100 g of dried plant material [29]. Roldan-Gutierrez et al. used an ultrasonic processor for the pre-treatment of thyme leaves, but the extraction of EO was further performed by steam distillation and superheated water methods [30].

2. Materials and Methods

2.1. Materials

Fresh thyme (stems and leaves) was purchased from Hofigal (Bucharest, Romania). The fresh leaves and stems were chopped into pieces of 1–2 cm, then dried in an air flow-heating oven (Memmert UNE 500 Universal Oven, Memmert GmbH + Co. KG, Schwabach, Germany) at 60 °C to a constant weight. Part of the dried vegetal material was used as such, and another part was ground using an electric grinder and screened to a particle size under 0.1 cm. The ground thyme was dosed in samples of 100 g (in sealed plastic container) and stored at 4–5 °C until extraction of EO. The water content of the dried thyme leaves was 7.7% (w), being determined with the help of a moisture analyzer (PMB 202 Moisture Analyzer, Adam Equipment Co., Ltd., Bletchley, UK). For the GC-MS quantification of thyme EO, the following standards were used: thymol, γ-terpinene, p-cymene, α-terpinene, and β-pinene from Merck (Merck & Co., Inc., Darmstadt, Germany).

2.2. Methods

2.2.1. Essential Oil Extraction Procedure

The extraction of EO from thyme was performed by MWHD using a multimode microwave oven (Plazmatronika, Plazmatronika NT SP Z.O.O., Wroclaw, Poland), which has a frequency of 2450 MHz. This apparatus is described in our previous work, Calinescu et al. [17]. During the experiment, temperature, time, and power were controlled via an operating console. Steam produced in the reactor carrying the thyme EO was directed to a modified Neo Clevenger trap with a 10 mL graduated tube. The extractions were carried out at different ratios of solvent to plant: 8/1, 10/1, and 12/1 (v/w). The particle size of vegetal material was varied, as follows: pieces of 1–2 cm or particles under 0.1 cm. The solvent used was distilled water and the mixture was subjected to extraction until no EO was obtained (80 min). The protocols for the MWHD method are described in our previous work [17]. Comparative extraction of thyme EO by conventional hydro-distillation (CHD) method was also performed. The conventional extraction was carried out for 180 min (after this time no EO was obtained) at a solvent to plant ratio of 12/1 (*v*/*w*). The separated EO was kept at 4 °C until GC-MS analysis. The extraction yield was expressed as grams of EO per 100 grams of dry matter (g EO/100 g DM).

2.2.2. Ultrasound Pre-Treatment of Plant Material

Before MWHD, the vegetal material mixed with the solvent (distilled water) was subjected to ultrasound pre-treatment for 30 min. The pre-treatment was performed using either an ultrasound bath (ES375H Bench Top Ultrasonic Tank, Hilsonic Ultrasonic Cleaners, Birkenhead, UK) with a volume of 3 L, a power of 120 W, and a frequency of 40 kHz, either an ultrasonic processor (Vibracell VCX-750, Sonics & Materials, Inc., Newtown, CT, USA) with a power of 750 W, a frequency of 20 kHz, and a titanium horn with a diameter of 13 mm. The sonication of the extraction mixture, using the ultrasonic processor, was applied with a duty cycle of 5 s on and 5 s off for an amplitude of 50% or 70%. The sonication for the ultrasonic bath was applied continuously. Control samples without any pre-treatment were directly subjected to MWHD or CHD.

2.2.3. GC-MS Analysis of the Thyme Essential Oil

The separated EO was analyzed by GC-MS method. The GS-MS apparatus, a Thermo Electron Corporation Focus GC system (Thermo Fisher Scientific S.p.A., Milan, Italy), is described in our previous work Calinescu et al. [17]. The main constituents of thyme EO (thymol, γ-terpinene, *p*-cymene, α-terpinene, and β-pinene) were identified according to retention times of known standards. Identification of the other constituents of thyme EO was performed by comparing the samples' spectral peaks with spectra from Wiley database. The main constituents of thyme EO were chosen based on preliminary tests that were part of a research project (PN-III-P2-PED-2019-2118, project: "IMUNOSTIM", no. 381PED/2020). Absolute amounts of the individual components were calculated based on GC peak areas, and were expressed as milligrams per 100 grams of dry matter (mg/100 g DM).

2.2.4. Statistical Analysis

All measurements were performed in triplicate, and the data were expressed with standard deviation (SD) as mean value ± SD for triplicate of samples ($n = 3$). All the results achieved at different levels of process factors were subjected to univariate one-way ANOVA and multivariate principal component analysis (PCA). Statistical analysis of the data was performed using the multiple comparison Duncan's post hoc tests in order to determine the significant statistical differences between the averages of the main components of two or more independent groups. To evaluate the strength of linear correlations between dependent variables, the Pearson correlation coefficient (r) was used [31]. The differences were considered statistically significant at $p < 0.05$. Statistical analysis was conducted using XLSTAT Version 2019.1 (Lumivero, Addinsoft, New York, NY, USA).

3. Results

3.1. Microwave-Assisted Hydro-Distillation (MWHD) vs. Conventional Hydro-Distillation (CHD)

The first step to optimize the extraction process of thyme EO consists of kinetics study of the control samples obtained by MWHD and CHD (Figure 1). As shown in Figure 1, the time required to achieve the reflux temperature (the time after the EO starts to be collected into the Clevenger apparatus) for MWHD (10 min) is lower compared with CHD (60 min). This can be explained by the volumetric heating of microwaves compared with the conventional one, which occurs by convection [27]. The MWHD leads to a rapid increase in EO content in the first 40 min (which corresponds to 83% of the total yield), achieving a maximum amount after 70 min (1.73 ± 0.03 g EO/100 g DM). For CHD, the EO content increases slowly, obtaining 1.65 ± 0.06 g EO/100 g DM after 170 min. Further, the experiments were performed by MWHD.

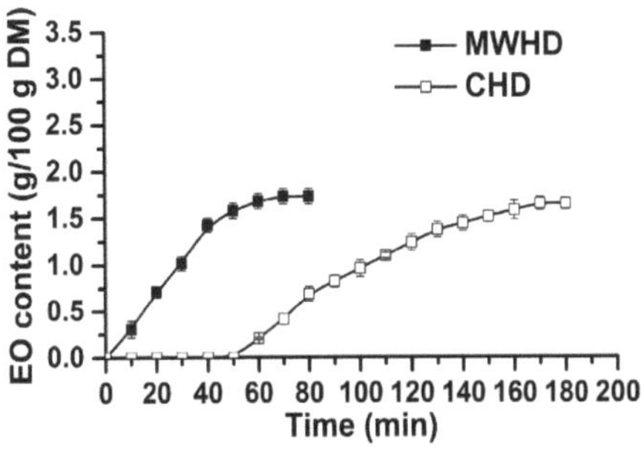

Figure 1. Kinetics study of microwave-assisted hydro-distillation (MWHD) vs. conventional hydro-distillation (CHD) for a solvent to plant ratio of 12/1 (v/w) and a particle size of 1–2 cm.

The thyme EO was analyzed and identified by GC-MS. The results, for both methods (MWHD and CHD), are shown in Table 1.

Table 1. Chemical composition of thyme essential oil (EO) for microwave-assisted hydro-distillation (MWHD) and conventional hydro-distillation (CHD) for a solvent to plant ratio of 12/1 (v/w) and a particle size of 1–2 cm. The letters ^{a} and ^{b} within table show the significant difference between groups analyzed by ANOVA ($p < 0.05$) and Duncan's post hoc t-tests.

RT (min)	Compound	CAS	BP (°C)	Component Content (mg/100 g DM)	
				MWHD	CHD
9.043	α-Pinene	80-56-8	155	13 ± 0.2^{b}	15 ± 0.7^{a}
9.205	Camphene	79-92-5	159	16 ± 0.2^{b}	26 ± 1.2^{a}
9.941	Sabinene	3387-41-5	163	7 ± 0.1^{b}	11 ± 0.5^{a}
10.039	***β-Pinene***	***127-91-3***	***165***	***33 ± 0.3^{b}***	***41 ± 1.8^{a}***
10.568	***α-Terpinene***	***99-86-5***	***173***	***57 ± 1.1^{b}***	***68 ± 2.9^{a}***
10.613	***p-Cymene***	***99-87-6***	***177***	***88 ± 3.0^{b}***	***110 ± 4.8^{a}***
10.783	Eucalyptol	470-82-6	176	6 ± 0.1^{b}	11 ± 0.5^{a}
11.245	***γ-Terpinene***	***99-85-4***	***183***	***523 ± 37.4^{a}***	***573 ± 24.9^{a}***
13.071	Terpinen-4-ol	562-74-3	219	-	14 ± 0.6^{a}
14.657	***Thymol***	***89-83-8***	***232***	***969 ± 28.9^{a}***	***689 ± 29.9^{b}***
16.724	β-(E)-Caryophyllene	87-44-5	254	4 ± 0.1^{b}	11 ± 0.5^{a}
17.616	γ-Cadinene	39029-41-9	271	3 ± 0.1^{b}	4 ± 0.2^{a}

RT—retention time; CAS—CAS registry number; BP—boiling point.

For both methods, approximately 12 compounds were identified. As shown in Table 1, the main constituents of thyme EO are thymol, γ-terpinene, *p*-cymene, α-terpinene, and β-pinene. These compounds constitute between 93% and 97% of the total amount of resulted EO. Considering the EO composition, there is only a slight difference between the two extraction methods. Thymol is extracted more efficiently by MWHD, while γ-terpinene is extracted more efficiently by CHD. This difference can be explained by the bioconversion of γ-terpinene in thymol, which is influenced by different factors, one of them being the extraction method [32]. The same behavior can be noticed for *p*-cymene, α-terpinene, and β-pinene, which are extracted more efficiently by CHD. Krause et al. proposed a biosynthetic pathway of thymol, starting from the cyclization of geranyl diphosphate forming γ-terpinene. The latter is oxidized by the P450 monooxygenases enzyme to cyclohexadienol intermediates. These products are unstable being dehydrogenated by a dehydrogenase/reductase enzyme to allylic ketone intermediates. Further, through keto-enol tautomerisms, thymol is formed [33].

3.2. Influence of Solvent to Plant Ratio on the Extraction of Thyme Essential Oil

The next step to optimize the EO extraction from thyme was to study the influence of solvent to plant ratio. To establish the solvent volume required to achieve as much EO as possible, different ratios of solvent to plant material were used. Moreover, to avoid degradation of thyme leaves due to direct microwave irradiation, an adequate amount of water is required. As shown in Figure 2a, the amount of EO increases with the increase in the water volume. The most efficient solvent to plant ratio for the extraction of thyme EO by MWHD is 12/1 (*v*/*w*).

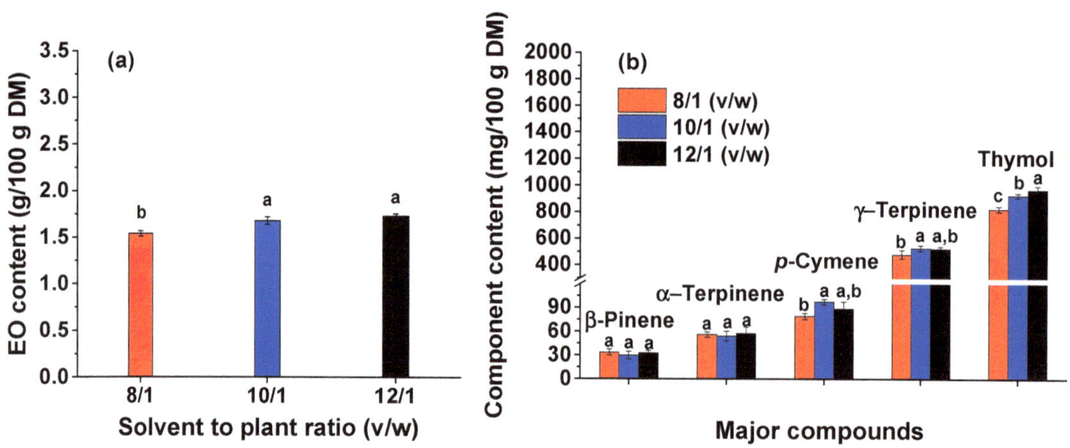

Figure 2. Influence of solvent to plant ratio on the extraction of thyme EO by microwave-assisted hydro-distillation (MWHD) for a particle size of 1–2 cm: (**a**) essential oil (EO) content; (**b**) major compounds identified by GC-MS. Different letters (a–c) within graph show the significant difference between groups analyzed by ANOVA ($p < 0.05$) and Duncan's post hoc *t*-tests.

The resulted EO, for all solvent to plant ratios, was analyzed by GC-MS. The main compounds of thyme EO identified by GC-MS are shown in Figure 2b. It can be noticed that the EO component content is similar for all three ratios and the amount of thymol is directly proportional with the solvent to plant ratio. Thus, a high amount of water is required to entrain the constituents with high boiling points. The ANOVA analysis showed that by increasing the solvent to plant ratio from 8/1 to 10/1, the EO content increases significantly ($p < 0.05$). By the further increase in the solvent to plant ratio, the EO content increases non-significantly (Figure 2a). Regarding the thymol extraction, the ANOVA analysis showed that by increasing the solvent to plant ratio, the thymol concentration increased significantly ($p < 0.05$, Figure 2b).

Although the amount of EO is similar for both 10/1 and 12/1 ratios, further experiments were carried out for a 12/1 (*v*/*w*) solvent to plant material ratio since the amount of the targeted compound (thymol) is higher and the specific energy (see Table 5) is lower compared with the other ratios used (969 ± 28.90 mg thymol/100 g DM for 12/1 ratio compared to 926 ± 16.93 and 824 ± 20.67 mg thymol/100 g DM for 10/1 and 8/1 ratios, respectively).

3.3. Influence of Ultrasound Equipment on the Extraction of Thyme Essential Oil

Prior to the extraction of thyme EO by MWHD, the extraction mixture was subjected to ultrasound pre-treatment for 30 min using two types of equipment: an ultrasound bath (ES375H Bench Top Ultrasonic Tank, Hilsonic Ultrasonic Cleaners, Birkenhead, UK) and an ultrasonic processor (Vibracell VCX-750, Sonics & Materials, Inc., Newtown, CT, USA).

It can be noticed (Figure 3a) that the ultrasound pre-treatment of the extraction mixture leads to higher amounts of EO compared with the experiments without pre-treatment, for both the ultrasound bath (approximatively 6% higher—1.83 ± 0.09 g EO/100 g DM) and the horn (approximatively 21% higher—2.10 ± 0.03 g EO/100 g DM). This could be due to the cavitation phenomenon, which can stimulate the disruption of cell walls and increase the mass transfer rate. In the ultrasonic bath, the cavitation occurs uncontrollably. The ultrasound energy is unequally dispersed in the bath and has low intensity. On the other hand, for the horn, the cavitation has a high localized intensity, and implicitly, the sonication process is more efficient [34]. Therefore, the higher amount of EO achieved when the ultrasound pre-treatment is carried out using the Vibracell equipment (an increase of approximately 15%) could be due to these differences between ultrasonic baths and horns. In addition, the ultrasound pre-treatment of the extraction mixture has a beneficial effect on the extraction time, achieving a maximum amount of EO in only 50 min compared with the extraction without pre-treatment when a maximum amount is achieved after 70 min. Further, the experiments were performed using the Vibracell ultrasonic processor to pre-treat the extraction mixture.

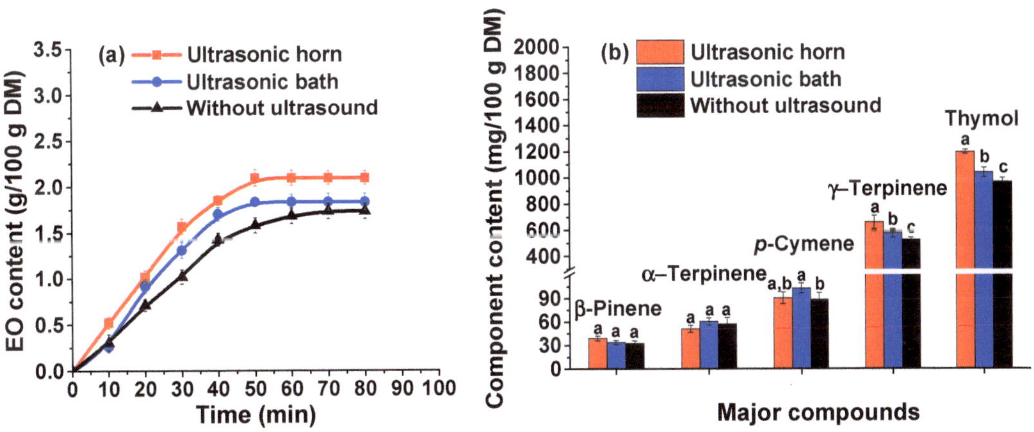

Figure 3. Influence of ultrasound pre-treatment on the extraction of thyme EO by microwave-assisted hydro-distillation (MWHD) for a solvent to plant ratio of 12/1 (v/w), a particle size of 1–2 cm, and an amplitude of 50%: (**a**) essential oil (EO) content; (**b**) major compounds identified by GC-MS. Different letters (a–c) within graph show the significant difference between groups resulted from ANOVA ($p < 0.05$) and Duncan's post hoc t-tests.

The GC-MS analyses of the EO obtained by consecutive use of ultrasound and microwave irradiation are shown in Figure 3b. It can be noticed that the major constituents of the EO are similar for all three methods (the extraction with ultrasound pre-treatment using both ultrasonic horn and bath and the extraction without pre-treatment). As shown in Figure 3b, the thymol is extracted more efficiently when the ultrasound pre-treatment is carried out using the ultrasonic probe (1196 ± 15.35 and 1040 ± 35.62 mg/100 g DM for ultrasonic probe and bath, respectively). Instead, *p*-cymene is obtained in lower amounts (90 ± 7.28 and 103 ± 6.49 mg/100 g DM for ultrasonic probe and bath, respectively). This could be due to the transformation of some compounds during ultrasound irradiation, such as isomerization, oxidation, and degradation of dimmers and polymers [35]. Also, in water, the cavitation phenomenon can lead to the formation of free radicals (H•, OH•) which can cause the transformation of the EO constituents [36]. The ANOVA analysis showed that pre-treatment with the ultrasonic horn led to a significant increase in the thymol content, while using the ultrasonic bath the content of *p*-cymene increased significantly ($p < 0.05$). The opposite behavior of *p*-cymene when the pre-treatment is performed using the ultrasonic bath could be due to the low intensity of the cavitation in such equipment. Since the ultrasound energy for the horn can be focused on specific sample area, the constituents with lower boiling points can be rapidly released, and implicitly, they can be degraded afterwards.

3.4. Influence of Ultrasound Amplitude on the Extraction of Thyme Essential Oil

The influence of ultrasound amplitude on the extraction process efficiency was also studied. Amplitudes of 50 and 70% were chosen. As shown in Figure 4a, the amount of thyme EO is directly proportional to the ultrasound amplitude. The EO content increases by 6% compared with the pre-treatment performed at an amplitude of 50% (2.10 ± 0.03 and 2.22 ± 0.06 g EO/100 g DM for an amplitude of 50 and 70%, respectively). The extraction mechanism involves the diffusion of the solvent through the cell walls and, implicitly, their rinsing, since the walls are disrupted. These phenomena can be enhanced by ultrasonic cavitation [37]. Thus, an efficient ultrasound pre-treatment of the extraction mixture can increase the EO yield.

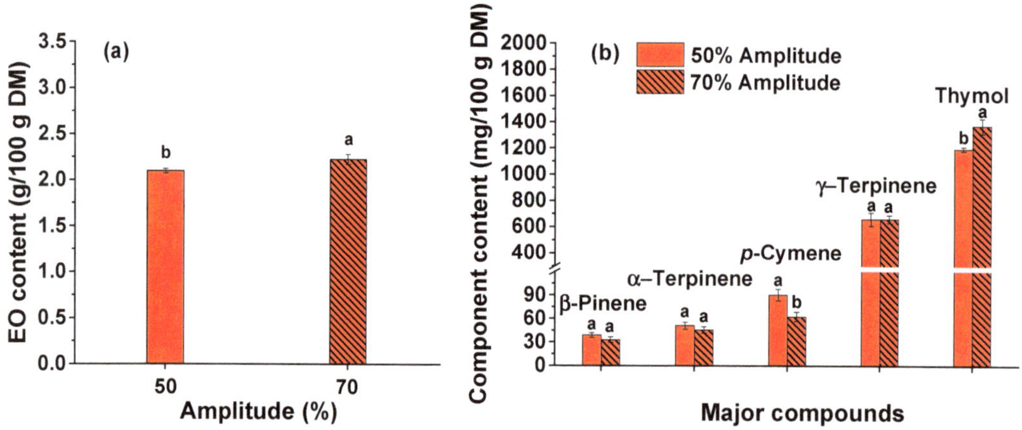

Figure 4. Influence of ultrasound amplitude on the extraction of thyme EO by microwave-assisted hydro-distillation (MWHD) for a solvent to plant ratio of 12/1 (v/w) and a particle size of 1–2 cm using the Vibracell ultrasonic processor: (**a**) essential oil (EO) content; (**b**) major compounds identified by GC-MS. Different letters (a,b) within graph show the significant difference between groups from one-way ANOVA ($p < 0.05$) and Duncan's post hoc t-tests.

The thyme EO was also analyzed and identified by GC-MS. The results for both ultrasound amplitudes are shown in Figure 4b. As in the previous experiments (see Figures 2b and 3b, and Table 1) the main constituents of thyme EO (thymol, γ-terpinene, p-cymene, α-terpinene, and β-pinene) remain unchanged. As shown in Figure 4b, the targeted compound, thymol, is extracted more efficiently when an amplitude of 70% is used (1196 ± 15.35 and 1370 ± 58.78 mg/100 g DM for an amplitude of 50 and 70%, respectively). Statistical analysis confirmed that by increasing the ultrasonic amplitude from 50 to 70%, the EO and thymol contents increase significantly ($p < 0.05$). However, the content of p-cymene is inversely proportional with the ultrasound power, achieving higher amounts for an amplitude of 50%. This can be due to an accentuated transformation of some constituents at high ultrasound powers.

3.5. Influence of Thyme Leaf Size on the EO Extraction

In addition to external glands, which can be easily destroyed by sonication, the EO can be found in the internal secretory structures of the vegetal materials. This can lead to a mass transfer resistance. A strategy to overcome this drawback is to mill the herb in order to increase the surface area. Thus, more cells will be directly exposed to the cavitation phenomenon, thus enhancing the mass transfer of the targeted compounds from vegetal matrix to the solvent [37]. As shown in Figure 5a, the particle sizes of the thyme leaves influence to a large extent the EO content. For a particle size under 0.1 cm, higher amounts of EO are achieved for both ultrasound pre-treatment (for both amplitudes applied, i.e., 50 and 70%) and extraction without pre-treatment. Using smaller particles, the EO content is 23% higher when an ultrasound pre-treatment (with an amplitude of 70%—2.67 ± 0.06 g EO/100 g DM) of the extraction mixture is applied, compared with the extraction without pre-treatment (2.18 ± 0.07 g EO/100 g DM). The ANOVA analysis showed that EO content increased significantly ($p < 0.05$) by decreasing the plant material particles size.

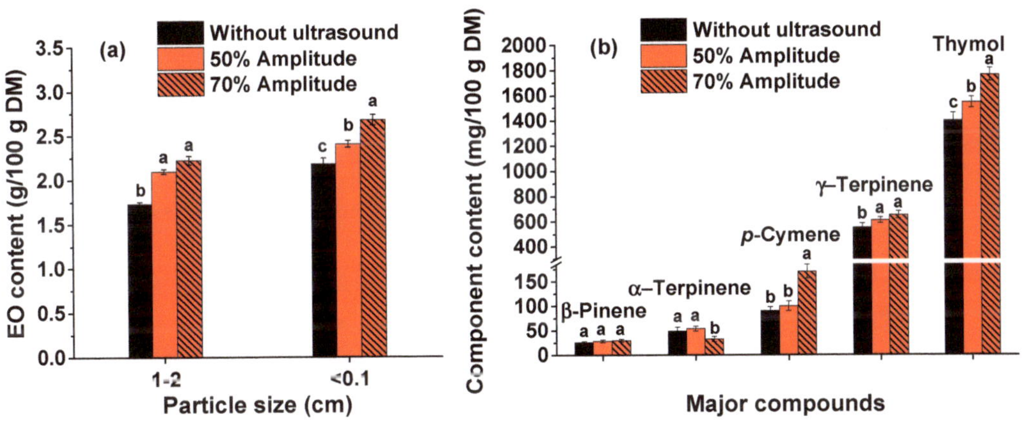

Figure 5. Influence of leaf particle size on the extraction of thyme EO by microwave-assisted hydrodistillation (MWHD) for a solvent to plant ratio of 12/1 (*v*/*w*): (**a**) essential oil (EO) content; (**b**) major compounds identified by GC-MS for a particle size under 0.1 cm. Letters (a–c) within graph show the significant difference between groups' ANOVA ($p < 0.05$) and Duncan's post hoc *t*-tests.

These strategies (the ultrasound pre-treatment using an ultrasonic probe and milling the plant material to a particle size under 0.1 cm) lead to a higher extraction yield compared with other studies. For example, Kowalski et al., [29] after applying ultrasonic pre-treatment using an ultrasound bath, followed by conventional steam distillation method, achieved an extraction yield of 2.15 g EO/100 g DM.

The major component content of thyme EO for a granulation of the leaves under 0.1 cm is similar with the one for a particle size of 1–2 cm. However, the amount of the targeted compound (thymol) is higher for the herb milled to a granulation under 0.1 cm (see Figures 3b, 4b and 5b). This behavior can be noticed for all three methods: the extraction with ultrasound pre-treatment applying both 50 and 70% amplitudes and the extraction without pre-treatment (1542 ± 42.54 and 1764 ± 52.68 mg/100 g DM for an amplitude of 50 and 70%, respectively, and 1396 ± 62.7 mg/100 g DM for the extraction without pre-treatment). As shown in Figure 5b, the thymol content is higher when an ultrasound pre-treatment with an amplitude of 70% is applied, while the α-terpinene content is lower (54 ± 4.91 and 33 ± 5.27 mg/100 g DM for an amplitude of 50 and 70%, respectively). As shown in Figure 4b, this behavior of α-terpinene can be also observed for the extraction with thyme leaves of a 1–2 cm particle size (51 ± 4.5 and 46 ± 4.02 mg/100 g DM for an amplitude of 50 and 70%, respectively). Indeed, according to ANOVA analysis, the decrease at an amplitude of 70% is insignificant (see Figure 4b), but the higher decrease when smaller particles are used can be explained by a more considerable exposure of ultrasound energy. Moreover, the *p*-cymene content is higher for an amplitude of 70% compared with 50% when the plant is milled (Figure 5b—100 ± 9.9 and 170 ± 14.06 mg/100 g DM for an amplitude of 50 and 70%, respectively), while for the extraction performed with leaves of 1–2 cm particle size, an opposite behavior is observed (Figure 4b—90 ± 7.28 and 63 ± 5.99 mg/100 g DM for an amplitude of 50 and 70%, respectively). This could be due to a possible oxidation of some components (such as α-terpinene and γ-terpinene) yielding *p*-cymene [38]. However, further research is required for possible oxidation under combined ultrasound and microwave treatments.

3.6. Principal Component Analysis

To evaluate the relation between the extraction method and the composition of the EO, principal component analysis (PCA) was performed. For the multivariate analysis, the chemical compositions of the thyme EO obtained by different extraction methods were determined. The PCA results show two eigenvalues higher than 1, i.e., those corresponding to PC1 (3.55) and PC2 (1.04). The main components of thyme EO in the plane formed by these first two PCs explain 90.68% of the variability, including 70.21% on the first axis and 20.47% on the second axis. The coordinates of variables (factor loadings) on the factor-plane PC1−PC2 are shown in Table 2, with the significant levels marked in bold. The projections of cases (factor scores) on the factor-plane PC1−PC2 are summarized in Table 3.

Table 2. Factor loadings.

Variables	PC1	PC2
β-Pinene	**−0.966**	0.045
α-Terpinene	**−0.888**	−0.201
p-Cymene	−0.150	**0.985**
γ-Terpinene	**−0.930**	−0.079
Thymol	**0.950**	−0.063

Table 3. Factor scores.

Method	Method Description	PC1	PC2
1	MWHD without ultrasound pre-treatment, solvent to plant ratio of 8/1, leaf particle size of 1–2 cm	−2.382 −3.041 −1.723	0.174 0.252 0.095
2	MWHD without ultrasound pre-treatment, solvent to plant ratio of 10/1, leaf particle size of 1–2 cm	−1.506 −2.039 −1.075	0.710 0.830 0.821
3	MWHD without ultrasound pre-treatment, solvent to plant ratio of 12/1, leaf particle size of 1–2 cm	−1.477 −1.941 −0.653	0.196 0.433 −0.170
4	MWHD + ultrasonic bath pre-treatment, solvent to plant ratio of 12/1, leaf particle size of 1–2 cm	−1.388 −1.663 −1.113	0.539 0.672 0.406
5	MWHD + ultrasonic horn pre-treatment (50% amplitude), solvent to plant ratio 12/1, leaf particle size 1–2 cm	−1.267 −1.339 −0.271	−0.454 −0.264 −0.698
6	MWHD + ultrasonic horn pre-treatment (70% amplitude), solvent to plant ratio of 12/1, leaf particle size of 1–2 cm	0.601 0.472 0.895	−1.865 −1.727 −2.018
7	MWHD without ultrasound pre-treatment, solvent to plant ratio of 12/1, leaf particle size < 0.1 cm	1.846 1.737 2.101	−0.559 −0.486 −0.754
8	MWHD + ultrasonic horn pre-treatment (50% amplitude), solvent to plant ratio of 12/1, leaf particle size < 0.1 cm	1.798 1.699 2.169	−0.479 −0.384 −0.915
9	MWHD + ultrasonic horn pre-treatment (70% amplitude, solvent to plant ratio of 12/1, leaf particle size < 0.1 cm	3.120 3.111 3.328	1.964 2.181 1.502

The PCA bi-plot and the correlation matrix are shown in Figure 6 and Table 4, respectively. The significant values of correlation coefficients (r) are highlighted in bold at a significance level $\alpha = 0.05$ (two-tailed test).

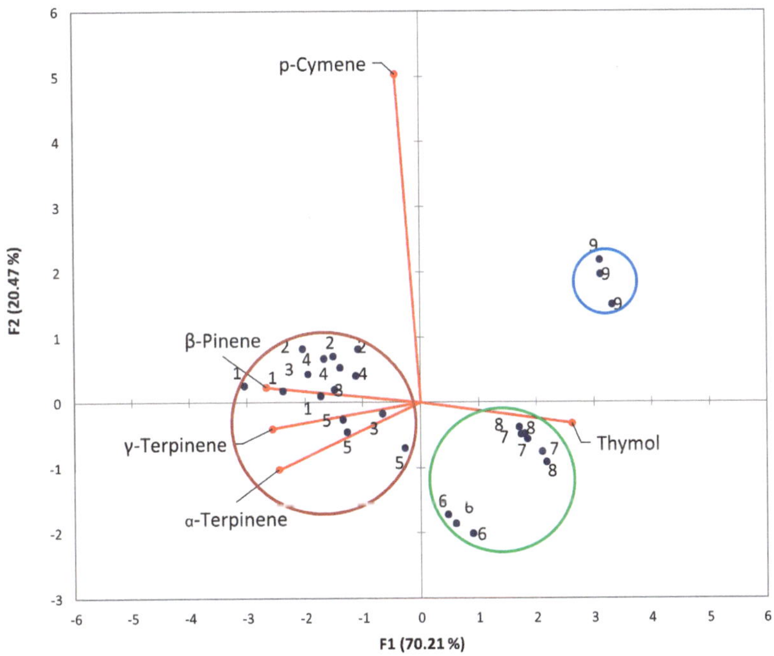

Figure 6. Projections of variables (β-Pinene, α-Terpinene, p-Cymene, γ-Terpinene, Thymol) and methods (1–9) on the factor-plane PC1–PC2.

Table 4. Correlation matrix.

Variables	β-Pinene	α-Terpinene	p-Cymene	γ-Terpinene	Thymol
β-Pinene	**1**	**0.797**	0.179	**0.908**	**−0.890**
α-Terpinene	**0.797**	**1**	−0.034	**0.742**	**−0.815**
p-Cymene	0.179	−0.034	**1**	0.055	−0.194
γ-Terpinene	**0.908**	**0.742**	0.055	**1**	**−0.832**
Thymol	**−0.890**	**−0.815**	−0.194	**−0.832**	**1**

The PCA bi-plot presented in Figure 6 and correlation matrix shown in Table 4 indicate the subsequent aspects:

- *p*-Cymene is weakly correlated with thymol, β-pinene, α-terpinene, and γ-terpinene ($-0.034 \leq r \leq 0.179$);
- Thymol is highly inversely correlated with β-pinene, α-terpinene, and γ-terpinene ($-0.890 \leq r \leq -0.815$);
- β-Pinene is highly directly correlated with α-terpinene, and γ-terpinene ($0.797 \leq r \leq 0.908$);
- The EO obtained by methods 6–9 (highlighted using blue and green circles) had a higher content of thymol and lower contents of β-pinene, α-terpinene, and γ-terpinene compared with the EO obtained by methods 1–5 (highlighted using red circle—discrimination on PC1);

- The EO obtained by method 9 (highlighted using blue circle) had a higher content of *p*-cymene compared with the EO obtained by methods 6–8 (highlighted using green circle—discrimination on PC2).
- Moreover, the highest amount of thymol was obtained by method 9.

3.7. Energy Considerations

During thyme leaf pre-treatment and extraction of EO, the microwave and ultrasound powers were recorded for all experiments. The power input for the electrical heater was measured at the heater power supply using a Wattmeter. Using the above-mentioned recorded values, the total energy introduced into the system was determined. The specific energy was also calculated using the following equation:

$$E_S = E_{total}/m_{EO} \text{ [kJ/g of EO]}, \quad (1)$$

where E_s is the specific energy [kJ/g of EO], E_{total} is the total energy introduced into the system [kJ], and m_{EO} is the amount of EO obtained by each extraction method [g].

The total energy introduced into the system and the specific energy for each method are shown in Table 5. The energy consumption for MWHD is 1104 kJ. The ultrasound pre-treatment step increases this quantity of energy concordantly with Table 5. For example, the highest increase in energy is given by the ultrasonic bath, the overall process energy increasing from 1104 kJ (for the extraction without pre-treatment) to 1896 kJ (for the extraction with pre-treatment using the ultrasound bath). Regarding the pre-treatment performed using the ultrasonic horn, the energy increase is much lower, between 47 and 82 kJ, as shown in Table 5.

For scaling up the microwave and ultrasound technologies, an important parameter to evaluate is the specific energy. The specific energy for MWHD (when thyme leaves with a particle size of 1–2 cm are used) decreases with the increase in the solvent to plant material ratio (see Table 5). Also, it can be noticed that the pre-treatment performed with the ultrasonic bath is not energetically efficient. In this case, the specific energy is 62% higher as compared with the extraction without pre-treatment. However, as compared with MWHD, using the ultrasonic probe to pre-treat the thyme leaves leads to a specific energy that is 14 and 16% lower for an amplitude of 50 and 70%, respectively. This difference is lower for thyme leaves with a particle size under 0.1 cm (a specific energy that is 6 and 13% lower for an amplitude of 50 and 70%, respectively, as compared with the extraction without pre-treatment).

The plant material grinding step implies energy consumption. In order to achieve a particle size under 0.1 cm (starting from 1–2 cm size) 20 s of milling were required. Thus, the grinding step will add 3 kJ to the energy of the overall process of EO extraction. However, this quantity of energy is insignificant, with the specific energy for the experiments performed with a thyme leaf particle size under 0.1 cm increasing by only 0.25%.

The aim of the current study was to develop a procedure appropriate to a small scale with scale-up possibilities. The reduced cost of combined ultrasound and microwave extraction is evidently beneficial for the proposed methods related to energy and time. The specific energy necessary to perform these extraction methods is 638.2 kJ/g of EO for MWHD and 534.5 kJ/g of EO for MWHD + ultrasonic horn pre-treatment. The extraction time is reduced compared with the conventional extraction. To scale-up these methods, Sairem has already commercialized different industrial microwave-assisted equipment [39] and also the ultrasound pre-treatment can be applied at industrial scale. Relating to the environmental effect, the level of the CO_2 fingerprint was calculated based on the supposition that 800 g CO_2 will be released for each 1 kWh obtained by combustion of fossil fuel [40]. It was observed that for the conventional extraction, the CO_2 emission (553 g CO_2/g of EO) is higher compared with MWHD (142 g CO_2/g of EO) and MWHD + ultrasonic horn pre-treatment at an amplitude of 70% (119 g CO_2/g of EO). According to these calculations, the combined ultrasound pre-treatment with MWHD is an environmentally friendly method.

Table 5. Energy consumption during thyme leaf pre-treatment and EO extraction.

	Extraction Methods	Total Energy (kJ)	Specific Energy (kJ/g of EO)
Leaf particle size of 1–2 cm	CHD	4104	2487.3
	MWHD without ultrasound pre-treatment (solvent to plant material ratio of 8/1)	1104	716.9
	MWHD without ultrasound pre-treatment (solvent to plant material ratio of 10/1)	1104	657.1
	MWHD without ultrasound pre-treatment (solvent to plant material ratio of 12/1)	1104	638.2
	MWHD + ultrasonic bath pre-treatment	1896	1036.1
	MWHD + ultrasonic horn pre-treatment (50% amplitude)	1151.1	548.2
	MWHD + ultrasonic horn pre-treatment (70% amplitude)	1186.7	534.5
Leaf particle size < 0.1 cm	MWHD without ultrasound pre-treatment (solvent to plant material ratio of 12/1)	1104	508.8
	MWHD + ultrasonic horn pre-treatment (50% amplitude)	1151.2	479.6
	MWHD + ultrasonic horn pre-treatment (70% amplitude)	1186.7	442.8

4. Conclusions

The purpose of this study was to investigate the influence of the ultrasound pre-treatment of thyme leaves, before MWHD, on the thymol content and on the EO extraction yield. Comparative extractions, without pre-treatment, by both MWHD and CHD were also performed. The influence of several parameters (solvent to plant ratio, particle size of the vegetal material, ultrasound equipment to pre-treat the extraction mixture, and amplitude of the ultrasonic processor) on the extraction of thyme EO by MWHD was investigated. The composition of the EO resulted from all methods was analyzed by GS-MS, the major constituents being thymol, γ-terpinene, *p*-cymene, α-terpinene, and β-pinene. The results showed that the ultrasound pre-treatment of the extraction mixture enhances the EO content, which can be extracted from thyme leaves. Moreover, by combining the strategies of ultrasound pre-treatment using an ultrasonic probe and milling the plant material to a particle size under 0.1 cm, a maximum yield of 2.67 ± 0.06 g EO/100 g DM was achieved. These strategies were also favorable for the targeted constituent, thymol, leading to a maximum content of 1764 ± 52.68 mg/100 g DM. However, the ultrasound pre-treatment requires additional energy to the MWHD. The energy consideration study showed that the ultrasound pre-treatment using an ultrasonic probe leads to a lower specific energy (442.8 kJ/g of EO) compared with the extraction without pre-treatment (508.8 kJ/g of EO). On the contrary, the energy consumption of the ultrasonic bath was high, leading to a higher specific energy compared with the extraction without pre-treatment, although the EO content was higher. This means that the use of an ultrasonic probe is a better and greener choice to pre-treat vegetal materials.

Author Contributions: Conceptualization, A.I.G. and I.P.; methodology, A.I.G. and I.P.; software, I.P. and O.C.P.; validation, A.I.G. and I.P.; statistical analysis, O.C.P.; formal analysis, C.G.C.-N., L.M. and M.L.G.; investigation, C.G.C.-N., L.M. and M.L.G.; data curation, A.I.G., O.C.P. and I.P.; writing—original draft preparation, C.G.C.-N., L.M. and M.L.G.; writing—review and editing, I.P.; supervision, A.I.G. All authors have read and agreed to the published version of the manuscript.

Funding: This research received no external funding.

Data Availability Statement: Data is contained within the article.

Conflicts of Interest: The authors declare no conflict of interest.

References

1. Prasanth Reddy, V.; Ravi Vital, K.; Varsha, P.V.; Satyam, S. Review on *Thymus vulgaris* Traditional Uses and Pharmacological Properties. *Med. Aromat Plants* **2014**, *3*, 1000164. [CrossRef]
2. Shalaby, A.S.; Razin, A.M. Dense Cultivation and Fertilization for Higher Yield of Thyme (*Thymus vulgaris* L.). *J. Agron. Crop Sci* **1992**, *168*, 243–248. [CrossRef]
3. Nieto, G. A Review on Applications and Uses of Thymus in the Food Industry. *Plants* **2020**, *9*, 961. [CrossRef]
4. Sheorain, J.; Mehra, M.; Thakur, R.; Grewal, S.; Kumari, S. In vitro anti-inflammatory and antioxidant potential of thymol loaded bipolymeric (tragacanth gum/chitosan) nanocarrier. *Int. J. Biol. Macromol.* **2019**, *125*, 1069–1074. [CrossRef] [PubMed]
5. Biswal, A.K.; Vashisht, I.; Khan, A.; Sharma, S.; Saha, S. Synthesis, characterization and antibacterial activity of thymol-loaded polylactic acid microparticles entrapped with essential oils of varying viscosity. *J. Mater. Sci.* **2019**, *54*, 9745–9758. [CrossRef]
6. Parkatzidis, K.; Chatzinikolaidou, M.; Koufakis, E.; Kaliva, M.; Farsari, M.; Vamvakaki, M. Multi-photon polymerization of bio-inspired, thymol-functionalized hybrid materials with biocompatible and antimicrobial activity. *Polym. Chem.* **2020**, *11*, 4078–4083 [CrossRef]
7. Venturini, T.P.; Rossato, L.; Chassot, F.; De Azevedo, M.I.; Al-Hatmi, A.M.S.; Santurio, J.M.; Alves, S.H. Activity of cinnamaldehyde, carvacrol and thymol combined with antifungal agents against *Fusarium* spp. *J. Essent. Oil Res.* **2021**, *33*, 502–508. [CrossRef]
8. Felici, M.; Tugnoli, B.; Ghiselli, F.; Massi, P.; Tosi, G.; Fiorentini, L.; Piva, A.; Grilli, E. In vitro anticoccidial activity of thymol, carvacrol, and saponins. *Poult. Sci.* **2020**, *99*, 5350–5355. [CrossRef]
9. Videla, E.A.; Giayetto, O.; Fernandez, M.E.; Chacana, P.A.; Marin, R.H.; Nazar, F.N. Immediate and transgenerational effects of thymol supplementation, inactivated Salmonella and chronic heat stress on representative immune variables of Japanese quail. *Sci. Rep.* **2020**, *10*, 18152. [CrossRef]
10. Alam, M.M.; Malebari, A.M.; Syed, N.; Neamatallah, T.; Almalki, A.S.A.; Elhenawy, A.A.; Obaid, R.J.; Alsharif, M.A. Design, synthesis and molecular docking studies of thymol based 1,2,3-triazole hybrids as thymidylate synthase inhibitors and apoptosis inducers against breast cancer cells. *Bioorg. Med. Chem.* **2021**, *38*, 116136. [CrossRef]
11. Munekata, P.E.S.; Alcantara, C.; Zugcic, T.; Abdelkebir, R.; Collado, M.C.; Garcia-Perez, J.V.; Jambrak, A.R.; Gavahian, M.; Barba, F.J.; Lorenzo, J.M. Impact of ultrasound-assisted extraction and solvent composition on bioactive compounds and in vitro biological activities of thyme and rosemary. *Food Res. Int.* **2020**, *134*, 109242. [CrossRef] [PubMed]
12. Ou, M.-C.; Liu, Y.-H.; Sun, Y.-W.; Chan, C.-F. The Composition, Antioxidant and Antibacterial Activities of Cold-Pressed and Distilled Essential Oils of *Citrus paradisi* and *Citrus grandis* (L.) Osbeck. *Evid.-Based Complement. Altern. Med.* **2015**, *2015*, 804091. [CrossRef] [PubMed]
13. Kant, R.; Kumar, A. Advancements in steam distillation system for oil extraction from peppermint leaves. *Mater. Today Proc.* **2021**, *47*, 5794–5799. [CrossRef]
14. Rombaut, N.; Tixier, A.-S.; Bily, A.; Chemat, F. Green extraction processes of natural products as tools for biorefinery. *Biofuels Bioprod. Biorefin.* **2014**, *8*, 530–544. [CrossRef]
15. Peng, X.; Feng, C.; Wang, X.; Gu, H.; Li, J.; Zhang, X.; Zhang, X.; Yang, L. Chemical composition and antioxidant activity of essential oils from barks of *Pinus pumila* using microwave-assisted hydrodistillation after screw extrusion treatment. *Ind. Crops Prod.* **2021**, *166*, 113489. [CrossRef]
16. Miljanovic, A.; Bielen, A.; Grbin, D.; Marijanovic, Z.; Andlar, M.; Rezic, T.; Roca, S.; Jerkovic, I.; Vikic-Topic, D.; Dent, M. Effect of Enzymatic, Ultrasound, and Reflux Extraction Pretreatments on the Chemical Composition of Essential Oils. *Molecules* **2020**, *25*, 4818. [CrossRef]
17. Calinescu, I.; Gavrila, A.I.; Ivopol, M.; Ivopol, G.C.; Popescu, M.; Mircioaga, N. Microwave assisted extraction of essential oils from enzymatically pretreated lavender (*Lavandula angustifolia* Miller). *Cent. Eur. J. Chem.* **2014**, *12*, 829–836. [CrossRef]
18. Liu, Z.; Li, H.; Cui, G.; Wei, M.; Zou, Z.; Ni, H. Efficient extraction of essential oil from *Cinnamomum burmannii* leaves using enzymolysis pretreatment and followed by microwave-assisted method. *Food Sci. Technol.* **2021**, *147*, 111497. [CrossRef]
19. Leila, M.; Ratiba, D.; Al-Marzouqi, A.-H. Experimental and mathematical modelling data of green process of essential oil extraction: Supercritical CO_2 extraction. *Mater. Today Proc.* **2021**, *49*, 1023–1029. [CrossRef]
20. Guo, J.; Yang, R.; Gong, Y.; Hu, K.; Hu, Y.; Song, F. Optimization and evaluation of the ultrasound-enhanced subcritical water extraction of cinnamon bark oil. *Food Sci. Technol.* **2021**, *147*, 111673. [CrossRef]
21. Yingngam, B.; Brantner, A.; Treichler, M.; Brugger, N.; Navabhatra, A.; Nakonrat, P. Optimization of the eco-friendly solvent-free microwave extraction of *Limnophila aromatica* essential oil. *Ind. Crops Prod.* **2021**, *165*, 113443. [CrossRef]
22. Asofiei, I.; Calinescu, I.; Gavrila, A.I.; Ighigeanu, D.; Martin, D.; Matei, C. Microwave Hydrodiffusion and Gravity, a Green Method for the Essential Oil Extraction from Ginger—Energy Considerations. *Univ. Politeh. Buchar. Sci. Bull. Ser. B Chem. Mater. Sci.* **2017**, *79*, 81–92.
23. Ferreira, D.F.; Lucas, B.N.; Voss, M.; Santos, D.; Mello, P.A.; Wagner, R.; Cravotto, G.; Barin, J.S. Solvent-free simultaneous extraction of volatile and non-volatile antioxidants from rosemary (*Rosmarinus officinalis* L.) by microwave hydrodiffusion and gravity. *Ind. Crops Prod.* **2020**, *145*, 112094. [CrossRef]
24. Chen, F.; Liu, S.; Zhao, Z.; Gao, W.; Ma, Y.; Wang, X.; Yan, S.; Luo, D. Ultrasound pre-treatment combined with microwave-assisted hydrodistillation of essential oils from *Perilla frutescens* (L.) Britt. leaves and its chemical composition and biological activity. *Ind. Crops Prod.* **2020**, *143*, 111908. [CrossRef]

25. Zhao, Y.; Wang, P.; Zheng, W.; Yu, G.; Li, Z.; She, Y.; Lee, M. Three-stage microwave extraction of cumin (*Cuminum cyminum* L.) Seed essential oil with natural deep eutectic solvents. *Ind. Crops Prod.* **2019**, *140*, 111660. [CrossRef]
26. Calinescu, I.; Vinatoru, M.; Ghimpeteanu, D.; Lavric, V.; Mason, T.J. A new reactor for process intensification involving the simultaneous application of adjustable ultrasound and microwave radiation. *Ultrason. Sonochem.* **2021**, *77*, 105701. [CrossRef]
27. Lee, C.S.; Binner, E.; Winkworth-Smith, C.; John, R.; Gomes, R.; Robinson, J. Enhancing natural product extraction and mass transfer using selective microwave heating. *Chem. Eng. Sci.* **2016**, *149*, 97–103. [CrossRef]
28. Vinatoru, M.; Mason, T.J.; Calinescu, I. Ultrasonically assisted extraction (UAE) and microwave assisted extraction (MAE) of functional compounds from plant materials. *TrAC Trends Anal. Chem.* **2017**, *97*, 159–178. [CrossRef]
29. Kowalski, R.; Wawrzykowski, J. Effect of ultrasound-assisted maceration on the quality of oil from the leaves of thyme *Thymus vulgaris* L. *Flavour Fragr. J.* **2009**, *24*, 69–74. [CrossRef]
30. Roldan-Gutierrez, J.M.; Ruiz-Jimenez, J.; Luque de Castro, M.D. Ultrasound-assisted dynamic extraction of valuable compounds from aromatic plants and flowers as compared with steam distillation and superheated liquid extraction. *Talanta* **2008**, *75*, 1369–1375. [CrossRef]
31. Egri, D.; Pârvulescu, O.C.; Ion, V.A.; Raducanu, C.E.; Calcan, S.I.; Badulescu, L.; Madjar, R.; Orbeci, C.; Dobre, T.; Mot, A.; et al. Vine Pruning-Derived Biochar for Agronomic Benefits. *Agronomy* **2022**, *12*, 2730. [CrossRef]
32. Zarshenas, M.M.; Samani, S.M.; Petramfar, P.; Moein, M. Analysis of the essential oil components from different *Carum copticum* L. samples from Iran. *Pharmacogn. Res.* **2014**, *6*, 62–66. [CrossRef]
33. Krause, S.T.; Liao, P.; Crocoll, C.; Boachon, B.; Forster, C.; Leidecker, F.; Wiese, N.; Zhao, D.; Wood, J.C.; Buell, C.R.; et al. The biosynthesis of thymol, carvacrol, and thymohydroquinone in Lamiaceae proceeds via cytochrome P450s and a short-chain dehydrogenase. *Proc. Natl. Acad. Sci. USA* **2021**, *118*, e2110092118. [CrossRef] [PubMed]
34. Panda, D.; Manickam, S. Cavitation Technology-The Future of Greener Extraction Method: A Review on the Extraction of Natural Products and Process Intensification Mechanism and Perspectives. *Appl. Sci.* **2019**, *9*, 766. [CrossRef]
35. Pingret, D.; Fabiano-Tixier, A.-S.; Chemat, F. Degradation during application of ultrasound in food processing: A review. *Food Control* **2013**, *31*, 593–606. [CrossRef]
36. Ince, N.H.; Tezcanli, G.; Belen, R.K.; Apikyan, P.G. Ultrasound as a catalyzer of aqueous reaction systems: The state of the art and environmental applications. *Appl. Catal. B* **2001**, *29*, 167–176. [CrossRef]
37. Vinatoru, M. An overview of the ultrasonically assisted extraction of bioactive principles from herbs. *Ultrason. Sonochem.* **2001**, *8*, 303–313. [CrossRef]
38. Asikainen, M.; Jauhiainen, O.; Aaltonen, O.; Harlin, A. Continuous catalyst-free aromatization of γ-terpinene using air as an oxidant. *Green Chem.* **2013**, *15*, 3230. [CrossRef]
39. Chemat, F.; Cravotto, G. (Eds.) *Microwave-Assisted Extraction for Bioactive Compounds*; Springer: Berlin/Heidelberg, Germany, 2013.
40. Bernard, J. Les centrales thermiques. *Sci. Vie* **2001**, *214*, 68–69.

agronomy

MDPI

Article

Biological Activities of Lamiaceae Species: Bio-Guided Isolation of Active Metabolites from *Salvia officinalis* L.

Eleni Tsitsigianni [1,2], Ekaterina-Michaela Tomou [1], Chara Almpani [2], Michail Ch. Rallis [2] and Helen Skaltsa [1,*]

[1] Department of Pharmacognosy & Chemistry of Natural Products, School of Health Sciences, Faculty of Pharmacy, National and Kapodistrian University of Athens, Panepistimiopolis, Zografou, 15771 Athens, Greece; eleni.tsitsigianni@gmail.com (E.T.); ktomou@pharm.uoa.gr (E.-M.T.)

[2] Section of Pharmaceutical Technology, School of Health Sciences, Faculty of Pharmacy, National & Kapodistrian University of Athens, Zografou, 15771 Athens, Greece; ch.almpani@gmail.com (C.A.); rallis@pharm.uoa.gr (M.C.R.)

* Correspondence: skaltsa@pharm.uoa.gr

Abstract: Lamiaceae family includes various medicinal and aromatic plants used in cosmetics and pharmaceutical industries. The present study aimed to investigate in vitro the cytotoxic, photoprotective and antioxidant activities of ten Lamiaceae taxa; *Melissa officinalis* subsp. *altissima* (Sm.) Arcang., *Rosmarinus officinalis* L., *Salvia officinalis* L., *Sideritis cypria* Post, *S. euboea* Heldr., *S. perfoliata* L. subsp. *perfoliata*, *S. scardica* Griseb., *S. sipylea* Boiss., *Stachys iva* Griseb., and *Thymus vulgaris* L. The aqueous extract of *Salvia officinalis* was bio-guided fractionated to obtain the main bioactive metabolites, which were evaluated for the aforementioned effects and their wound-healing potential. In total, five compounds were isolated and identified through NMR spectra, namely salvianic acid A, rosmarinic acid, salvianolic acid K, luteolin-3′-O-β-D-glucuronide and hispidulin-7-O-β-D-glucuronide. All the compounds were photoprotective and non cytotoxic, while no statistically significant oxidative stress reduction was obtained. Regarding the wound-healing potential, salvianolic acid K was the most promising candidate. Overall, this study suggests photoprotective natural agents from various Lamiaceae species, widely found in Greece, and provides a better insight into *Salvia officinalis* and its bioactive constituents.

Keywords: *Salvia officinalis*; Lamiaceae; sage; photoprotection; cytotoxicity; oxidative stress; wound healing; bio-guided isolation; phenolics; skin health

1. Introduction

Skin is the largest organ of the human body and contributes to protecting the organism against injuries, infections, and toxic substances, as well as from ultraviolet radiation (UV) [1]. Skin aging is a multifactorial biological process affected by genetic, hormonal, metabolic, and environmental factors [2]. Extrinsic aging or photoaging could be caused by chronic exposure to solar UV radiation, resulting in gradual loss of elasticity and changes of skin connective tissue with subsequent sagging and appearance of crevices, wrinkles, dryness, and brittleness. In addition, skin aging is accompanied by impaired wound healing and skin pigmentation alterations [2,3].

Solar UV radiation is divided into three categories based on their wavelength: (i) long wave UV-A (320–380 nm), medium wave UV-B (280–320 nm), and short wave UV-C (180–280 nm) [4,5]. Of them, approximately 40% of UV-A and 10% of UV-B radiation reaches the earth's surface and their absorptivity is not the same since the radiation's penetration depth is relative to the wavelength. As a result, UV-A could penetrate deeper into the dermis affecting most of the skin cells and leading to skin photoaging and carcinogenic transformation, whereas UV-B could be absorbed in the epidermis by various cellular biochromes (mainly cutaneous pigments, lipids, proteins, and nucleic acids) and could reach the papillary dermis, causing erythema or sunburn [5,6]. Furthermore, UV-B could directly cause DNA damage that leads to the formation of cyclopyrimidine dimers (CPDs) and other photoproducts like pyrimidine-(6-4)-pyrimidone photoproducts (6-4PPs), responsible for skin cancer, oxidative stress, and photoaging [7,8]. However, UV radiation could also possess beneficial effects on human skin. In particular, it is mentioned that it could stimulate wound healing, exhibit antibacterial activity, and be used for the treatment of various skin conditions such as psoriasis, atopic dermatitis, vitiligo, and eczema [5,9,10].

Plant extracts and natural compounds have demonstrated important effects as photoprotective agents through different mechanisms of action such as inhibition of scavenging reactive oxygen species (ROS), stimulating the production of melanin, and absorbing UV radiation [5]. Specifically, plant polyphenols (e.g., phenolic compounds and flavonoids) exhibit photoprotective activity which could be attributed to the chromophore groups of their structures which absorb radiation, reducing its penetration to the skin and its harmful consequences [5,11,12].

Lamiaceae, being one of the largest families, includes approximately 236 genera and more than 7100 species [13]. Many plants of this widespread family have aroused significant interest due to their pharmacological properties and their use in the food industry and cosmetics [14]. Genus *Salvia* L. comprises over 900 species worldwide [15] and traditionally its species have been widely used to treat various diseases such as respiratory disorders and skin inflammation [16]. According to the European Medicines Agency (EMA), preparations from leaves of *Salvia officinalis* L., commonly known as sage, are traditional herbal medicines for four indications including the relief of minor skin inflammations [16]. Many studies have reported a broad range of pharmacological properties of *S. officinalis* such as antioxidant and anti-inflammatory activities [17–20]. These effects have been attributed to its rich phytochemical content, including terpenoids, flavonoids, and phenolic acids [20–22].

The present study aimed to investigate the biological activities of different extracts/infusions from ten Lamiaceae taxa [*Melissa officinalis* subsp. *altissima* (Sm.) Arcang., *Rosmarinus officinalis* L., *Salvia officinalis* L., *Sideritis cypria* Post, *S. euboea* Heldr., *S. perfoliata* L. subsp. *perfoliata*, *S. scardica* Griseb., *S. sipylea* Boiss., *Stachys iva* Griseb., and *Thymus vulgaris* L.]. The main objective was the bio-guided fractionation of the most potent plant with a rich phytochemical profile to explore its compounds for their cytotoxic, photoprotective, antioxidant, and wound-healing properties.

2. Materials and Methods

2.1. General Experimental Procedures

1D and 2D-NMR spectra were recorded in CD_3OD on Bruker DRX 400 instrument at 295 K. Chemical shifts were given in ppm (δ) and were referenced to the solvent signals at 3.31/49.0 ppm. COSY (COrrelation SpectroscopY), HSQC (Heteronuclear Single Quantum Correlation), and HMBC (Heteronuclear Multiple Bond Correlation) experiments were performed using standard Bruker microprograms. Column chromatography (CC): Sephadex LH-20 (Pharmacia) and Vacuum Liquid Chromatography (VLC, Merck, Art. 9385, Darmstadt, Germany). Preparative–thin-layer chromatography (Prep-TLC) plates were pre-coated with silica gel (Merck, Art. 5721, Darmstadt, Germany). Fractionation was always monitored by TLC silica gel 60 F-254 (Merck, Art. 5554, Darmstadt, Germany) with visualization under UV (254 and 366 nm) and spraying with the vanillin-sulfuric acid reagent. All obtained extracts/infusions, fractions, and isolated compounds were evaporated to dryness in a vacuum under low temperature and then put in activated desiccators with P_2O_5 until their weights had stabilized.

2.2. Plant Samples

Aerial parts of 10 different taxa belonging to the Lamiaceae family were provided (*Melissa officinalis* subsp. *altissima* (Sm.) Arcang., *Rosmarinus officinalis* L., *Salvia officinalis* L., *Sideritis cypria* Post, *Sideritis euboea* Heldr., *Sideritis perfoliata* L. subsp. *perfoliata*, *Sideritis scardica* Griseb., *Sideritis sipylea* Boiss., *Stachys iva* Griseb., and *Thymus vulgaris* L.).

A detailed description of the used plant materials is shown in Table 1. All samples were authenticated and voucher specimens of wild and cultivated populations were kept at a personal herbarium of the Department of Pharmacognosy and Chemistry of Natural Products, Faculty of Pharmacy, NKUA. *M. officinalis* subsp. *altissima*; *R. officinalis*; *Salvia officinalis*; *T. vulgaris*, as well as *Sideritis sipylea* were authenticated by Assoc. Prof. Th. Constantinidis and K. Goula (voucher specimen numbers: Dimas & Skaltsa 01–04 and Lytra & Skaltsa 01, respectively). *Sideritis perfoliata* subsp. *perfoliata* were provided by the Cypriot National Centre of Aromatic Plants in trays and *S. cypria* from the Cypriot National Agricultural Department and seeds of the plant are kept at the Agricultural Research Institute, national gene bank (Accession number: ARI02415). *S euboea*; *S. scardica* and *St. iva* were provided by ELGO Dimitra and authenticated by Dr. P. Chatzopoulou (codes 19–17; GRC017 and 99/2015, respectively).

Table 1. Origin, extract-preparation, and abbreviation of investigated plant samples.

Plant Sample	Origin	Extract	Abbreviation
Melissa officinalis subsp. *altissima* (Sm.) Arcang.	wild—Crete	EtOAc/Aqueous	MOE/MOW
Rosmarinus officinalis L.	wild—Mt. Pelion	EtOAc/Aqueous	ROE/ROW
Salvia officinalis L.	wild—Kozani	EtOAc/Aqueous	SOE/SOW
Sideritis cypria Post	cultivated—Cyprus	MeOH/Infusion	SCM/SCI
Sideritis euboea Heldr.	cultivated—ELGO Dimitra	MeOH/Infusion	SEM/SEI
Sideritis perfoliata L. subsp. *perfoliata*	cultivated—Cyprus	MeOH/Infusion	SPM/SPI
Sideritis scardica Griseb.	cultivated—ELGO Dimitra	MeOH/Infusion	SSM/SSW
Sideritis sipylea Boiss.	wild—Samos Island	MeOH/Infusion	SSMe/SSI
Stachys iva Griseb.	cultivated—ELGO Dimitra	MeOH/Infusion	SIM/SII
Thymus vulgaris L.	wild—Mt. Pelion	EtOH/Aqueous	TVE/TVW

2.3. Preparation of Extracts and Infusions

Air-dried aerial parts of each plant (100 g) were cut into small pieces and then extracted at room temperature with different solvents based on their phytochemical content. Specifically, *M. officinalis* subsp. *altissima*, *R. officinalis*, and *S. officinalis* were extracted with ethyl acetate (EtOAc; 2 × 24 h) and water (2 × 24 h), successively. *T. vulgaris* was extracted with ethanol (EtOH; 2 × 24h) and then successively with water (2 × 24 h).

S. cypria, *S. euboea*, *S. perfoliata* subsp. *perfoliata*, *S. scardica*, *S. sipylea*, *St. iva* were extracted with methanol (MeOH; 2 × 24 h). Moreover, their infusions were independently prepared based on the monograph of the European Medicines Agency [23]. Four grams (4 g) of aerial parts of the taxa were given into 200 mL boiling distilled water for 5 min, were filtered and the solvent was removed under reduced pressure.

All obtained extracts and infusions were concentrated to dryness and were stored as solids at −20 °C until instrumental and biological analyses.

2.4. Chromatographic Separation and Isolation of the Active Metabolites from S. officinalis Aqueous Extract

Part of the aqueous extract (3.5 g) of *S. officinalis* was subjected to vacuum column chromatography (VLC, 10.0 cm × 3.0 cm) on silica gel, eluting with solvent mixtures of increasing polarity (EtOAc:MeOH:H_2O). Ten fractions (SOW_A-SOW_J) were obtained and tested for their cytotoxicity, photoprotective activity, and their potential to reduce oxidative stress. Afterward, fractions SOW_E, SOW_F, and SOW_G were selected for further phytochemical study. Fraction SOW_E (163.5 mg; eluted with EtOAc:MeOH:H_2O 75:25:2.5) was submitted in CC on Sephadex LH-20 (16.0 cm × 2.0 cm) with isocratic elution (MeOH 100%), yielding compounds rosmarinic acid (**2**; 11.0 mg) and luteolin-3′-O-β-D-glucuronide (**4**; 1.5 mg). Fraction SOW_F (93.1 mg; eluted with EtOAc:MeOH:H_2O 70:30:3) was subjected to CC on Sephadex LH-20 (16.0 cm × 2.0 cm) with isocratic elution (MeOH 100%) and 27 fractions were received. The fractions SOW_FH (10.2 mg), SOW_FI (6.6 mg), and SOW_FK (4.0 mg) were separately submitted in prep-TLC over silica gel with

EtOAc:MeOH:H_2O 6.5:1.5:1.0 to afford compounds salvianic acid A (**1**; Rf: 0.36; 2.7 mg), hispidulin-7-O-β-D-glucuronide (**5**; Rf: 0.19; 1.0 mg), and luteolin-3′-O-β-D-glucuronide (**4**; Rf: 0.42; 3.0 mg), respectively. Fraction SOW_G (112.9 mg) was submitted in CC on Sephadex LH-20 (16.0 cm × 2.0 cm) with isocratic elution (MeOH 100%) and gave compounds salvianic acid A (**1**; 3.3 mg) and salvianolic acid K (**3**; 3.9 mg).

2.5. Activity of Plant Extracts/Infusions, Fractions, and Isolated Compounds in Fibroblasts

2.5.1. Materials and Equipment

A scale with precision to four decimal places from Mettler Toledo B154 College Digital Analytical Balance Scale was used. The plates were 96 and 24-well Corning 96-Well Plates—Sigma-Aldrich and the Filament flow chamber was from Abductor Telstar PV100, Spain. An Axiovert 25 ZEISS (Schwabach, Switzerland) inverted microscope and a Fluostar Galaxy BMG Microplate Photometer (Ortenberg, Germany) were used. The Laboratory Oven was a Memmert (Schwabach, Germany) and the liquid nitrogen freezing cell container was a 34 XT Taylor-Wharton (Cambridge Scientific, Merck KGaA, Darmstadt, Germany). The plate shaker was an MS1 Minishakersmall orbital shaker, (Vortex–IKA, Ludwigsburg, Germany). Also, Camera Canon PowerShot G5 Xzoom lens (7.2–28.8 mm). Haemocytometer measuring Neubauer cells and an irradiation lamp Astralux Type UVA MED, UK were used. Fluorescence images were obtained using a Canon PC1049, 16X (Tokyo, Japan) camera with Boligor (Tokyo, Japan) and Carl Zeiss 426126 and MC 80 DX-1.0X (Oberkochen, Germany) adapters for Axiovert 25 microscope, with fluorescent light MBQ 52ac LEJ (Jena, Germany) emitter.

DMEM 1X, high glucose (Dulbecco's Modified Eagle Medium, Taufkirchen, Germany), PBS (Phosphate Buffered Saline, Cologne, Germany), Antibiotic Antimycotic solution (100×), Trypsin-EDTA 0.05% were purchased from Biosera (Cholet, France), while Sodium Lauryl Sulphate (SLS), Chlorpromazine (CPZ) were from Serva (Heidelberg, Germany). FBS (Fetal Bovine Serum) was from PanBiotech (Aidenbach, Germany) and Neutral Red Solution was from Apollo Scientific (Stockport, UK). Ethanol Absolute and Glacial Acetic Acid, as well as the water for injection, were from Millipore–Sigma (Merck KGaA, Darmstadt, Germany). 6-chloromethyl-2′,7′-dichlorodihydrofluorescein diacetate, acetyl ester (CM-H2DCFDA) was purchased from Gibco-Life Technologies (ThermoFisher Scientific, New York, NY, USA). NIH/3T3 Fibroblasts were kindly donated by Dr D. Kletsas, Laboratory of Cell Proliferation and Ageing, Demokritos, Greece.

2.5.2. Cytotoxicity Assay in NIH/3T3 Fibroblasts

Cell viability was assessed on NIH/3T3 fibroblasts. The proliferation rates of NIH/3T3 after treatment with plant extracts/infusions (Table 1) as well as fractions and isolated compounds of *S. officinalis* were determined by Neutral Red (NR) Uptake assay. The assay relies on the ability of living cells to incorporate and bind neutral red, a weak cationic dye, in lysosomes. Cells were maintained in the Dulbecco's Modified Essential Medium (DMEM) high glucose, supplemented with 10% fetal bovine serum (FBS) and 1X Penicillin-Streptomycin. The cells were seeded in 96-well flat-bottomed microplates at a density of 10,000 cells/well in serum-containing medium and incubated at 37 °C, 5% carbon dioxide. After 24 h of incubation to ensure cell attachment, medium was changed to DMEM (5% FBS) containing graded concentrations of plant extracts/infusions (from 10^{-5} to 1.0 mg/mL), while the fractions and the isolated compounds were evaluated at concentrations ranging from 10^{-4} to 10^{-2} mg/mL and 10^{-7} to 10^{-2} mg/mL, respectively. Cultures incubated with the corresponding vehicle served as negative controls, whereas cultures incubated with serial dilutions of Sodium Lauryl Sulfate (SLS, 10^{-3}–100 mg/mL) served as positive control. After 24 h incubation, the morphology of cells was examined under microscope. The cells were washed with DPBS. Neutral Red medium was added, followed by incubation at 37 °C in a humidified atmosphere of 5% CO_2 for 3 h. After incubation, the NR medium was replaced with NR desorb solution (ETOH/acetic acid). The microtitre plate was shaken on a microtitre plate shaker for 10 min until NR was extracted from the cells and formed a homogeneous solution. The optical density was measured in a microplate reader at the wavelength of 540 nm.

Cell viability was calculated as follows:

$$\% \text{ cell viability} = (\text{Absorbance sample}/\text{Absorbance control}) \times 100,$$

where Absorbance control is the absorbance of cells treated DMEM (5%FBS) and Absorbance sample is the absorbance of cells treated with the test sample.

Samples with cell viability of less than 70% were considered cytotoxic.

2.5.3. Photoprotection Assay in NIH/3T3 Fibroblasts

The photoprotective potential of plant extracts/infusions (Table 1), fractions, and the isolated compounds was studied in a UVA-induced phototoxicity assay (5 J/cm^2). The selected UV-A dose was slightly cytotoxic for the NIH/3T3 fibroblast cell line, in order to reveal the possible cytoprotective efficacy of the extracts/infusions. NIH/3T3 fibroblasts were seeded in 96-well clear-bottomed white microplates at a density of 10,000 cells/well in DMEM 10% FBS and were left to adhere for 24 h. Two plates per plant extract were pre-incubated for 1 h, using the concentrations described in the cytotoxicity assay (Section 2.5.2). Chlorpromazine hydrochloride was used as positive control (10^{-2}–10^{3} μg/mL). One of the two plates was irradiated for 50 min, whereas the other plate was kept in the dark. In both plates, the treatment buffer was replaced with fresh culture medium and cell viability was determined by NRU after an 18–24 h incubation as previously described.

2.5.4. Intracellular Reactive Oxygen Species (ROS) Assay

The NIH/3T3 fibroblasts were plated in 96-well clear-bottomed black microplates at a density of 10,000 cells/well in DMEM 10% FBS and were left to adhere for 24 h. To assess the antioxidant capacity of the plant samples, cells were incubated for 1 h in DPBS with different concentrations of plant extracts/infusions (10^{-5}–1.0 mg/mL) or fractions (10^{-7}–10^{-2} mg/mL) or isolated compounds (10^{-7}–10^{-2} mg/mL). The cells were then irradiated with a low UV-A dose of 2.5 J/cm^2 within 25 min, which causes non cytotoxic effects (unpublished data). H_2O_2, a known oxidative stress inducer, is used as positive control at concentrations of 10^{-3}–100 μM. Oxidative stress was evaluated using CM-H2DCFDA ester (5-(and-6)-chloromethyl-2′,7′-dichlorodihydrofluorescein diacetate, acetyl ester), which reacts with free radicals (HO·, HOO·) or oxidizing molecules (H_2O_2 and ONOO-). The induced fluorescence is due to the existence of reactive oxygen compounds. Briefly, after the irradiation, the cells were washed with DPBS, CM-H2DCFDA ester was added in each well at a 3 μM and the cells were incubated for 45 min. The medium was replaced by DPBS, and, after 15 min, fluorescence emission was determined at 520 nm, following excitation at 485 nm in a microplate reader, and fluorescence images were obtained.

2.6. Scratch Assay

The isolated compounds were tested for their healing effect by evaluating their NIH/3T3 fibroblasts migration ability. The procedure involves growing a cell monolayer to confluence in a 12-well assay plate. The 'wound', a cell-free zone in the monolayer, was created by a sterile micropipette yellow tip. The recolonization of the scratched region was monitored. The incision is followed by PBS wash in order to remove any dead cells and cellular debris. The isolated compounds of *S. officinalis* were added at two different concentrations (10^{-4} and 10^{-3} mg/mL), in DMEM (0.2% FBS) so that the cells are synchronized and cell proliferation is inhibited. DMEM (15% FBS) was used as a positive control and DMEM (0.2% FBS) as a control. Images of the incision closure were acquired at 0 and 24 h at 4× magnification. All images were analyzed using Image-J software. The reduction of the gap distance of the section was calculated by the following formula: % section closure = (A−B) * 100/A, where A = initial intersection distance in time (t_{oh}) and B = intersection distance in time (t_{24h}).

2.7. Statistical Processing of Results

GraphPad Prism 8 software (GraphPad Software, San Diego, CA, USA) was used for statistical analysis. Experiments were performed in triplicates. Data normality was checked for all the data with the Shapiro-Wilk test. Since data followed a normal distribution, statistical significance was tested with parametric methods (One-Way Anova, Post Hoc analysis LSD, *t*-test). The differences were considered statistically significant at $p < 0.05$ which is indicated in graphs by one asterisk.

3. Results and Discussion

3.1. In Vitro Cytotoxicity, Photoprotection, and Antioxidant Activity of Plant Extracts/Infusions

The plant extracts and infusions did not demonstrate any cytotoxic effect at the concentrations from 10^{-5} to 0.1 mg/mL, apart from the ethyl acetate extract of *R. officinalis* (ROE), which showed a cytotoxic activity at a concentration of 0.1 mg/mL (Figures S1–S4).

The ethanol, ethyl acetate, and methanol extracts showed cytotoxicity at 1 mg/mL, the highest concentration tested, with cell viability ranging in the mean between 45–60% (Figures S1 and S2), with the exceptions of the ethyl acetate extract of *M. officinalis* subsp. *altissima* (MOE) and methanol extract of *S. perfoliata* subsp. *perfoliata* (SPM) which did not show any cytotoxicity. All aqueous extracts and infusions (Figures S3 and S4) did not exhibit any cytotoxicity with the exceptions of *R. officinalis* aqueous extract (ROW) and the infusions of *S. sipylea* (SSI) and *S. euboea* (SEI), which were cytotoxic at the highest tested concentration (1 mg/mL).

The photoprotective effect of the plant samples was explored at a concentration range of 10^{-5} to 1.0 mg/mL (Figures S5–S8). UV-A irradiation of NIH/3T3 fibroblasts with 5 J/cm^2 caused an approximately 30% decrease in fibroblast viability (IRR control), while the non-irradiated cells (NIRR control) viability was set to 100%. The photosensitive substance, chlorpromazine (positive control), had an IC_{50} value of 1.0 μg/mL on cell viability after irradiation. *S. officinalis* ethyl acetate extract (SOE) showed important photoprotective activity at the concentration range from 10^{-2} to 1.0 mg/mL with cell viability ranging from 80 to 97%. In addition, the ethanol extract of *T. vulgaris* (TVE) demonstrated a photoprotective effect at all tested concentrations (cell viability ranges: 80–95%). Among the methanol extracts, those of *S. euboea* (SEM) and *S. perfoliata* subsp. *perfoliata* (SPM) had a photoprotective effect at a concentration range from 10^{-2} to 1.0 mg/mL and from 10^{-5} to 1.0 mg/mL, respectively. The aqueous extracts of *S. officinalis* (SOW) and *T. vulgaris* (TVW) unveiled the most important photoprotective effect at the concentration of 0.1 mg/mL with viability rates ranging from 85 to 94%. In addition, the aqueous extract of *R. officinalis* (ROW) showed moderate photoprotective activity at the lowest concentration, while greater activity was observed at higher concentrations (from 10^{-4} to 1.0 mg/mL). The infusions of *S. euboea* (SEI) and *S. perfoliata* subsp. *perfoliata* (SPI) demonstrated significant activity with cell viability ranging from 87 to 100%. It is noteworthy that both infusions showed similar viability rates to non-irradiated cells (NIRR control) at the two highest concentrations (0.1 and 1.0 mg/mL) (Figure S8).

Considering the potential of the plant extracts and infusions to reduce oxidative stress (Figures S9–S12), the aqueous extracts and infusions did not significantly reduce oxidative stress with the exception of *S. sipylea* infusion (SSI).

Overall, the diverse results among the tested plant samples in each biological assay could be also attributed to the different phytochemical constituents which were included in the used plants and extracts. Previous studies have reported a broad range of bioactive metabolites from these plants, belonging to triterpenoids (*M. officinalis* subsp. *altissima, R. officinalis, Sideritis taxa, Stachys taxa*, and *T. vulgaris*), diterpenoids (*R. officinalis, S. officinalis, Sideritis taxa, Stachys taxa*), iridoids (*Sideritis taxa, St. iva*), flavonoids (*R. officinalis, S. officinalis, Sideritis taxa, St. iva*, and *T. vulgaris*), phenylethanoid glycosides (*Sideritis taxa* and *St. iva*), and phenolic acids (*M. officinalis* subsp. *altissima, R. officinalis, S. officinalis, Sideritis taxa, St. iva*, and *T. vulgaris*) [21,22,24–36].

3.2. Bio-Guided Investigation of *S. officinalis* Aqueous Extract

Taking into consideration the overall outcomes from the plant extracts and infusions, the aqueous extract of *S. officinalis* (SOW) was chosen for further bio-guided isolation of the main responsible bioactive metabolites due to its rich phytochemical profile and photoprotective and non cytotoxic activity. This extract was fractionated and the obtained fractions (SOW_A-SOW_J) were evaluated for their cytotoxicity, photoprotective activity, and their potential to reduce oxidative stress. The fractions were initially tested for their in vitro cytotoxicity (Figure 1) at a concentration range from 10^{-4} to 0.01 mg/mL. None of the fractions were cytotoxic at all tested concentrations. All the fractions showed photoprotective activity at the tested concentrations, with the exception of SOW_H, SOW_I, and SOW_J fractions (Figure 2). It should be mentioned that a statistically significant photoprotective effect was exhibited by fractions SOW_E at the concentration of 10^{-2} mg/mL, SOW_F at the two highest concentrations (10^{-3} and 10^{-2} mg/mL) and SOW_G at all concentrations with the highest cell viability reaching 130 and 140% respectively, causing a 60 to 70% increase in viability compared to irradiated cells (IRR control, Figure 2). Regarding their antioxidant capacity, statistically significant effects were observed by the fractions SOW_C (at 10^{-5} and 10^{-3} mg/mL), SOW_F (at 10^{-3} and 0.01 mg/mL), and SOW_H (at 10^{-5} and 10^{-3} mg/mL) (Figure 3).

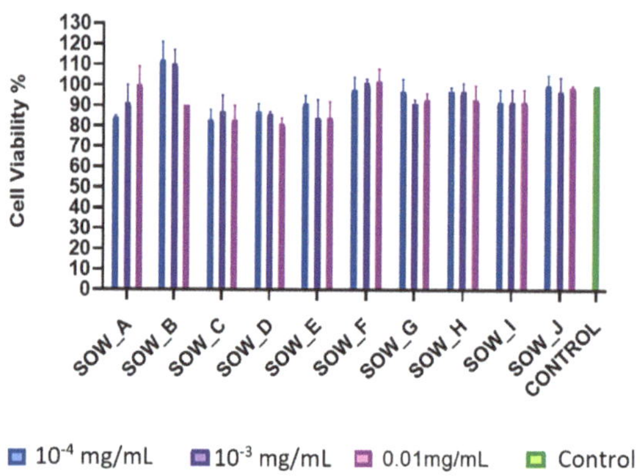

Figure 1. Cytotoxic activity of *S. officinalis* fractions in NIH/3T3 fibroblasts after 24 h of treatment. Bars ± SD.

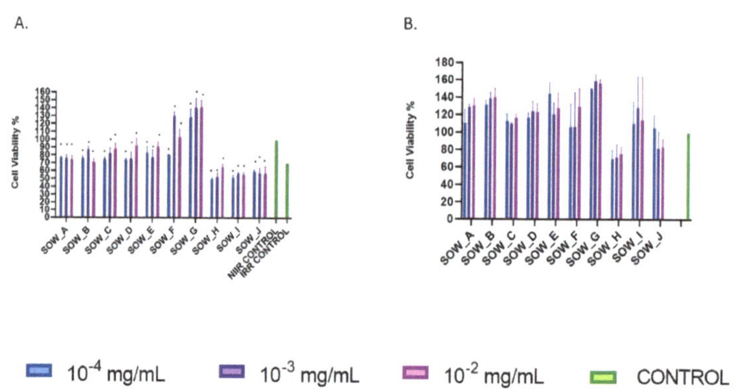

Figure 2. Photoprotective activity of *S. officinalis* fractions. UV-A-induced cytotoxicity was assessed on mouse skin fibroblasts (NIH/3T3) pre-treated with the indicated fractions for 1 h (**A**). Fraction—induced cytotoxicity after 1 h of treatment (**B**). PBS-treated cultures served as controls. UV-A untreated control (NIRR) set to 100%. Bars ± SD. * $p < 0.05$.

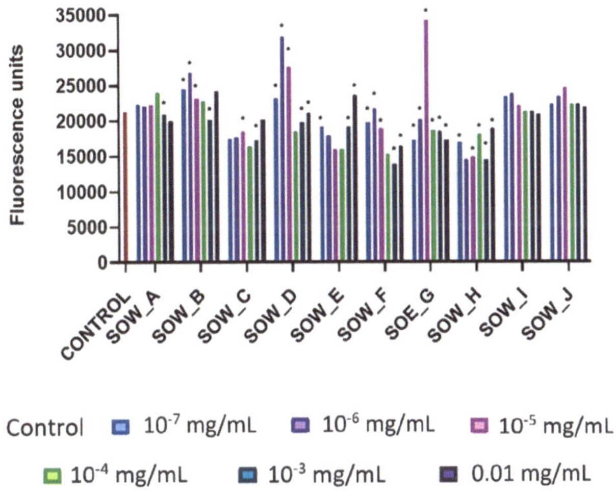

Figure 3. Intracellular antioxidant activity of *S. officinalis* fractions. 2′,7′-dichlorofluorescein (DCF)-fluorescence of mouse skin fibroblasts (NIH/3T3) treated with the indicated fractions for 1 h was used for the assessment of UVA-stimulated reactive oxygen species (ROS) levels. Vehicle (PBS)—treated irradiated cells were used as negative controls. Bars ± SD. * $p < 0.05$.

Based on the overall results, fractions SOW_E, SOW_F, and SOW_G were selected for further phytochemical study in order to isolate the most bioactive constituents, using different chromatographic techniques. In total, five compounds were obtained and identified through 1D- and 2D-NMR spectra, including three phenolic derivatives: salvianic acid A (**1**) [37], rosmarinic acid (**2**) [38], and salvianolic acid K (**3**) [38], and two flavone glucuronides namely luteolin-3′-O-β-D-glucuronide (**4**) [39] and hispidulin-7-O-β-D-glucuronide (**5**) [40]. Compounds **2**–**4** were previously found in *S. officinalis* [21,24,38]. To the best of our knowledge, although compounds **1** and **5** were reported in the genus *Salvia* [21,41,42], they were isolated for the first time from this species. In general, phenolic acids derivatives are commonly found in the genus *Salvia* [24]. Among them, rosmarinic acid is the most principal caffeic acid dimer, while salvianolic acids A–K are widely found in the genus *Salvia*. Furthermore, luteolin glycosides and their glucuronides seemed to be more common compared to those of apigenin in *Salvia taxa* [21].

The isolated compounds were evaluated for their cytotoxicity, photoprotective activity, and their potential to reduce oxidative stress. It should be mentioned that hispidulin-7-O-β-D-glucuronide (**5**) was not tested in the in vitro assays due to the fact that it was isolated in a very low amount. So, compounds **1**–**4** were tested in NIH/3T3 fibroblasts for 24 h at concentrations from 10^{-7} to 0.01 mg/mL. No statistically significant cytotoxic activity was observed compared to the control (Figure 4).

All isolated compounds exhibited statistically significant photoprotective activity in the tested concentrations (10^{-7} to 10^{-2} mg/mL) (Figure 5). These results are in accordance with the outcomes from the photoprotective effects of the fractions (Figure 2). It is noteworthy to mention that luteolin-3′-O-β-D-glucuronide (**4**) at a concentration of 0.01 mg/mL showed a cell viability value of approximately 120% ($p < 0.05$) compared to the control. The significant photoprotective effects of rosmarinic acid have been previously explored [43,44]. Among the different salvianolic acids, the photoprotection of salvianolic acid B has been reported [45]. Based on our knowledge this is the first report on the photoprotective activity of salvianic acid A, salvianolic acid K, and luteolin-3′-O-β-D-glucuronide.

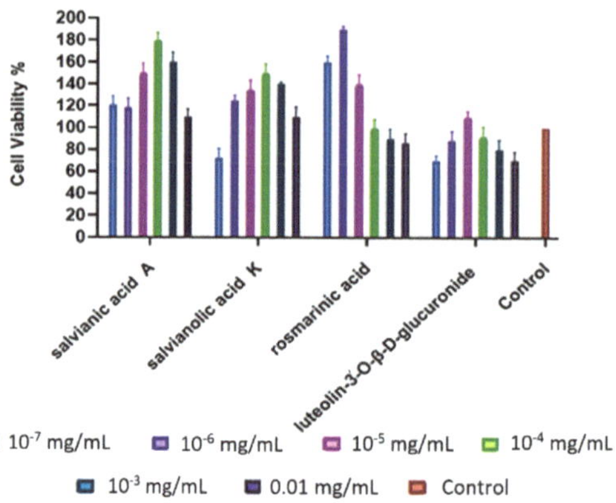

Figure 4. Cytotoxic activity of the isolated compounds (**1–4**) from *S. officinalis* in NIH/3T3 fibroblasts after 24 h of treatment. Bars ± SD.

Regardless of the potential of the isolated compounds to reduce oxidative stress, none of them demonstrated statistically significant changes in oxidative stress levels compared to the control group at the tested concentrations (Figure 6). Previous studies have mentioned that rosmarinic acid and salvianolic acids are responsible compounds for the antioxidant activity of *Salvia* species [21,24,46,47]. Although the initial fractions had been shown to reduce the oxidative stress in specific concentrations (Figure 3), it was not observed any effect from their isolated constituents. This might be attributed to the synergistic effects of the compounds in the fractions.

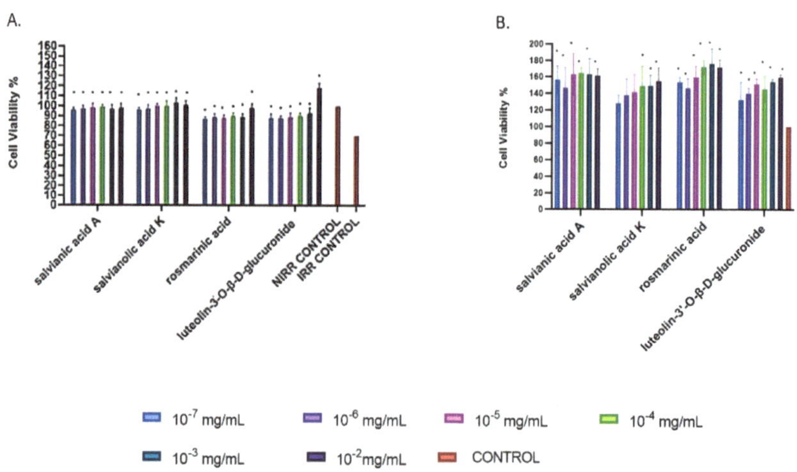

Figure 5. Photoprotective activity of the isolated compounds (**1–4**) from *S. officinalis*. UV-A-induced cytotoxicity was assessed on mouse skin fibroblasts (NIH/3T3) pre-treated with the indicated compounds for 1 h (**A**). Compound—induced cytotoxicity after 1 h of treatment (**B**). PBS-treated cultures served as controls. UV-A untreated control (NIRR) set to 100%. Bars ± SD. * $p < 0.05$.

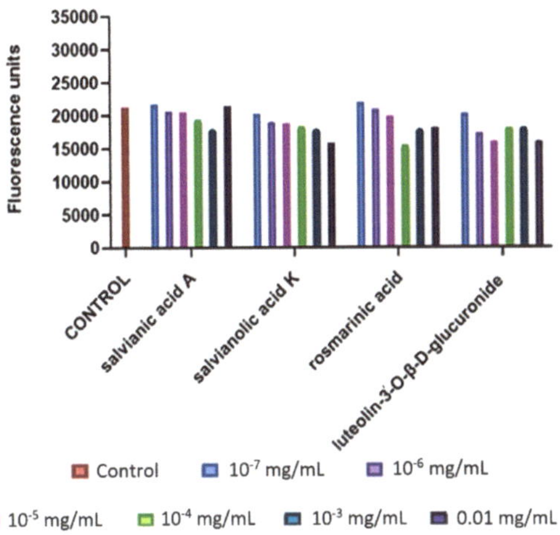

Figure 6. Intracellular antioxidant activity of the isolated compounds (**1–4**) from *S. officinalis*. 2′,7′-dichlorofluorescein (DCF)-fluorescence of mouse skin fibroblasts (NIH/3T3) treated with the indicated compounds for 1 h was used for the assessment of UVA-stimulated reactive oxygen species (ROS) levels. Vehicle (PBS)—treated irradiated cells were used as negative controls. Bars ± SD.

3.3. *Wound-Healing Potential of the Isolated Compounds in Scratch Assay*

The wound healing activity of isolated compounds (**1-4**) is presented in Figure 7. Mild migration was observed in the control group at 37%, while in the positive control (FBS 15%) 80% cell migration and proliferation were detected after 24 h of scratch formation. Among the isolated compounds, salvianolic acid K (**3**) showed the greatest, statistically significant wound healing activity at the concentration of 10^{-3} mg/mL, enhancing cell migration by 70%, in a similar way to positive control (80%). Previous studies have reported various pharmacological activities of this compound [46,48,49], however, this is the first time that a wound-healing activity of salvianolic acid K is reported. Among the tested compounds, rosmarinic acid (**2**) has already been reported as an important wound-healing agent [50].

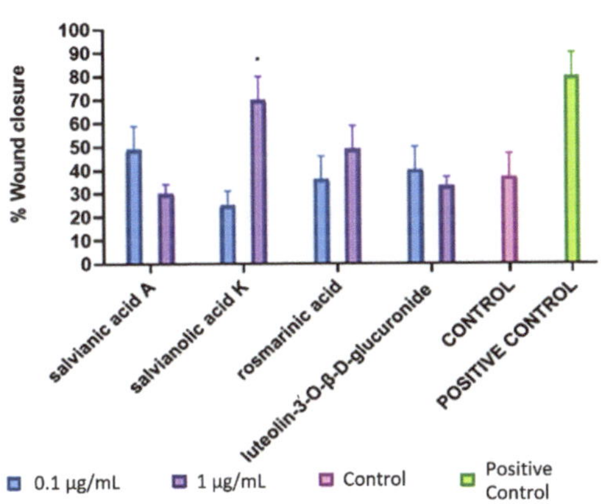

Figure 7. Wound-healing activity of the isolated compounds (**1–4**) from *S. officinalis* in NIH/3T3 fibroblasts. The data are expressed as mean of three experiments. Bars ± SD. * $p < 0.05$.

4. Conclusions

This study explored the cytotoxic and photoprotective effects, as well as the potential to reduce oxidative stress in fibroblasts of ten Lamiaceae taxa. *S. officinalis* aqueous extract was bio-guided fractionated leading to the isolation and identification of the main responsible bioactive metabolites, namely salvianic acid A, rosmarinic acid, salvianolic acid K, luteolin-3′-O-β-D-glucuronide, and hispidulin-7-O-β-D-glucuronide. All the isolated compounds exhibited significant photoprotective activity in the tested concentration while none of them statistically significantly decreased oxidative stress. Salvianolic acid K wound healing effect was much enhanced, appearing as a good candidate for in vivo wound healing studies.

Overall, the present study unveiled the importance of Lamiaceae plants in skin health. Furthermore, it could provide a better insight into the traditional applications of the genus *Salvia*, aiming to validate ethnomedicinal use and identify the responsible metabolites.

Supplementary Materials: The following supporting information can be downloaded at: https://www.mdpi.com/article/10.3390/agronomy13051224/s1, Figure S1. Cytotoxic activity of ethanol and ethyl acetate extracts of *S. officinalis* (SOE), *M. officinalis* subsp. *altissima* (MOE), *R. officinalis* (ROE), and *T. vulgaris* (TVE) in mouse skin fibroblasts (NIH/3T3) after 24 h of treatment; Figure S2. Cytotoxic activity of methanol extracts of *S. scardica* (SSM), *S. sipylea* (SSMe), *S. cypria* (SCM), *St. iva* (SIM), *S. euboea* (SEM), and *S. perfoliata* subsp. *perfoliata* (SPM) in mouse skin fibroblasts (NIH/3T3) after 24 h of treatment; Figure S3. Cytotoxic activity of aqueous extracts of *S. officinalis* (SOW), *T. vulgaris* (TVW), *M. officinalis* subsp. *altissima* (MOW), and *R. officinalis* (ROW) in mouse skin fibroblasts (NIH/3T3) after 24 h of treatment; Figure S4. Cytotoxic activity of infusions of *S. cypria* (SCI), *S. scardica* (SSW), *S. sipylea* (SSI), *S. euboea* (SEI), *S. perfoliata* subsp. *perfoliata* (SPI), and *St. iva* (SII) in mouse skin fibroblasts (NIH/3T3) after 24 h of treatment; Figure S5. Photoprotective activity of ethanol and ethyl acetate extracts of *S. officinalis* (SOE), *M. officinalis* subsp. *altissima* (MOE), *R. officinalis* (ROE), and *T. vulgaris* (TVE). UV-A-induced cytotoxicity was assessed on mouse skin fibroblasts (NIH/3T3) pre-treated with the indicated extracts for 1 h (A). Extract—induced cytotoxicity after 1 h of treatment (B); Figure S6. Photoprotective activity of methanol extracts of *S. scardica* (SSM), *S. sipylea* (SSMe), *S. cypria* (SCM), *St. iva* (SIM), *S. euboea* (SEM), and *S. perfoliata* subsp. *perfoliata* (SPM). UV-A-induced cytotoxicity was assessed on mouse skin fibroblasts (NIH/3T3) pre-treated with the indicated extracts for 1 h (A). Extract-induced cytotoxicity after 1 h of treatment (B); Figure S7. Photoprotective activity of aqueous extracts of *S. officinalis* (SOW), *T. vulgaris* (TVW), *M. officinalis* subsp. *altissima* (MOW), and *R. officinalis* (ROW). UV-A-induced cytotoxicity was assessed on mouse skin fibroblasts (NIH/3T3) pre-treated with the indicated extracts for 1 h (A). Extract-induced cytotoxicity after 1 h of treatment (B); Figure S8. Photoprotective activity of infusions of *S. cypria* (SCI), *S. scardica* (SSW), *S. sipylea* (SSI), *S. euboea* (SEI), *S. perfoliata* subsp. *perfoliata* (SPI), and *St. iva* (SII). UV-A-induced cytotoxicity was assessed on mouse skin fibroblasts (NIH/3T3) pre-treated with the indicated extracts for 1 h (A). Extract-induced cytotoxicity after 1 h of treatment (B); Figure S9. Intracellular antioxidant activity of ethanol and ethyl acetate extracts of *S. officinalis* (SOE), *M. officinalis* subsp. *altissima* (MOE), *R. officinalis* (ROE), and *T. vulgaris* (TVE); Figure S10. Intracellular antioxidant activity of methanol extracts of *S. scardica* (SSM), *S. sipylea* (SSMe), *S. cypria* (SCM), *St. iva* (SIM), *S. euboea* (SEM), and *S. perfoliata* subsp. *perfoliata* (SPM); Figure S11. Intracellular antioxidant activity of aqueous extracts of *S. officinalis* (SOW), *T. vulgaris* (TVW), *M. officinalis* subsp. *altissima* (MOW), and *R. officinalis* (ROW); Figure S12. Intracellular antioxidant activity of infusions of *S. cypria* (SCI), *S. scardica* (SSW), *S. sipylea* (SSI), *S. euboea* (SEI), *S. perfoliata* subsp. *perfoliata* (SPI), and *St. iva* (SII).

Author Contributions: Conceptualization, M.C.R. and H.S.; methodology, M.C.R., E.-M.T. and H.S.; chemical investigation, E.T. and E.-M.T.; biological investigation: E.T.; statistical analysis, E.T.; writing—original draft preparation, E.T., E.-M.T. and C.A.; writing—review and editing, M.C.R. and H.S.; supervision, M.C.R. and H.S. All authors have read and agreed to the published version of the manuscript.

Funding: This research received no external funding.

Data Availability Statement: Not applicable.

Acknowledgments: The authors wish to thank N. Tzortzakis and A. Chrysargyris (Cyprus University of Technology) for providing the cultivated plant materials *S. cypria* Post and *S. perfoliata* subsp. *perfoliata* from Cyprus, as well as P. Chatzopoulou (Hellenic Agricultural Organization-ELGO Dimitra) for the cultivated plant materials of *Sideritis euboea* Heldr. *Sideritis scardica* Griseb., and *Stachys iva* Griseb. In addition, the authors wish to thank Th. Constantinidis and K. Goula (Section of Ecology and Systematics, Department of Biology, National and Kapodistrian University of Athens) for the identification of wild populations. E.T. thanks Maria Kostaki, for the initial training in the in vitro assays.

Conflicts of Interest: The authors declare no conflict of interest.

References

1. Arda, O.; Göksügür, N.; Tüzün, Y. Basic histological structure and functions of facial skin. *Clin. Dermatol.* **2014**, *32*, 3–13. [CrossRef]
2. Zouboulis, C.C.; Ganceviciene, R.; Liakou, A.I.; Theodoridis, A.; Elewa, R.; Makrantonaki, E. Aesthetic aspects of skin aging, prevention, and local treatment. *Clin. Dermatol.* **2019**, *37*, 365–372. [CrossRef]
3. Ibrahim, N.; Wong, S.; Mohamed, I.; Mohamed, N.; Chin, K.-Y.; Ima-Nirwana, S.; Shuid, A. Wound Healing Properties of Selected Natural Products. *Int. J. Environ. Res. Public. Health* **2018**, *15*, 2360. [CrossRef]
4. Svobodová, A.; Psotová, J.; Walterová, D. Natural phenolics in the prevention of UV-induced skin damage. A review. *Biomed. Pap. Med. Fac. Univ. Palacky. Olomouc Czech Repub.* **2003**, *147*, 137–145. [CrossRef]
5. Solano, F. Photoprotection and Skin Pigmentation: Melanin-Related Molecules and Some Other New Agents Obtained from Natural Sources. *Molecules* **2020**, *25*, 1537. [CrossRef] [PubMed]
6. Cavinato, M.; Jansen-Dürr, P. Molecular mechanisms of UVB-induced senescence of dermal fibroblasts and its relevance for photoaging of the human skin. *Exp. Gerontol.* **2017**, *94*, 78–82. [CrossRef] [PubMed]
7. Cavinato, M.; Waltenberger, B.; Baraldo, G.; Grade, C.V.C.; Stuppner, H.; Jansen-Dürr, P. Plant extracts and natural compounds used against UVB-induced photoaging. *Biogerontology* **2017**, *18*, 499–516. [CrossRef] [PubMed]
8. Hernandez, D.F.; Cervantes, E.L.; Luna-Vital, D.A.; Mojica, L. Food-derived bioactive compounds with anti-aging potential for nutricosmetic and cosmeceutical products. *Crit. Rev. Food Sci. Nutr.* **2021**, *61*, 3740–3755. [CrossRef] [PubMed]
9. Hernández, A.R.; Vallejo, B.; Ruzgas, T.; Björklund, S. The Effect of UVB Irradiation and Oxidative Stress on the Skin Barrier-A New Method to Evaluate Sun Protection Factor Based on Electrical Impedance Spectroscopy. *Sensors* **2019**, *19*, 2376. [CrossRef]
10. Mohania, D.; Chandel, S.; Kumar, P.; Verma, V.; Digvijay, K.; Tripathi, D.; Choudhury, K.; Mitten, S.K.; Shah, D. Ultraviolet Radiations: Skin Defense-Damage Mechanism. *Adv. Exp. Med. Biol.* **2017**, *996*, 71–87. [PubMed]
11. Pérez-Sánchez, A.; Barrajón-Catalán, E.; Herranz-López, M.; Castillo, J.; Micol, V. Lemon balm extract (*Melissa officinalis*, L.) promotes melanogenesis and prevents UVB-induced oxidative stress and DNA damage in a skin cell model. *J. Dermatol. Sci.* **2016**, *84*, 169–177. [CrossRef] [PubMed]
12. Kostyuk, V.; Potapovich, A.; Albuhaydar, A.R.; Mayer, W.; De Luca, C.; Korkina, L. Natural Substances for Prevention of Skin Photoaging: Screening Systems in the Development of Sunscreen and Rejuvenation Cosmetics. *Rejuvenation Res.* **2018**, *21*, 91–101. [CrossRef]
13. Aghakhani, F.; Kharazian, N.; Lori Gooini, Z. Flavonoid constituents of Phlomis (Lamiaceae) species using liquid chromatography mass spectrometry. *Phytochem. Anal.* **2018**, *29*, 180–195. [CrossRef] [PubMed]
14. Singh, P.; Pandey, A.K. Prospective of essential oils of the genus *Mentha* as biopesticides: A review. *Front. Plant Sci.* **2018**, *9*, 1295. [CrossRef] [PubMed]
15. Walker, J.B.; Sytsma, K.J.; Treutlein, J.; Wink, M. *Salvia* (Lamiaceae) is not monophyletic: Implications for the systematics, radiation, andecological specializations of *Salvia* and tribe Mentheae. *Am. J. Bot.* **2004**, *91*, 1115–1125. [CrossRef] [PubMed]
16. Committee on Herbal Medicinal Products. *Assessment Report on Salvia officinalis L., Folium and Salvia officinalis L., Aetheroleum*; EMA/HMPC/150801/2015; Committee on Herbal Medicinal Products: London, UK, 2016.
17. Lamaison, J.L.; Petitjean-Freytet, C.; Carnat, A. Lamiacées médicinales à propriétés antioxydantes, sources potentielles d'acide rosmarinique [Medicinal Lamiaceae with antioxidant properties, a potential source of rosmarinic acid]. *Pharm. Acta Helv.* **1991**, *66*, 185–188.
18. Dapkevicius, A.; Venskutonis, R.; van Beek, T.A.; Linssen, J.P.H. Antioxidant activity of extracts obtained by different isolation procedures from some aromatic herbs grown in Lithuania. *J. Sci. Food Agric.* **1998**, *77*, 140–146. [CrossRef]
19. Hohmann, J.; Zupkó, I.; Rédei, D.; Csányi, M.; Falkay, G.; Máthé, I.; Janicsák, G. Protective effects of the aerial parts of *Salvia officinalis*, *Melissa officinalis* and *Lavandula angustifolia* and their constituents against enzyme-dependent and enzyme-independent lipid peroxidation. *Planta Med.* **1999**, *65*, 576–578. [CrossRef]
20. Bonesi, M.; Loizzo, M.R.; Acquaviva, R.; Malfa, G.A.; Aiello, F.; Tundis, R. Anti-inflammatory and Antioxidant Agents from *Salvia* Genus (Lamiaceae): An Assessment of the Current State of Knowledge. *Antiinflamm. Antiallergy Agents Med. Chem.* **2017**, *16*, 70–86. [CrossRef]
21. Lu, Y.; Foo, L.Y. Polyphenolics of *Salvia*-a review. *Phytochemistry* **2002**, *59*, 117–140. [CrossRef]
22. Topçu, G. Bioactive triterpenoids from *Salvia* species. *J. Nat. Prod.* **2006**, *69*, 482–487. [CrossRef] [PubMed]
23. Committee on Herbal Medicinal Products. *European Union Herbal Monograph on Sideritis scardica Griseb.; Sideritis clandestina (Bory & Chaub.) Hayek; Sideritis raeseri Boiss. & Heldr.; Sideritis syriaca L., Herba*; EMA/HMPC/39455/2015; Committee on Herbal Medicinal Products: London, UK, 2016.
24. Lu, Y.; Foo, L.Y. Flavonoid and phenolic glycosides from *Salvia officinalis*. *Phytochemistry* **2000**, *55*, 263–267. [CrossRef] [PubMed]
25. González-Burgos, E.; Carretero, M.E.; Gómez-Serranillos, M.P. *Sideritis* spp.: Uses, chemical composition and pharmacological activities—A review. *J. Ethnopharmacol.* **2011**, *135*, 209–225. [CrossRef] [PubMed]

26. Fraga, B.M. Phytochemistry and chemotaxonomy of *Sideritis* species from the Mediterranean region. *Phytochemistry* **2012**, *76*, 7–24. [CrossRef]
27. Tundis, R.; Peruzzi, L.; Menichini, F. Phytochemical and biological studies of *Stachys* species in relation to chemotaxonomy: A review. *Phytochemistry* **2014**, *102*, 7–39. [CrossRef] [PubMed]
28. Andrade, J.M.; Faustino, C.; Garcia, C.; Ladeiras, D.; Reis, C.P.; Rijo, P. *Rosmarinus officinalis* L.: An update review of its phytochemistry and biological activity. *Future Sci. OA* **2018**, *4*, FSO283. [CrossRef]
29. Mailis, T.; Skaltsa, H. Polar Constituents of *Salvia willeana* (Holmboe) Hedge, Growing Wild in Cyprus. *Plants* **2018**, *7*, 18 [CrossRef]
30. Sharifi-Rad, J.; Ezzat, S.M.; El Bishbishy, M.H.; Mnayer, D.; Sharopov, F.; Kılıç, C.S.; Neagu, M.; Constantin, C.; Sharifi-Rad, M.; Atanassova, M.; et al. *Rosmarinus* plants: Key farm concepts towards food applications. *Phytother. Res.* **2020**, *34*, 1474–1518. [CrossRef]
31. Żyżelewicz, D.; Kulbat-Warycha, K.; Oracz, J.; Żyżelewicz, K. Polyphenols and Other Bioactive Compounds of *Sideritis* Plants and Their Potential Biological Activity. *Molecules* **2020**, *25*, 3763. [CrossRef]
32. Shimada, A.; Ueno, H.; Inagaki, M. Glutaminase inhibitory activities of pentacyclic triterpenes isolated from *Thymus vulgaris* L *Nat. Prod. Res.* **2021**, *36*, 2864–2868. [CrossRef]
33. Tomou, E.M.; Barda, C.; Skaltsa, H. Genus *Stachys*: A Review of Traditional Uses, Phytochemistry and Bioactivity. *Medicines* **2020**, *7*, 63. [CrossRef]
34. Dimas, D.; Tomou, E.-M.; Karamanou, C.; Sfiniadakis, I.; Siakavella, K.I.; Liakopoulou, A.; Hatziantoniou, S.; Rallis, M.; Skaltsa, H *Melissa officinalis* ssp. *altissima* extracts: A therapeutic approach targeting psoriasis in mice. *J. Ethnopharmacol.* **2019**, *246*, 112208.
35. Patil, S.M.; Ramu, R.; Shirahatti, P.S.; Shivamallu, C.; Amachawadi, R.G. A systematic review on ethnopharmacology, phytochemistry and pharmacological aspects of *Thymus vulgaris* Linn. *Heliyon* **2021**, *7*, e07054. [CrossRef]
36. Pritsas, A.; Tomou, E.M.; Tsitsigianni, E.; Papaemmanouil, C.D.; Diamantis, D.A.; Chatzopoulou, P.; Tzakos, A.G.; Skaltsa, H. Valorisation of stachysetin from cultivated *Stachys iva* Griseb. as anti-diabetic agent: A multi-spectroscopic and molecular docking approach. *J. Biomol. Struct. Dyn.* **2021**, *39*, 6452–6466. [CrossRef] [PubMed]
37. Yahara, S.; Satoshiro, M.; Nishioka, I.; Nagasawa, T.; Oura, H. Isolation and characterization of phenolic compounds from Coptidis Rhizoma. *Chem. Pharm. Bull.* **1985**, *33*, 527–531. [CrossRef]
38. Lu, Y.; Foo, L.Y. Rosmarinic acid derivatives from *Salvia officinalis*. *Phytochemistry* **1999**, *51*, 91–94. [CrossRef]
39. Heitz, A.; Carnat, A.; Fraisse, D.; Carnat, A.-P.; Lamaison, J.-L. Luteolin 3′-glucuronide, the major flavonoid from *Melissa officinalis* subsp. *officinalis*. *Fitoterapia* **2000**, *71*, 201–202. [CrossRef]
40. Zaabat, N.; Hay, A.E.; Michalet, S.; Skandrani, I.; Chekir-Ghedira, L.; Dijoux-Franca, M.G.; Akkal, S. Chemical Composition, Antioxidant, Genotoxique and Antigenotoxic Potentials of *Phlomis bovei* De Noé Aerial Parts. *Iran. J. Pharm. Res.* **2020**, *19*, 282–291.
41. Martins, N.; Barros, L.; Santos-Buelga, C.; Henriques, M.; Silva, S.; Ferreira, I.C.F.R. Evaluation of bioactive properties and phenolic compounds in different extracts prepared from *Salvia officinalis* L. *Food Chem.* **2015**, *170*, 378–385. [CrossRef]
42. Xu, J.; Wei, K.; Zhang, G.; Lei, L.; Yang, D.; Wang, W.; Han, Q.; Xia, Y.; Bi, Y.; Yang, M.; et al. Ethnopharmacology, phytochemistry, and pharmacology of Chinese *Salvia* species: A review. *J. Ethnopharmacol.* **2018**, *225*, 18–30. [CrossRef]
43. Psotova, J.; Svobodova, A.; Kolarova, H.; Walterova, D. Photoprotective properties of *Prunella vulgaris* and rosmarinic acid on human keratinocytes. *J. Photochem. Photobiol. B.* **2006**, *84*, 167–174. [CrossRef] [PubMed]
44. Sánchez-Campillo, M.; Gabaldon, J.A.; Castillo, J.; Benavente-García, O.; Del Baño, M.J.; Alcaraz, M.; Vicente, V.; Alvarez, N.; Lozano, J.A. Rosmarinic acid, a photo-protective agent against UV and other ionizing radiations. *Food Chem. Toxicol.* **2009**, *47*, 386–392. [CrossRef] [PubMed]
45. Sun, Z.; Park, S.Y.; Hwang, E.; Zhang, M.; Jin, F.; Zhang, B.; Yi, T.H. Salvianolic Acid B Protects Normal Human Dermal Fibroblasts Against Ultraviolet B Irradiation-Induced Photoaging Through Mitogen-Activated Protein Kinase and Activator Protein-1 Pathways. *Photochem. Photobiol.* **2015**, *91*, 879–886. [CrossRef]
46. Lu, Y.; Foo, Y.L. Antioxidant activities of polyphenols from sage (*Salvia officinalis*). *Food Chem.* **2001**, *75*, 197–202. [CrossRef]
47. Xiao, Z.; Liu, W.; Mu, Y.P.; Zhang, H.; Wang, X.N.; Zhao, C.Q.; Chen, J.M.; Liu, P. Pharmacological Effects of Salvianolic Acid B Against Oxidative Damage. *Front. Pharmacol.* **2020**, *11*, 572373. [CrossRef] [PubMed]
48. Kasimu, R.; Tanaka, K.; Tezuka, Y.; Gong, Z.-N.; Li, J.-X.; Basnet, P.; Namba, T.; Kadota, S. Comparative Study of Seventeen *Salvia* Plants: Aldose Reductase Inhibitory Activity of Water and MeOH Extracts and Liquid Chromatography-Mass Spectrometry (LC-MS) Analysis of Water Extracts. *Chem. Pharm. Bull.* **1998**, *46*, 500–504. [CrossRef]
49. Ayoub, I.M.; George, M.Y.; Menze, E.T.; Mahmoud, M.; Botros, M.; Essam, M.; Ashmawy, I.; Shendi, P.; Hany, A.; Galal, M.; et al. Insights into the neuroprotective effects of *Salvia officinalis* L. and *Salvia microphylla* Kunth in the memory impairment rat model. *Food Funct.* **2022**, *13*, 2253–2268. [CrossRef]
50. Küba, M.C.; Türkoğlu, A.; Oğuz, A.; Tuncer, M.C.; Kaya, Ş.; Başol, Ö.; Bilge, H.; Tatlı, F. Comparison of local rosmarinic acid and topical dexpanthenol applications on wound healing in a rat experimental wound model. *Folia Morphol.* **2021**, *80*, 618–624. [CrossRef]

agronomy

MDPI

Article

A Comparative Evaluation of Aquaponic and Soil Systems on Yield and Antioxidant Levels in Basil, an Important Food Plant in Lamiaceae

Maryam A. O. K. Albadwawi, Zienab F. R. Ahmed *, Shyam S. Kurup, Mohammed A. Alyafei and Abdul Jaleel *

Department of Integrative Agriculture, College of Agriculture and Veterinary Medicine, United Arab Emirates University, Al Ain P.O. Box 15551, United Arab Emirates

* Correspondence: zienab.ahmed@uaeu.ac.ae (Z.F.R.A.); abdul.jaleel@uaeu.ac.ae (A.J.)

Abstract: Greenhouse and aquaponic productions of basil (*Ocimum basilicum* L.) are well established, but the comparison between these two methods is not studied in detail. A study was conducted to evaluate the performance of basil in both aquaponic and soil systems under greenhouse conditions. The plants were raised in aquaponic beds with ornamental fish and a set of plants was raised in soil pots under a greenhouse setup. The studied parameters were morphological, biochemical and antioxidant levels. In order to analyze the stress effects of an aquaponic system on plant defense mechanism, two different antioxidant enzymes (catalase and peroxidase) were analyzed. Water quality parameters were monitored during the entire study period. Based on the results, there was a significant increase in growth parameters in the aquaponic system when compared to the conventional greenhouse cultivation of basil plants. The photosynthetic parameters showed a decline in the aquaponic system, but the biochemical parameters showed an enhancement in the aquaponic system of growing basil plants. The antioxidants exhibited a significant increase in the aquaponic system, which suggests a water stress effect on the plants induced by the aquaponic growing system. From the results of this study, it can be concluded that the aquaponic system is the best suitable method for basil production in the UAE condition.

Keywords: *Ocimum*; ornamental fish; aquaponics; antioxidants; sustainable production

1. Introduction

Water management is the one of the main challenges worldwide, especially in countries such as the United Arab Emirates, in which it poses a problem which needs immediate attention. The present system of conventional agriculture is no longer feasible in this aspect, as it requires a lot of water, which is very precious in this country.

As an alternative way of modern agriculture, aquaponics is considered as a system that can use the water-containing residues from fish production for producing plants in order to save water and produce clean products in a shorter time [1]. The principle of aquaponics is that it uses the waste from fish and utilizes it in the form of nutrients required for normal plant growth [2]. In this aspect, an aquaponics system can conserve precious water, especially in countries such as the UAE, where the water supply is limited, and there are very few arable lands. Additionally, the environmental impact is lower compared to the conventional farming systems. With aquaponic technology, a sustainable fish and vegetable production can be achieved with the simultaneous conservation of precious natural resources in the country.

The UAE is moving toward sustainable food security in the near future, and for that, a multi-faceted agro-ecological intensification of food production and a decoupling from unsustainable resource use is of utmost importance. Techniques such as aquaponics perform a great role in achieving the food security of modern times [3]. In countries such as the UAE, where the climatic and land characteristics are not adapted for conventional farming, this type of low water consumption agricultural systems is more promising [4]. Additionally, the usage of synthetic fertilizers can partly or completely be eliminated in aquaponic systems, and the use of pesticides is likewise reduced in this type of aquaponic systems, as the system relies on fish waste to provide nutrients to the plants [5].

The working principle of aquaponics is quite simple, as it involves only three kinds of living organisms: fish, beneficial bacteria and plants. The interrelations between them are highly complex and interdependent [6]. Additionally, the issue of toxic components such as ammonia created in the system from fish and food waste is resolved by the action of bacteria by the process of conversion of ammonia into nitrite and then into nitrate, which is non-toxic and at the same time beneficial to plants [7]. In a recirculating aquaponics system (RAS), the water is circulated through the fish tanks and grow beds, and the waste of water can be highly reduced [8]. It can produce fish and plants in high quantity and quality simultaneously.

When selecting the fish to be used in an aquaponic system, the main point to be noted is the resistance of the fish species to various environmental and water conditions [9]. The availability and usefulness of the species is also to be considered. The major fish species used in any aquaponics system is Nile Tilapia (*Oreochromis niloticus*), which can be used to run the plant-based system very successfully [10]. The studies related to the usage of ornamental fish in aquaponics systems is relatively scanty [11]. In this study, we selected koi fish (*Cyprinus rubrofuscus* var. "koi"), which is an important ornamental species loved by the hobbyists and aquaculturists throughout the world. Koi fish belong to the family Cyprinidae and are relatively resistant to different situations in fish tanks [12]. The resistance of this fish to various levels of water parameters and its ability to grow and reproduce in significant numbers is the main reason for our selection for it in aquaponics. This fish is comparatively less disease-susceptible and also fast growing, making it an easy species when cultivated together with plants [13].

Ocimum basilicum, popularly known as basil, is a very popular culinary herb that is mainly used in food production [14]. Growing basil is also very common among farmers. Basil has repellent or pathogen-killing properties such as nematicidal, antimicrobial, fungistatic and insecticidal effects [15]. Considering these benefits and effects, basil has multiple interesting uses in the food industries, perfume industry, traditional medicine industry as well as pharmaceutical industry. *O. basilicum* is a very common plant that is grown in aquaponics or more specifically in coupled aquaponic conditions [14]. Basil leaves develop an unmistakable and distinctive scent via peculiar oil glands, and basil is therefore considered as a very useful herb [15]. Apart from being a very common and important ingredient in the kitchen, basil has several essential properties, and is therefore used in fresh and dry form in almost all countries and cultures. Basil is suitable for soil as well as soilless cultivation [16].

Basil production through aquaponics is well established [17]. Basil has high growth potential and is therefore especially suitable for aquaponics. Inevitably, the demand for basil is considerably high among both hydroponics and aquaponics producers. Studies indicate that owing to its characteristics, basil is the most used herb for different hydroponics and aquaponics experiments [18]. Basil produces 1.8 kg per meter square under aquaponic production, and only 0.6 kg per meter square in soil cultivation [19]. Thus, aquaponic production is considered to be more efficient and environmentally friendly than soil cultivation [20]. Basil is among one of the top priorities for farmers all over the world due to the high value that it carries as a cash crop [21–23].

Even though previous studies show that basil growth and production is higher in the case of aquaponic systems compared to soil cultivation, researchers did not focus adequately on comparing the level of antioxidants in the basil grown in both cases. This research considers and seeks to fulfil this research gap by conducting a comparative evaluation of yield as well as levels of antioxidants in basil grown in aquaponics and soil systems.

The soil system of growing basil or *O. basilicum* is one of the most profound and rich ways of producing basil [24]. Basil does its best work when growing with the help of rich soil that has an accurate pH level for its growth. A perfect drainage system, a soil that has the required pH level and a place where sunlight reaches for about six to eight hours every day, will give any farmer basil of high quality. Basil, in its most basic nature, is a tropical herb that will grow and thrive easily in moist and nutrition-filled soil [24–26].

With the use of the aquaponic crop-production system, basil can be grown in a way that is environmentally friendly; no soil, less water and high-quality crop yield [27]. The crops produced with the help of the aquaponic system are not affected by any shift in the environment and sudden shifts in the climate [28]. The aquaponic system takes the help of fish, water tanks, plants and bacteria to grow products that are of high quality [29]. The quality is maintained through proper checkups and maintenance of the system.

Previous studies showed that basil has a lot of characteristics and properties that make it one of the most common herbs used extensively in the kitchen as well as in several other business sectors [15,18,20]. Nevertheless, while previous researchers found that basil production and yield is higher in the case of aquaponic systems than in soil systems, no information is available on the level of antioxidants present in basil grown in both systems. The purpose of this research was to investigate the efficiency of aquaponic systems in comparison to soil systems, particularly in the context of yield and level of antioxidants in basil. In other words, this research seeks to identify and understand how basil grown in soil systems differs from basil grown in aquaponic systems.

2. Materials and Methods

2.1. Experiment Location

The experiment was carried out in in a polycarbonate greenhouse in Falaj Hazza campus unit of aquaponics of the College of Agriculture and Veterinary Medicine, the UAEU, in Al Ain city, 160 km East of Abu Dhabi, the capital city of the United Arab Emirates, co-ordinate latitude and longitude of 24.2191° N and 55.7146° E, with ambient lighting, average daily temperature 19.2–26.4 °C and relative humidity 40–60%.

2.2. Seed Collection and Germination

The certified seeds of *Ocimum basilicum* were obtained from agriculture material supply companies in Al Ain. Seeds of *O. basilicum* were sown in the nursery in foam plates under slight media layer and kept in the greenhouse conditions. Water spray was performed two times per day. Plants were germinated after 4–6 days. The plants were sown along with the commencement of fish culture. Normal cultivation practices were followed for soil-based plants in the greenhouse using agriculture sandy soil.

2.3. Experimental Setup

The seedlings were grown in moisturized Rockwool cubes (2.0 cm × 2.0 cm × 2.0 cm), placed in plastic containers and sprayed with water every day. The well-grown seedlings were transplanted in a Rockwool slabs (100 cm × 20 cm × 2.5 cm) and connected to the hydroponics system, with water circulated from the fish tanks after filtration (Figures S1 and 1).

2.4. Fish Selection and Growth Conditions

A total of 50 fingerlings per m^3 of *Cyprinus rubrofuscus* var. "koi" were added in each fish tank (500 L) with an average weight of 5 g. The fish were fed as per 2.5% bodyweight with 36% protein commercial fish diet from Arabian Agricultural Services Company (ARASCO), Saudi Arabia. They were fed three times a day.

Figure 1. Different stages during the experimental study. (**a**) Early stages of growth in green house, (**b**) growth in aquaponic bed, (**c**) sample collection, (**d**) laboratory analysis.

2.5. Plant Growth Conditions and Parameters Analyzed

Basil was harvested on 40 and 70 days after sowing (DAS). Group of plants were randomly harvested on each harvesting time. Basil characteristics of each harvest was evaluated by measuring the length from leaf to root, shoot length, root length, total weight, shoot weight, leaf weight and average leaf number per plant. The plant height (from the soil level to the tip of the shoot) was measured and the values are expressed in cm. The plant root length of longest root was measured (from the point of first cotyledonary node to the tip) and expressed in cm. The total number of fully developed leaves were counted and expressed as number of leaves per plant.

The total leaf area of fully developed leaves was measured using a leaf area meter (LICOR Photo Electric Area Meter, Model LI-3100, Lincoln, NE, USA) and expressed in cm^2 per plant. Fresh weight was determined by using an electronic balance and the values were expressed in grams. The plants were dried for 48 h at 60 °C in a hot air oven after taking fresh weight. After drying, the weight was measured and the values were expressed in grams.

2.6. Estimation of Photosynthetic Pigments

Chlorophyll and carotenoid of the leaf were extracted and assessed by the method of Arnon [30]. Five hundred milligrams of fresh leaf were ground with 10 mL of 80% acetone at 4 °C and centrifuged at 2500 rpm for 10 min at 4 °C. The extract was transferred to a graduated tube and made up to 10 mL with 80% acetone and assayed immediately. The absorbance was read at 645, 663 and 480 nm with a spectrophotometer (U-2001-Hitachi) against 80% acetone as blank. Chlorophyll content was calculated using the formula of Arnon [30] and expressed in milligram/gram of fresh weight (FW).

Anthocyanin [31] and xanthophyll [32] from the basil were extracted and estimated and the results were expressed in milligrams/gram FW.

2.7. Biochemical Analysis

Protein was estimated according to the method of Bradford [33]. Five hundred milligrams of fresh plant material were ground with 10 mL of 20% trichloro acetic acid (TCA), homogenized and centrifuged for 15 min at 800× *g*. A total of 5 mL of 0.1 N NaOH was added to the pellet after the supernatant was discarded, to solubilize the protein, and the solution was centrifuged at 800× *g* for 15 min. The supernatant was composed of 10 mL with 0.1 N NaOH, and then used for the estimation of the protein content. The protein solution containing 10–100 µg protein in a volume of 0.1 mL was pipetted into 12 × 100 mm test tubes. Five milliliters of Bradford's reagent were added to the test tube and the contents were mixed by vortexing. The absorbance at 595 nm was measured after 2 min with 3 mL cuvette against a reagent blank, prepared by adding 0.1 mL of 0.1 N NaOH and 5 mL of Bradford's reagent. A standard curve was prepared using BSA V fraction, which was used to determine the protein content.

Total phenols were estimated according to the method by Malick and Singh [34]. Five hundred milligrams of fresh plant tissue were ground in a pestle and mortar with 10 mL of 80% ethanol. The homogenate was centrifuged at 10,000 rpm for 20 min. The supernatant was evaporated, and then the residue was dissolved with 5 mL of distilled water and used as extract. A volume of 0.5 mL of Folin–Ciocalteau reagent was added to 2 mL of the extract. After 3 min, 2 mL of 20% Na_2CO_3 solution was added and mixed thoroughly. The mixture was kept in boiling water for exactly one min and after cooling; the absorbance was read at 650 nm. The total phenols were determined using a standard curve prepared with different concentrations of gallic acid.

2.8. Non-Enzymatic Antioxidant Contents

Ascorbic acid content was assayed as described by Omaye et al. [35]. One gram of fresh material was ground in a pestle and mortar with 5 mL of 10 per cent TCA, and the extract was centrifuged at 3500 rpm for 20 min. The pellet was re-extracted twice with 10 per cent TCA and 10 mL of supernatant was used for estimation. To 0.5 mL of extract, 1 mL of DTC reagent (2,4-Dinitrophenyl hydrazine-Thiourea-$CuSO_4$ reagent) was added and mixed thoroughly. The tubes were incubated at 37 °C for 3 h, and to this 0.75 mL of ice-cold 65 per cent H_2SO_4 was added. The tubes were then allowed to stand at 30 °C for 30 min. The resulting color was read at 520 nm in spectrophotometer (U-2001-Hitachi). The ascorbic acid content was determined using a standard curve prepared with ascorbic acid and the results are expressed in mg g^{-1} fresh weight.

α-Tocopherol content was estimated as described by Backer et al. [36]. Five hundred milligrams of fresh tissue was homogenized with 10 mL of a mixture of petroleum ether and ethanol (2:1.6 *v*/*v*) and the extract was centrifuged at 10,000 rpm for 20 min. The supernatant was used for estimation of α-tocopherol. To one mL of extract, 0.2 mL of 2 per cent 2,2-dipyridyl in ethanol was added and mixed thoroughly and kept in the dark for 5 min. The resulting red color was diluted with 4 mL of distilled water and mixed well. The resulting color in the aqueous layer was measured at 520 nm. The α-tocopherol content was calculated using a standard graph made with known amount of α-tocopherol and expressed in mg g^{-1} fresh weight.

2.9. Antioxidant Enzymes

Peroxidase and catalase crude enzyme extract were prepared according to a method by Hwang et al. [37]. One gram of plant tissue was homogenized with 10 mL of ice-cold buffer 50 mM sodium phosphate containing 1 mM PMSF. The homogenate was strained through two layers of cheesecloth and then centrifuged at 12,500× g for 20 min at 4 °C. The supernatant content was 10 mL with 50 mM of sodium phosphate buffer, used as the source of enzymes. The enzyme protein was determined using a method by Bradford [33] for expressing the specific activity of all the enzymes.

Catalase activity was assayed as a method described by Chandlee and Scandalios [38]. Frozen material (0.5 g) was homogenized in 5 mL of ice-cold 50 mM sodium phosphate buffer (pH 7.5) containing 1 mM PMSF (phenyl methyl sulfonyl fluoride). The homogenate was centrifuged at 4 °C for 20 min at 12,500× g. The supernatant was used for enzyme assay. The assay mixture contained 0.4 mL of 15 mM H_2O_2, 2.6 mL of 50 mM potassium phosphate buffer (pH 7.0) and 0.04 mL of enzyme extract. The decomposition of H_2O_2 was followed by the decline in absorbance at 240 nm. The enzyme activity was expressed in units of 1 mM of H_2O_2 reduction per minute per mg protein.

Peroxidase was assayed according to a method by Kumar and Khan [39]. The assay mixture of peroxidase contained 2 mL of 0.1 M phosphate buffer (pH 6.8), 1 mL of 0.01 M pyrogallol, 1 mL of 0.005 M H_2O_2 and 0.5 mL of enzyme extract. The solution was incubated for 5 min at 25 °C, after which the reaction was terminated by adding 1 mL of 2.5 N H_2SO_4. The amount of purpurogallin formed was determined by measuring the absorbance at 420 nm against a blank (the extract after the addition of 2.5 N H_2SO_4 at zero time). The activity was expressed in unit mg^{-1} protein. One unit is defined as the change in the absorbance by 0.1 min−1 mg^{-1} protein.

2.10. Microelements and Macroelements

The analysis was conducted using ICP-OES-Agilent Technologies [40]. Samples were prepared accurately by weighing 0.5 g in the microwave digestion vessels. The digestion process began by adding 10 mL of concentrated nitric acid (HNO_3) and 2 mL of hydrochloric acid (HCL) to the sample. The vessels were capped and placed in the microwave digestion system. The analysis was conducted using ICP-OES-Agilent Technologies, 710. The percentages of different elements in this sample were determined by the corresponding standard calibration curves obtained by using standard AR grade solutions of the elements, for example K, Mg, Ca, Na, Fe, Mn, Zn, P and S.

2.11. Statistical Analysis

The experiment followed a complete randomized design. The data were analyzed by SAS (SAS Institute Inc., 2000, Cary, NC, USA). Comparisons between the aquaponic (soilless) and soil systems at the harvest times of 40 and 70 days were conducted using the t-test at $p \leq 0.05$. The values are mean ± SD for seven replicates from each cultivation system.

3. Results

3.1. Morphological Parameters

The plants in aquaponic and soil system exhibited an increase in height with age, as they started growing since the first DAS, and almost reached their maximum height at 70 DAS. However, when compared to the aquaponic system at 40 DAS (69.42 cm), the greenhouse-grown plants exhibited (65.55 cm) a decreased height. A similar trend was noticed in the case of 70 DAS, when aquaponic-grown plants measured 98.05 cm, while the greenhouse-grown plants measured only 84.21 cm in height (Table 1).

The root length increased with age in both systems. The root length decreased in the soil system when compared to aquaponic-grown plants (Table 1). The average root length of basil plants was 27.98 cm in the aquaponic system, but it was further reduced in soil-system-grown plants to 23.21 cm at 40 DAS. Additionally, at 70 DAS, the root length of the basil plants in the aquaponic and greenhouse systems were 59.21 cm and 54.36 cm, respectively.

Table 1. Different growth parameters of *Ocimum basilicum* in aquaponic and soil (greenhouse) systems at 40 and 70 DAS.

Growth Parameters	DAS	Soilless (aquaponic) System	Soil (Greenhouse) System
Total plant height (cm)	40	69.42 ± 6.22 [a]	65.55 ± 6.11 [b]
	70	98.05 ± 4.35 [a]	84.21 ± 9.55 [b]
Root length (cm)	40	27.98 ± 4.15 [a]	23.21 ± 2.11 [b]
	70	59.21 ± 3.98 [a]	54.36 ± 5.23 [b]
Leaves (No.)	40	107.00 ± 20.19 [a]	99.88 ± 10.03 [b]
	70	174 ± 12.55 [a]	148 ± 15.19 [b]
Leaf area $(cm)^2$	40	3.78 ± 0.22 [a]	3.08 ± 0.70 [a]
	70	3.74 ± 0.31 [a]	3.66 ± 0.38 [a]
Total fresh weight (g)	40	47.80 ± 5.31 [a]	45.89 ± 3.33 [b]
	70	84.31 ± 5.89 [a]	79.25 ± 6.31 [b]
Total dry weight (g)	40	4.96 ± 0.51 [a]	3.77 ± 0.19 [b]
	70	7.11 ± 0.22 [a]	6.33 ± 0.35 [b]
Root dry weight (g)	40	0.77 ± 0.03 [a]	0.44 ± 0.01 [b]
	70	1.2 ± 00.02 [a]	0.99 ± 0.022 [b]

Values are given as mean ± SD of seven samples in each group. Values that do not share a common superscript (a,b) differ significantly at $p \leq 0.05$ using *t*-test. DAS—Days after sowing.

The number of leaves increased with age in aquaponic- and greenhouse-grown plants. However, the number of leaves was reduced in the greenhouse system when compared to the aquaponic system at both 40 and 70 DAS. In the aquaponic system, the number of leaves at 40 DAS was 107.00, while in the soil system it was reduced to 99.88. Similarly, at 70 DAS, the aquaponic exhibited 174 leaves when compared with 148 in the soil system.

The total leaf area showed a similar trend under the greenhouse system when compared to aquaponic basil plants, representing 3.78 cm and 3.74 cm in the aquaponic system at 40 and 70 DAS, respectively. In greenhouse plants, it represented 3.08 and 3.66, respectively, at 40 and 70 DAS. The fresh weight was lower in greenhouse-grown plants when compared to aquaponic plants. This is due to the higher shoot and root growth of aquaponic plants, which contributed to the increased fresh and dry weights of the basil plants in the system.

3.2. Photosynthetic Pigments

The chlorophyll 'a' content of the basil leaves was extracted and estimated from the randomly selected leaves at 40 and 70 DAS (Figure 2). Contrary to the growth parameters, the chlorophyll pigment contents in basil showed an increasing trend under the greenhouse cultivation system. In the case of chlorophyll 'a', it represented 0.48 mg/g FW and 0.55 mg/g FW at 40 DAS in the aquaponic and green house systems, respectively. At 70 DAS, the content had increased and represented 0.37 mg/g FW and 0.49 mg/g FW, respectively, in the aquaponic and greenhouse systems. At 40 DAS, the chlorophyll 'b' content represented 0.68 mg/g FW and 0.82 mg/g FW in aquaponic and greenhouse plants, respectively, and it is evident that chlorophyll 'b' was enhanced in the greenhouse system (Figure 2). The total chlorophyll content of the basil leaves increased in greenhouse-cultivated plants (Figure 2). The greenhouse-grown plants harvested at 40 DAS showed higher contents (0.137 mg/g FW) when compared to the aquaponic plants (0.116 mg/g FW). A similar trend was observed in the case of plants harvested at 70 DAS, where the soil-cultivated plants' chlorophyll content (1.15 mg/g FW) was higher than that of the soilless-cultivated plants (0.88 mg/g FW).

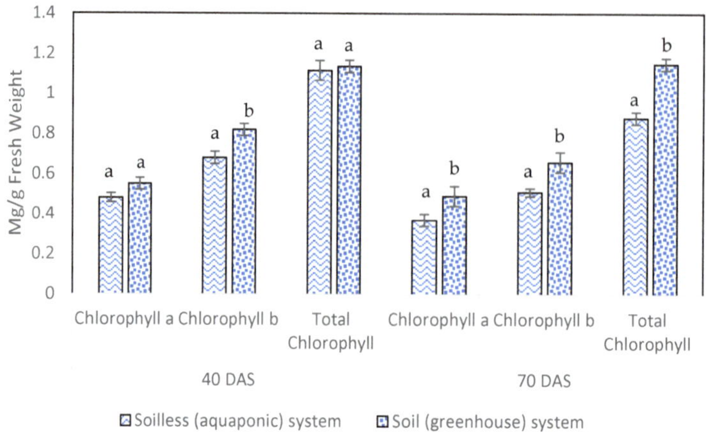

Figure 2. Chlorophyll contents of *Ocimum basilicum* in (aquaponic) and soil (greenhouse) systems at 40 and 70 DAS. Values are given as mean ± SD of seven samples in each group. Values that do not share a common superscript (a, b) differ significantly at $p \leq 0.05$ using t-test.

The anthocyanin content of basil plants increased with age in both production systems. Additionally, in comparison with the aquaponic system, the greenhouse system exhibited more anthocyanin content in the leaves. At 40 DAS, the basil plants exhibited an anthocyanin concentration of 0.562 mg/g FW in aquaponics- and 0.698 mg/g FW in greenhouse-cultivated plants. Similarly, the basil leaves from the aquaponic system at 70 DAS exhibited a 0.59 mg/g FW anthocyanin concentration and the greenhouse-system plants exhibited a 0.77 mg/g FW anthocyanin concentration. Overall, the highest anthocyanin contents were recorded in basil leaves grown in the greenhouse system at 70 DAS (Figure 3a).

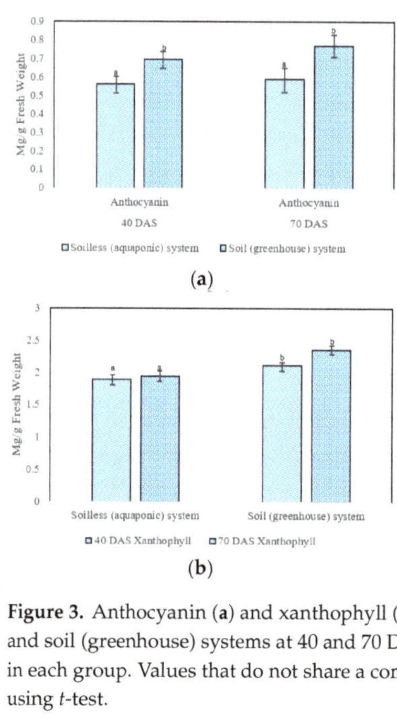

Figure 3. Anthocyanin (**a**) and xanthophyll (**b**) concentrations of *Ocimum basilicum* in (aquaponic) and soil (greenhouse) systems at 40 and 70 DAS. Values are given as mean ± SD of seven samples in each group. Values that do not share a common superscript (a, b) differ significantly at $p \leq 0.05$ using t-test.

The xanthophyll concentration (Figure 3b) at 40 DAS in basil plants was the lowest among the studied samples. As the age increased, the xanthophyll pigment concentration showed an increase in concentration. The highest concentration was recorded in basil plants grown in the greenhouse system at 70 DAS (2.36 mg/g FW) when compared to plants grown in the aquaponic system (1.95 mg/g FW).

3.3. Biochemical Parameters

The plants grown under the aquaponic system exhibited higher protein contents (36.25 μg mL^{-1}) when compared to the plants grown under the conventional greenhouse system (29.03 μg mL^{-1}) at 40 DAS. The protein content was high at 70 DAS as well in aquaponic-grown plants. However, the different samplings at 40 and 70 DAS did not influence the quantity of protein in both the aquaponic and greenhouse plants. The phenol concentration showed a remarkable increase in plants with age. The increase was clear at 40 and 70 DAS in both the systems. The phenol concentration at 40 DAS was 8.84 μg mL^{-1} and 7.01 μg mL^{-1} in aquaponic and soil systems, respectively. This increased to 10.11 and 9.98 μg mL^{-1}, respectively, at 70 DAS (Table 2).

Table 2. Biochemical parameters of *Ocimum basilicum* in soilless (aquaponic) and soil (greenhouse) systems at 40 and 70 DAS.

Biochemical Parameters	DAS	Soilless (aquaponic) System	Soil (Greenhouse) System
Protein (μg mL^{-1})	40	36.25 ± 12.7 ^{a}	29.03 ± 8.62 ^{b}
	70	36.90 ± 12.8 ^{a}	30.00 ± 8.7 ^{b}
Total phenol (mg GAEq g^{-1} DW)	40	8.84 ± 2.8 ^{a}	7.01 ± 2.0 ^{b}
	70	10.11 ± 3.6 ^{a}	9.98 ± 2.9 ^{b}

Values are given as mean ± SD of seven samples in each group. Values that do not share a common superscript (a, b) differ significantly at $p \leq 0.05$ using *t*-test.

3.4. Antioxidants

The ascorbic acid content of the basil plants increased with age in the soil- and aquaponic-growing systems. The aquaponic system significantly increased the ascorbic acid content of the plants as compared to the greenhouse plants; it represented 3.57 mg/g FW at 70 DAS when compared to 3.2 mg/g FW in the greenhouse system (Figure 4a). Like ascorbic acid, in basil plants, the α-tocopherol content increased with age in the soil- and aquaponic-growing systems. The aquaponic system significantly increased the ascorbic acid content of the plants as compared to the greenhouse plants, representing 6.66 mg/g FW at 70 DAS when compared to 5.02 mg/g FW in the greenhouse system (Figure 4b).

3.5. Antioxidant Enzyme Activities

The antioxidant enzyme activities, such as catalase and peroxidase, increased with time in soilless- and aquaponic-grown plants. Their activities were higher in aquaponic-grown plants when compared to soilless-grown plants. At 40 DAS, the content was 7.19 units/mg protein in aquaponic-grown plants and 4.227 in soil-grown plants. The difference in activities of the catalase enzyme at 40 and 70 DAS was significant between aquaponic and greenhouse plants (Figure 5a).

The peroxidase enzyme activity was higher in aquaponic plants when compared to greenhouse-grown plants. Their activity was 1.571 in aquaponic plants at 40 DAS and 1.161 in greenhouse-grown plants. Similarly, the activity was higher at 70 DAS as well in aquaponic plants when compared to greenhouse-grown basil plants (Figure 5b).

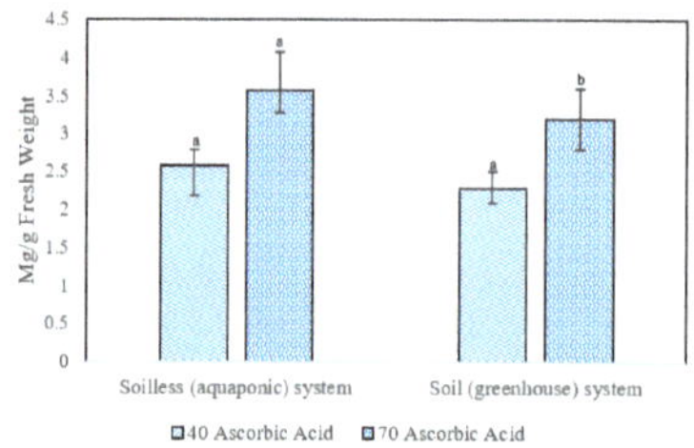

(**a**)

α-tocopherol

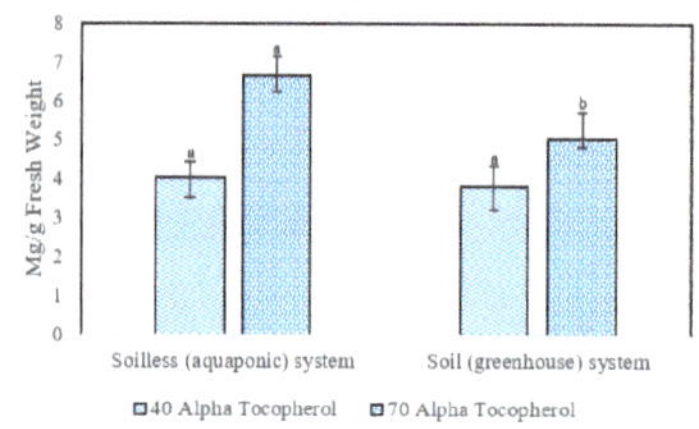

(**b**)

Figure 4. Ascorbic acid (**a**) and α-tocopherol (**b**) concentrations of *Ocimum basilicum* in (aquaponic) and soil (greenhouse) systems at 40 and 70 DAS. Values are given as mean ± SD of seven samples in each group. Values that do not share a common superscript (a, b) differ significantly at $p \leq 0.05$ using *t*-test.

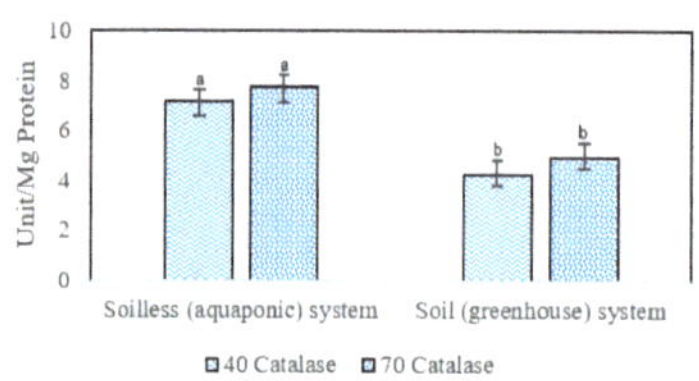

(**a**)

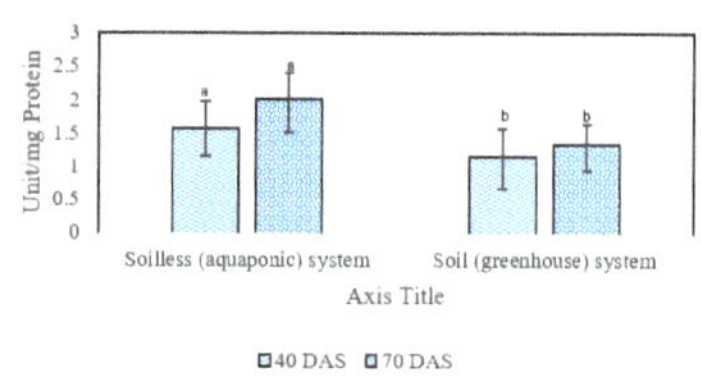

(**b**)

Figure 5. Catalase (**a**) and peroxidase (**b**) activities of *Ocimum basilicum* in soilless (aquaponic) and soil (greenhouse) systems at 40 and 70 DAS. Values are given as mean ± SD of seven samples in each group. Values that do not share a common superscript (a,b) differ significantly at $p \leq 0.05$ using *t*-test.

3.6. Macronutrients and Micronutrients

The analysis of elements was only conducted at the end of the experimental period (70 DAS). Some of the levels of macro- and micronutrients were higher in the soilless system when compared to the conventional greenhouse system (Table 3).

Table 3. Macronutrients (%) and micronutrients (mg kg^{-1}) in basil leaves under soilless (aquaponic) and soil systems at 70 DAS.

Nutrients	**Growth Systems**	
Macronutrients (%)	Aquaponics	Soil
Nitrogen (N)	6.32 ± 0.09 ^{a}	4.89 ± 0.11 ^{b}
Phosphorus (P)	1.70 ± 0.05 ^{a}	0.99 ± 0.0 ^{b}
Potassium (K)	0.69 ± 0.02 ^{a}	0.71 ± 0.02 ^{a}
Magnesium (Mg)	0.59 ± 0.01 ^{a}	0.38 ± 0.0 ^{b}
Calcium (Ca)	2.81 ± 0.11 ^{a}	2.87 ± 0.13 ^{a}
Sulphur (S)	0.29 ± 0.01 ^{a}	0.29 ± 0.01 ^{a}

Table 3. *Cont.*

Nutrients	Growth Systems	
Micronutrients (mg kg^{-1})		
Boron (B)	41.9 ± 1.70 [a]	34.9 ± 1.10 [b]
Copper (Cu)	13.4 ± 0.86 [a]	13.8 ± 0.64 [a]
Iron (Fe)	95.2 ± 4.10 [a]	97.2 ± 2.70 [b]
Manganese (Mn)	99.8 ± 10.50 [a]	87.7 ± 2.40 [b]
Sodium (Na)	88.6 ± 13.50 [a]	74.0 ± 6.10 [b]
Zinc (Zn)	51.1 ± 5.30 [a]	61.1 ± 1.80 [b]

Values are given as mean ± SD of seven samples in each group. Values that do not share a common superscript (a,b) differ significantly at $p \leq 0.05$ using *t*-test.

3.7. Water Parameters

The levels of water temperature, dissolved oxygen, pH, total dissolved solids (TDSs), electrical conductivity (EC), ammonia, nitrate and nitrite showed variations during different intervals in the aquaponic system. It is known that when ammonia increased, the nitrate and nitrite levels increased. The pH level also exhibited various levels during the period of the experiment.

The highest level of dissolved oxygen was recorded to be 5.22 mg/l on day 10 and the lowest was of 5.01 mg/l on day 1. The level of hydrogen pH was recorded to be the highest on day 10 at 6.99 and the lowest on day 30 at 6.42. The level of TDS was recorded throughout the analysis of 40 days from the recorded 310 ppm on the 1st day to 410 ppm on the 40th day. The average water parameters of EC resulted from the fluctuation throughout the observation of 40 days. It was recorded to be the highest on day 40 and the lowest on day 20. The percentage of ammonia was recorded to be the highest on day 40 at 0.60 mg/l and the lowest on days 1–10 at 0.10 mg/l. On day 20, the level of ammonia was recorded to be 0.12 mg/l, and on day 30 it was reduced to 0.11 mg/l. As per the results, the highest level of nitrate was recorded to be of 18.79 mg/l, and the lowest was level of nitrate was recorded to be 5.81 mg/l. The level of nitrite was recorded to be the highest on day one and the lowest on day 20 (Table 4).

Table 4. The average water parameters under aquaponic system during 10 day intervals.

Analysis Intervals	Temperature (°C)	Dissolved Oxygen (mg/L)	pH	TDS (ppm)	EC (mS/cm)	Ammonia (mg/L)	Nitrate (mg/L)	Nitrite (mg/L)
Day 1	24.01 ± 0.03	5.01 ± 0.03	6.85 ± 0.03	310 ± 1.100	0.62 ± 0.04	0.10 ± 0.03	5.81 ± 0.03	0.30 ± 0.02
Day 10	25.09 ± 0.03	5.22 ± 0.04	6.99 ± 0.03	380.11 ± 1.24	0.76 ± 0.04	0.10 ± 0.03	6.09 ± 0.04	0.10 ± 0.04
Day 20	24.02 ± 0.03	5.10 ± 0.04	6.87 ± 0.02	350.21 ± 0.81	0.70 ± 0.04	0.12 ± 0.03	11.99 ± 0.07	0.08 ± 0.04
Day 30	25.02 ± 0.02	5.03 ± 0.03	6.42 ± 0.04	390.10 ± 1.32	0.78 ± 0.02	0.11 ± 0.02	17.22 ± 0.06	0.10 ± 0.03
Day 40	25.24 ± 0.03	5.10 ± 0.03	6.99 ± 0.04	410.41 ± 1.05	0.82 ± 0.03	0.60 ± 0.02	18.79 ± 0.09	0.11 ± 0.03
Day 50	24.84 ± 0.02	5.20 ± 0.03	7.01 ± 0.05	420.47 ± 1.06	0.84 ± 0.04	0.61 ± 0.02	17.79 ± 0.09	0.12 ± 0.03
Day 60	26.01 ± 0.04	5.11 ± 0.03	6.91 ± 0.04	420.61 ± 1.15	0.84 ± 0.03	0.71 ± 0.03	14.11 ± 0.06	0.12 ± 0.03
Day 70	24.21 ± 0.03	5.10 ± 0.03	6.98 ± 0.05	410.11 ± 1.05	0.82 ± 0.02	0.62 ± 0.02	18.25 ± 0.08	0.14 ± 0.04

4. Discussion

Ocimum basilicum L., commonly known as basil, is a herbal plant used for medicinal purposes in the treatment of headaches, coughs as well as also kidney malfunctions. There are many major chemical components in the plant which are the basis for these medicinal properties. Basil is a source of antioxidant vitamins E, A and C [41]. These sources of antioxidants help to elevate the health system of humans.

4.1. Plant Growth Parameters

Primarily, the total height of the basil plant exhibited an increase with the consequences of age. This is based on the analysis of the first DAS of the aquaponics and soil systems.

The plant was able to reach its maximum height in seventy days. The result of plants grown on the soil system of the greenhouse showed a distinct difference from the aquaponic plant growth. The root length of the plants of both systems showed a similar result regarding height. In comparison with conventional basil cultivation, plant height in the present study was higher, which is contrary to the findings of a previous study by Pasch et al. [17], where they reported a reduction in growth of basil in different culture conditions with catfish (*Clarias gariepinus*) in decoupled aquaponics. Our results agree with the earlier studies of Mangmang et al. [42,43], where the inoculation effect of *Azospirillum brasilense* on basil significantly increased the growth and production under the aquaponics production system.

Roosta [44] showed a decrease in vegetative growth when studying the comparison of the vegetative growth, eco-physiological characteristics and mineral nutrient content of basil plants irrigated with different hydroponic:aquaponic solutions. In aquaponics, the plants exhibited high growth performance and basil leaf area, fresh herbage yield and root weight were increased by up to 27, 11 and 11%, respectively. An increase in biomass and height were reported in basil plants grown in decoupled aquaponics by Rodgers et al. [45].

4.2. Photosynthetic Pigment Contents

The chlorophyll contents of plants show defining results at both 40 and 70 DAS. Chlorophyll 'a' was subjected to show higher pigments in the greenhouse-grown basil than the aquaponic-grown basil. Chlorophyll 'b' also showed similar results to chlorophyll 'a', namely that the increase was seen more in the greenhouse plants than in the aquaponic plants. Similar to that of chlorophyll, the plants showed high levels of anthocyanin in greenhouse growth and lower levels in the aquaponic system. The contents of xanthophyll, the highest concentration, were recorded in plants grown in the greenhouse as well as in the aquaponic system.

There was visible chlorosis, as reported by Roosta [44], in basil plants under irrigation with different ratios of hydroponic:aquaponic solutions. That means the content was lesser in the aquaponic system of growth, which is concomitant with our studies. Leaf chlorosis can be considered as an indicator of less physiological functions under aquaponics [46,47].

There was no difference in chlorophyll content or leaf nutrients between the aquaponics and hydroponics of basil plants when grown in comparison to conventional systems [48]. Saha et al. [48] also reported that the pigment concentrations were unaltered in the aquaponic production of basil, whereas, contrary to our results, Ferrarezi and Bailey [49] reported an increase in chlorophyll contents in basil plants grown in aquaponics. Similarly, in one of the recent studies by Rodgers et al. [45], high chlorophyll index ratios in basil plants when grown in decoupled aquaponic setups were reported.

4.3. Biochemical Analysis

According to the results, the biochemical composition of the basil plants exhibited a higher content of proteins in the aquaponic system than in the conventional soil system. Additionally, the total phenol level was high in the aquaponic system and low in the soil system. As basil is considered a herbal medicinal plant, it is of importance that the high level of phenol provides antioxidant properties [50]. An increase in protein and other biochemical concentrations were reported earlier in aquaponic-grown basil by Yang and Kim [24,51].

4.4. Non-Enzymatic Antioxidant Contents

The antioxidants present in basil provide medical benefits. As per the analysis of the results, ascorbic acid contents in basil showed an increase in the aquaponics system in both the analyses at different DAS. The contents of ascorbic acid in greenhouse plants were lower. This can be based on the fact that the aquaponic system delivers more essential nutrients and helps build plant food in a better way than the greenhouse system [6]. Ascorbic acid is known as vitamin C, which is an essential antioxidant that works to protect firm cellular components from damage. These elements of basil tend to scavenge free radicals [52].

As per the results, α-tocopherol also increased in the aquaponic system as compared to the low α-tocopherol components in the soil system. The content of alpha-tocopherol is proven to provide antioxidant activity that helps in protecting the membrane components, similar to that in ascorbic acid [53]. We can assume that the plants under aquaponic growth are probably subjected to a type of water stress (flooding), which in turn increased the antioxidant content in the plants to fight against it [54].

4.5. Antioxidant Enzymes

Antioxidant activity in basil plants is based on several activities, such as catalase and peroxide. The results show that these activities tend to increase with the age factor of the plants in both systems. The activities are higher in aquaponic plants than that in soil systems. In the case of peroxidase, the activities are also higher in aquaponic systems.

These antioxidant enzyme activities help in eliminating superoxides and the hydrogen peroxide [55]. Additionally, the survival of many plants depends on their antioxidant activities. Overall, the plants under the aquaponic system experience a type of flood stress at the root level, which can be the reason behind the increased activities of antioxidant enzymes. Antioxidant enzyme activities are used to scavenge the potential free radicals which are produced due to the stress in the plants [56]. The increase in peroxidase in plants at 70 DAS can be correlated to the onset of flowering as reported earlier [57].

4.6. Microelements and Macroelements

The results are based on the macro- and micronutrients in basil plants in the analysis at 70 DAS in both aquaponic and soil systems. The major macronutrients present in basil plants are nitrogen, phosphorus, potassium, calcium and magnesium. The analysis showed that in an aquaponic system, the contents of nitrogen were higher than that of the soil system. The components of phosphorus also showed similar results as that of nitrogen. While the content of potassium and calcium exhibited a lower content in the aquaponic system and higher contents in the soil system, conversely, the macronutrients of sulphur exhibited the presence of equal levels in both the aquaponic and soil system at 70 DAS and after. As per the results, the levels of micronutrients were also shown. Boron, magnum and sodium exhibited higher content levels in the aquaponic system than in the soil system, while the level of copper, iron and zinc showed lower content levels in the aquaponic system and higher content levels in the soil system at 70 DAS.

Soil-system plants have lower levels of nutrient content. Aquaponics plants, with water beds of recirculating water, contain more nutrients, and this helps the basil plant to grow faster. The waste generated by fish in the aquaponics system tends to deliver essential nutrients for plant growth, such as nitrogen, calcium, magnesium and potassium [58]. As per the recent data, the information on aquaponic nutrient manipulation is scanty. Potassium and iron can be added to the system as main nutrient elements to be added as a supplement to the aquaponic solution as potassium hydroxide and iron chelates as a foliar spray [59].

4.7. Water Quality Parameters

The results of the water quality parameters are based on the analysis of different internal observations on the aquaponic system of basil plants. In the analysis, the temperature kept fluctuating throughout days 1–40 from 24 °C to 25 °C. The dissolved oxygen level was considered to be the highest on day 10 of the analysis

As per the literature, the accurate pH ranges are 6–9 for tilapia fish, 5.5–6 for plants and 7–8 for nitrifying bacteria, so we can conclude that pH 7 is considered an ideal compromise for aquaponics [60]. In this study, the EC in aquaponics resulted mainly from the daily nutrient release from the fish feed. The water temperature recommendation for basil is 20–25 °C in any system of cultivation [49].

5. Conclusions

The main aim of this experiment was to standardize the aquaponic production of basil plants with the help of ornamental fish, such as Koi fish, and to compare the growth and quality of basil plants with conventional production methods such as greenhouse cultivation in the UAE's climatic condition. Based on the obtained results, the growth parameters showed an increase under aquaponic production, while the photosynthetic pigments were higher under greenhouse cultivation. The biochemical parameters exhibited a significant enhancement in aquaponic-grown plants when compared to the standard greenhouse system. The non-enzymatic and enzymatic antioxidants exhibited a significant increase under the aquaponic system in basil plants. The conventional greenhouse system exhibited lower antioxidant levels. Micro- and macro-nutrient levels also exhibited varied responses under aquaponic cultivation. From the results of this study, it can be concluded that the basil crop can be grown in both aquaponic and greenhouse soil systems, but the quantity and quality of the crop can be increased when a soilless system is adopted. In a soilless aquaponic system, the height, fresh and dry weight of basil increased significantly when compared to conventional soil systems. This increase may be due to the additional fertilizers from the fish waste produced in the system.

Supplementary Materials: The following supporting information can be downloaded at: https://www.mdpi.com/article/10.3390/agronomy12123007/s1, Figure S1. Experimental set up showing tanks, growth trays and filters.

Author Contributions: Conceptualization, A.J. and Z.F.R.A.; methodology, A.J. and M.A.A.; validation, A.J., Z.F.R.A., S.S.K. and M.A.A.; formal analysis, M.A.O.K.A.; investigation, M.A.O.K.A.; data curation, M.A.O.K.A.; writing—original draft preparation, M.A.O.K.A., A.J. and Z.F.R.A.; writing—review and editing, M.A.O.K.A., A.J., S.S.K. and Z.F.R.A.; supervision, A.J. and M.A.A. All authors have read and agreed to the published version of the manuscript.

Funding: This research received no external funding.

Data Availability Statement: Not applicable.

Acknowledgments: The authors would like to thank the support from UAEU in the form of research support to Master Students from MS Horticulture Program. The assistance from staff of Falaj Hazza Experimental Station for aquaponic and greenhouse setups and collection of samples and E3, F1 labs, CAVM, UAEU are greatly acknowledged.

Conflicts of Interest: The authors declare no conflict of interest.

References

1. Blidariu, F.; Grozea, A. Increasing the economic efficiency and sustainability of indoor fish farming by means of aquaponics-review. *Anim. Sci. Biotechnol.* **2011**, *44*, 1–8.
2. Yildiz, H.Y.; Robaina, L.; Pirhonen, J.; Mente, E.; Domínguez, D.; Parisi, G. Fish Welfare in Aquaponic Systems: Its Relation to Water Quality with an Emphasis on Feed and Faeces—A Review. *Water* **2017**, *9*, 13. [CrossRef]
3. Khan, M.M.; Akram, M.T.; Janke, R.; Qadri, R.W.K.; Al-Sadi, A.M.; Farooque, A.A. Urban Horticulture for Food Secure Cities through and beyond COVID-19. *Sustainability* **2020**, *12*, 9592. [CrossRef]
4. Al Shamsi, K.B.; Compagnoni, A.; Timpanaro, G.; Cosentino, S.L.; Guarnaccia, P. A Sustainable Organic Production Model for "Food Sovereignty" in the United Arab Emirates and Sicily-Italy. *Sustainability* **2018**, *10*, 620. [CrossRef]
5. Joyce, A.; Timmons, M.; Goddek, S.; Pentz, T. Bacterial Relationships in Aquaponics: New Research Directions. In *Aquaponics Food Production Systems*; Springer: Cham, Switzerland, 2019; pp. 145–161.
6. Goddek, S.; Delaide, B.; Mankasingh, U.; Ragnarsdottir, K.V.; Jijakli, H.; Thorarinsdottir, R. Challenges of Sustainable and Commercial Aquaponics. *Sustainability* **2015**, *7*, 4199–4224. [CrossRef]
7. Zou, Y.; Hu, Z.; Zhang, J.; Xie, H.; Liang, S.; Wang, J.; Yan, R. Attempts to improve nitrogen utilization efficiency of aquaponics through nitrifies addition and filler gradation. *Environ. Sci. Pollut. Res.* **2016**, *23*, 6671–6679. [CrossRef]
8. Yep, B.; Zheng, Y. Aquaponic trends and challenges–A review. *J. Clean. Prod.* **2019**, *228*, 1586–1599. [CrossRef]
9. Pinho, S.M.; David, L.H.; Garcia, F.; Keesman, K.J.; Portella, M.C.; Goddek, S. South American fish species suitable for aquaponics: A review. *Aquac. Int.* **2021**, *29*, 1427–1449. [CrossRef]
10. Liang, J.-Y.; Chien, Y.-H. Effects of feeding frequency and photoperiod on water quality and crop production in a tilapia–water spinach raft aquaponics system. *Int. Biodeterior. Biodegrad.* **2013**, *85*, 693–700. [CrossRef]

11. Maucieri, C.; Forchino, A.A.; Nicoletto, C.; Junge, R.; Pastres, R.; Sambo, P.; Borin, M. Life cycle assessment of a micro aquaponic system for educational purposes built using recovered material. *J. Clean. Prod.* **2018**, *172*, 3119–3127. [CrossRef]
12. Hu, F.; Li, Y.; Wang, Q.; Zhu, B.; Wu, S.; Wang, Y.; Zeng, W.; Yin, J.; Liu, C.; Bergmann, S.M.; et al. Immersion immunization of koi (*Cyprinus carpio*) against cyprinid herpesvirus 3 (CyHV-3) with carbon nanotube-loaded DNA vaccine. *Aquaculture* **2021**, *539*, 736644. [CrossRef]
13. Adamek, M.; Oschilewski, A.; Wohlsein, P.; Jung-Schroers, V.; Teitge, F.; Dawson, A.; Gela, D.; Piackova, V.; Kocour, M.; Adamek, J.; et al. Experimental infections of different carp strains with the carp edema virus (CEV) give insights into the infection biology of the virus and indicate possible solutions to problems caused by koi sleepy disease (KSD) in carp aquaculture. *Vet. Res.* **2017**, *48*, 12. [CrossRef]
14. Mokat, D.N.; Kharat, T.D. Essential Oil Composition in Leaves of *Ocimum* Species Found in Western Maharashtra, India. *J. Essent. Oil Bear. Plants* **2022**, *25*, 1–8. [CrossRef]
15. Knaus, U.; Pribbernow, M.; Xu, L.; Appelbaum, S.; Palm, H.W. Basil (*Ocimum basilicum*) cultivation in decoupled aquaponics with three hydro-components (grow pipes, raft, gravel) and African catfish (*Clarias gariepinus*) production in Northern Germany. *Sustainability* **2020**, *12*, 8745. [CrossRef]
16. Rehman, R.; Hanif, M.A.; Mushtaq, Z.; Mochona, B.; Qi, X. Biosynthetic factories of essential oils: The aromatic plants. *Nat. Prod. Chem. Res.* **2016**, *4*, 227. [CrossRef]
17. Moncada, A.; Miceli, A.; Vetrano, F. Use of plant growth-promoting rhizobacteria (PGPR) and organic fertilization for soilless cultivation of basil. *Sci. Hortic.* **2021**, *275*, 109733. [CrossRef]
18. Pasch, J.; Ratajczak, B.; Appelbaum, S.; Palm, H.W.; Knaus, U. Growth of basil (*Ocimum basilicum*) in DRF, raft, and grow pipes with effluents of African catfish (*Clarias gariepinus*) in decoupled aquaponics. *AgriEngineering* **2021**, *3*, 92–109. [CrossRef]
19. Babmann, B.; Harbach, H.; Weißbach, S.; Palm, H.W. Effect of plant density in coupled aquaponics on the welfare status of African catfish, *Clarias gariepinus*. *J. World Aquac. Soc.* **2020**, *51*, 183–199. [CrossRef]
20. Hundley, G.C.; Navarro, F.K.S.P.; Filho, O.P.R.; Navarro, R.D. Integration of Nile tilapia (*Oreochromis niloticus* L.) production *Origanum majorana* L. and *Ocimum basilicum* L. using aquaponics technology. *Acta Sci. Technol.* **2018**, *40*, 35460. [CrossRef]
21. Obirikorang, K.A.; Sekey, W.; Gyampoh, B.A.; Ashiagbor, G.; Asante, W. Aquaponics for Improved Food Security in Africa: A Review. *Front. Sustain. Food Syst.* **2021**, *5*, 705549. [CrossRef]
22. Pandey, J.; Verma, R.K.; Singh, S. Suitability of aromatic plants for phytoremediation of heavy metal contaminated areas: A review. *Int. J. Phytoremed.* **2019**, *21*, 405–418. [CrossRef] [PubMed]
23. Bahl, J.R.; Singh, A.K.; Lal, R.K.; Gupta, A.K. High-yielding improved varieties of medicinal and aromatic crops for enhanced income. In *New Age Herbals*; Springer: Singapore, 2018; pp. 247–265.
24. Yang, T.; Kim, H.-J. Comparisons of nitrogen and phosphorus mass balance for tomato-, basil-, and lettuce-based aquaponic and hydroponic systems. *J. Clean. Prod.* **2020**, *274*, 122619. [CrossRef]
25. Tolay, I. The impact of different Zinc (Zn) levels on growth and nutrient uptake of Basil (*Ocimum basilicum* L.) grown under salinity stress. *PLoS ONE* **2021**, *16*, e0246493. [CrossRef] [PubMed]
26. Simon, J.E.; Morales, M.R.; Phippen, W.B.; Vieira, R.F.; Hao, Z. Basil: A source of aroma compounds and a popular culinary and ornamental herb. *Perspect. New Crops New Uses* **1999**, *16*, 499–505.
27. Quagrainie, K.K.; Flores, R.M.V.; Kim, H.-J.; McClain, V. Economic analysis of aquaponics and hydroponics production in the U.S. Midwest. *J. Appl. Aquac.* **2018**, *30*, 1–14. [CrossRef]
28. Danner, R.I.; Mankasingh, U.; Anamthawat-Jonsson, K.; Thorarinsdottir, R.I. Designing Aquaponic Production Systems towards Integration into Greenhouse Farming. *Water* **2019**, *11*, 2123. [CrossRef]
29. Ahmed, N.; Thompson, S.; Glaser, M. Global Aquaculture Productivity, Environmental Sustainability, and Climate Change Adaptability. *Environ. Manag.* **2019**, *63*, 159–172. [CrossRef]
30. Arnon, D.I. Copper enzymes in isolated chloroplasts. Polyphenoloxidase in *Beta vulgaris*. *Plant Physiol.* **1949**, *24*, 1. [CrossRef]
31. Beggs, C.J.; Wellmann, E. Analysis of light-controlled anthocyanin formation in coleoptiles of *Zea mays* L.: The role of UV-B, blue, red and far-red light. *Photochem. Photobiol.* **1985**, *41*, 481–486. [CrossRef]
32. Neogy, M.; Datta, J.K.; Mukherji, S.; Roy, A.K. Effect of aluminium on pigment content, Hill activity and seed yield in mungbean. *Indian J. Plant Physiol.* **2001**, *6*, 381–385.
33. Bradford, M.M. A rapid and sensitive method for the quantitation of microgram quantities of protein utilizing the principle of protein-dye binding. *Anal. Biochem.* **1976**, *72*, 248–254. [CrossRef] [PubMed]
34. Malik, C.P.; Singh, M.B. *Plant Enzymology and Histo-Enzymology*; Kalyani Publications: New Delhi, India, 1980.
35. Omaye, S.T.; Turnbull, J.D.; Sauberlich, H.E. Selected methods for the determination of ascorbic acid in animal cells, tissues, and fluids. In *Methods in Enzymology*; Academic Press: Amsterdam, The Netherlands, 1979; Volume 62, pp. 3–11.
36. Baker, H.; Frank, O.; Deangelis, B.; Feingold, S. Plasma tocopherol in man at various times after ingesting free or acetylated tocopherol. *Nutr. Rep. Int.* **1980**, *21*, 531–536.
37. Hwang, S.-Y.; Lin, H.-W.; Chern, R.-H.; Lo, H.F.; Li, L. Reduced susceptibility to waterlogging together with high-light stress is related to increases in superoxide dismutase and catalase activities in sweet potato. *Plant Growth Regul.* **1999**, *27*, 167–172. [CrossRef]
38. Chandlee, J.M.; Scandalios, J.G. Analysis of variants affecting the catalase developmental program in maize scutellum. *Theor. Appl. Genet.* **1984**, *69*, 71–77. [CrossRef] [PubMed]

39. Kumar, K.B.; Khan, P.A. Peroxidase and polyphenol oxidase in excised ragi (Eleusine corocana cv PR 202) leaves during senescence. *Indian J. Exp. Biol.* **1982**, *20*, 412–416.
40. USEPA United States Environmental Protection Agency. *Revision 5.4 Determination of Trace Elements in Waters and Wastes by Inductively Coupled Plasma–Mass Spectrometry*; Method 200.8; USEPA United States Environmental Protection Agency: Washington, DC, USA, 1994.
41. Juliani, H.R.; Simon, J.E. Antioxidant activity of basil. In *Trends in New Crops and New Uses*; ASHS Press: Alexandria, VA, USA, 2002; Volume 575.
42. AlShrouf, A. Hydroponics, aeroponic and aquaponic as compared with conventional farming. *Am. Acad. Sci. Res. J. Eng. Technol. Sci.* **2017**, *27*, 247–255.
43. MangMang, J.S.; Deaker, R.; Rogers, G. Inoculation effect of Azospirillum brasilense on basil grown under aquaponics production system. *Org. Agric.* **2016**, *6*, 65–74. [CrossRef]
44. Roosta, H.R. Comparison of the vegetative growth, eco-physiological characteristics and mineral nutrient content of basil plants in different irrigation ratios of hydroponic: Aquaponic solutions. *J. Plant Nutr.* **2014**, *37*, 1782–1803. [CrossRef]
45. Rodgers, D.; Won, E.; Timmons, M.B.; Mattson, N. Complementary Nutrients in Decoupled Aquaponics Enhance Basil Performance. *Horticulturae* **2022**, *8*, 111. [CrossRef]
46. Yang, T.; Kim, H.-J. Nutrient management regime affects water quality, crop growth, and nitrogen use efficiency of aquaponic systems. *Sci. Hortic.* **2019**, *256*, 108619. [CrossRef]
47. Mullis, D.; Reyes, J. The overall effects of soilless agricultural system on basil (*Ocimum basilicum* L.). *Ccamlr Sci.* **2019**, *26*, 300–307.
48. Saha, S.; Monroe, A.; Day, M.R. Growth, yield, plant quality and nutrition of basil (*Ocimum basilicum* L.) under soilless agricultural systems. *Ann. Agric. Sci.* **2016**, *61*, 181–186. [CrossRef]
49. Ferrarezi, R.S.; Bailey, D.S. Basil Performance Evaluation in Aquaponics. *HortTechnology* **2019**, *29*, 85–93. [CrossRef]
50. Shahrajabian, M.H.; Sun, W.; Cheng, Q. Chemical components and pharmacological benefits of Basil (*Ocimum basilicum*): A review. *Int. J. Food Prop.* **2020**, *23*, 1961–1970. [CrossRef]
51. Yang, T.; Kim, H.-J. Characterizing Nutrient Composition and Concentration in Tomato-, Basil-, and Lettuce-Based Aquaponic and Hydroponic Systems. *Water* **2020**, *12*, 1259. [CrossRef]
52. Ghandar, A.; Ahmed, A.; Zulfiqar, S.; Hua, Z.; Hanai, M.; Theodoropoulos, G. A decision support system for urban agriculture using digital twin: A case study with aquaponics. *IEEE Access* **2021**, *9*, 35691–35708. [CrossRef]
53. Liebler, D.C.; Kling, D.S.; Reed, D.J. Antioxidant protection of phospholipid bilayers by alpha-tocopherol. Control of alpha-tocopherol status and lipid peroxidation by ascorbic acid and glutathione. *J. Biol. Chem.* **1986**, *261*, 12114–12119. [CrossRef]
54. Yiu, J.-C.; Juang, L.-D.; Fang, D.Y.-T.; Liu, C.-W.; Wu, S.-J. Exogenous putrescine reduces flooding-induced oxidative damage by increasing the antioxidant properties of Welsh onion. *Sci. Hortic.* **2009**, *120*, 306–314. [CrossRef]
55. Wang, F.; Yang, Q.; Zhao, Q.; Li, X. Cold shock treatment with oxalic acid could alleviate chilling injury in green bell pepper by enhancing antioxidant enzyme activity and regulating proline metabolism. *Sci. Hortic.* **2022**, *295*, 110783. [CrossRef]
56. Ahmad, S.; Wang, G.-Y.; Muhammad, I.; Chi, Y.-X.; Zeeshan, M.; Nasar, J.; Zhou, X.-B. Interactive Effects of Melatonin and Nitrogen Improve Drought Tolerance of Maize Seedlings by Regulating Growth and Physiochemical Attributes. *Antioxidants* **2022**, *11*, 359. [CrossRef]
57. Al Shamsi, S.R.H.A.; Rabert, G.A.; Kurup, S.S.; Alyafei, M.A.M.; Jaleel, A. Biochemical Changes and Antioxidant Variations in Date Palm (*Phoenix dactylifera* L.) Varieties during Flower Induction and Development. *Plants* **2021**, *10*, 2550. [CrossRef] [PubMed]
58. Rakocy, J.; Shultz, R.C.; Bailey, D.S.; Thoman, E.S. Aquaponic production of tilapia and basil: Comparing a batch and staggered cropping system. In *South Pacific Soilless Culture Conference-SPSCC 648*; International Society for Horticultural Science: Leuven, Belgium, 2003; pp. 63–69.
59. Roosta, H.R.; Hamidpour, M. Effects of foliar application of some macro-and micro-nutrients on tomato plants in aquaponic and hydroponic systems. *Sci. Hortic.* **2011**, *129*, 396–402. [CrossRef]
60. Heise, J.; Müller, H.; Probst, A.J.; Meckenstock, R.U. Ammonium Removal in Aquaponics Indicates Participation of Comammox Nitrospira. *Curr. Microbiol.* **2021**, *78*, 894–903. [CrossRef] [PubMed]

agronomy

MDPI

Article

Chemical Composition and Insecticidal Properties of *Origanum vulgare* (Lamiaceae) Essential Oil against the Stored Product Beetle, *Sitophilus granarius*

Angelica Plata-Rueda [1,2], Marcelo Henrique Dos Santos [3], José Eduardo Serrão [4] and Luis Carlos Martínez [4,*]

1 Department of Entomology, Federal University of Viçosa, Viçosa 36570-000, Brazil
2 Faculty of Science, National University of Colombia, Bogotá 316-5000, Colombia
3 Department of Chemistry, Federal University of Viçosa, Viçosa 36570-000, Brazil
4 Department of General Biology, Federal University of Viçosa, Viçosa 36570-000, Brazil
* Correspondence: lc.martinez@outlook.com; Tel.: +55-31-3899-4012

Abstract: Although phosphides are utilized in stored pest control, efforts have been made to discover environmentally friendly insecticides. For insecticidal properties, essential oils (EOs) are considered to be novel alternatives for pesticide use. This study characterized the *Origanum vulgare* EO by gas chromatography–flame ionization detector (GC–FID) × gas chromatography–mass spectrometry (GC–MS) and assessed the insecticidal activities against *Sitophilus granarius*. Mortality, post-exposure survival, behavior, and respiration caused by this EO in *S. granarius* were investigated. The majority of the compounds were p-cymene, carvacrol, linalool, and thymol. In dose–mortality bioassays, the lethality of this EO (LD_{50} = 3.05 μg insect^{-1} and LD_{90} = 10.02 μg insect^{-1}) was confirmed in *S. granarius*. The survival rate was 99.9% in adults not treated with *O. vulgare* EOs, reducing to 44.9% and 10.3% in weevils treated with 3.05 μg insect^{-1} and 10.02 μg insect^{-1}, respectively. The *O. vulgare* EO alters the behavioral pattern in terms of walking distance and resting time, displaying repellency. Additionally, this EO reduced the gas exchange of weevils from 2.78 to 2.36 μL CO_2 h^{-1} at 3.05 μg insect^{-1}, after 3 h EO exposure. The results suggest that *O. vulgare* EOs affect different biological functions in the insect, and open new perspectives for controlling stored pests, representing a first step in the innovation of green pesticides.

Keywords: gas chromatography; repellency; respiration; terpenoids; toxicity; survivorship

1. Introduction

The weevil, *Sitophilus granarius* Linnaeus (Coleoptera: Curculionidae), is a devastating stored pest of grains, including *Avena sativa* (L.), *Hordeum vulgare* (L.), *Sorghum bicolor* (L.), *Triticum aestivum* (L.), and *Zea mays* (L.) (Poaceae), worldwide. *Sitophilus granarius* causes feeding damage to stored agricultural commodities [1], contaminates food with their molted exoskeleton and feces [2], and act as a vector of fungi [3], with a strong impact on market quality and access. Some methods to control *S. granarius* include temperature treatment [4], sun treatment [5], a controlled atmosphere [6], and the fumigation of synthetic chemicals [7]. In *S. granarius*, collateral effects encouraged by synthetic groups, such as organophosphates and phosphides, have been investigated [8]. However, negative consequences have developed as a result of insecticides (physiological and behavioral), including resistance [9], environmental pollution [10], and residual toxicity [8], which have limited the demand of chemical control. The search for new pest-control tactics can be accomplished to ensure the protection of stored products, considering the harmful effects of synthetic insecticides.

Plant essential oils (EOs) are proposed for the pest control of fields and food storage, and they display several insecticidal activities [11]. EOs alter the insect's digestion [12], causing repellency [13] and disrupting olfactory response [14]. Moreover, impacts on physiology cause growth anomaly [15], developmental impairment [16], oxygen deprivation [17], and energy depletion [18] in insects. EOs are a blend of phytochemicals, mainly alkaloids, flavonoids, and terpenes. The latter are the most abundant chemical group within the composition of EOs, and act as neurotoxins in insects, affecting acetylcholine [19], γ-aminobutyric acid [20], and octopaminergic receptors [21], while also inhibiting the electron–transport complex [22]. The EOs are administered on insects via contact (through the integumentary system) [23], inhalation (through the respiration system) [24], or they are administered orally (through the digestive system) [25].

The insecticidal properties of EOs may vary according to plant species, and their efficacy has been demonstrated in coleopteran grain pests [17,24,26]. For the insecticidal effects of plant EOs, preliminary investigations demonstrated that *Calendula incana* was toxic to *Necrobia rufipes* DeGeer (Cleridae) [26], *Carlina acaulis* to *Prostephanus truncatus* (Horn) (Bostrichidae) [27], and *Mentha spicata* to *Callosobruchus chinensis* Linnaeus (Bruchidae) [28], supporting the utilization of plant types in pest-suppression tactics. In this sense, EOs from Amarydillaceae [29], Annonaceae [22], Lauraceae [30], Meliaceae [17], and Poaceae [31] are the most hopeful for exerting toxicity on insects.

Oregano, *Origanum vulgare* Linnaeus (Lamiales: Lamiaceae), is a prominent plant rich in secondary metabolites and is used in medicine [32], the food industry [33], and agriculture [34]. *Origanum vulgare* EO is utilized for a continuous period as a natural tool to safeguard against several microorganisms of stored grains [35], with low animal toxicity and rapid degradation in the environment. Among the antimicrobial properties, this EO exhibits strong insecticidal effects against stored pests [25]. In particular, *O. vulgare* EO has been demonstrated, with promising results, to control several Coleopteran stored pests, such as *Acanthoscelides obtectus* (Say) (Bruchidae) [36], *Alphitobius diaperinus* (Panzer) (Tenebrionidae) [37], and *Trogoderma granarium* (Everts) (Dermestidae) [38]. However, *O. vulgare* EO has been not evaluated to manage *S. granarius* populations.

The objective in this research was to characterize the principal compounds of *O. vulgare* EO and to evaluate its effect on the mortality, survival, behavior, and respiration rate of *S. granarius*.

2. Materials and Methods

2.1. Weevils

Sitophilus granarius was obtained from a mass-rearing colony in the Institute of Applied Biotechnology for Agriculture (BIOAGRO) of the Federal University of Viçosa (UFV), Viçosa, Minas Gerais, Brazil. Adults were kept in plastic bottles (750 mL) at 27 ± 3 °C and 55 ± 25% relative humidity under a 12:12 h light/dark cycle. The weevils were fed *Triticum aestivum* Linnaeus (Poaceae) grains. Newly emerged (24 h-old) adults were utilized in the experiments.

2.2. Essential Oil

The organic *O. vulgare* EO, produced on an industrial scale by steam distillation (using a Clevenger-type apparatus), was purchased from Ferquina Industry and Commerce Ltda. (Catanduva, São Paulo, Brazil).

2.3. Gas Chromatography-Flame Ionization Detector (GC–FID) Analysis

Quantitative analysis of *O. vulgare* EO was made using a Shimadzu GC-17A Series instrument (Shimadzu Corporation, Japan), equipped with a capillary column (Supelco DB-5 30 m × 0.22 mm × 0.25 μm film) and coupled with a flame-ionization detector (FID). The operating conditions were the following: carrier gas, helium at a flow rate of 1.5 mL min^{-1}; injector temperature, 220 °C; detector temperature, 240 °C; column temperature to start at 40 °C (isothermal for 3 min), with a ramp of 3 °C min^{-1}, until reaching 240 °C, and held isothermally at 240 °C for 10 min; injection, 1 μL (1% *w*/*v* in dichloromethane, three times); split ratio, 1:10; and column pressure, 118 kPa. For each component identified, the amount was expressed in relative percentage, calculated by the normalization of chromatographic peak areas.

2.4. Gas Chromatography–Mass Spectrometry (GC–MS) Analysis

GC–MS analyses were made on a Shimadzu GCMS-QP5050A gas chromatograph equipped with a Rtx-5MS (Restek Corporation, Bellefonte, USA) capillary column (30 m × 0.25 mm i.d., 0.25 μm film thickness). The desorption was operated via the splitless mode (1:10 ratio), with a programmed temperature of 50 °C to 220 °C at 5 °C min^{-1}, and a final holding time of 240 min. An aliquot (1 μL) of this EO (in 1% *w/v* in dichloromethane) was injected three times, and the spectra were recorded in electron impact mode (ionization energy at 70 eV), with a range of 40–400 Da. The identified *O. vulgare* EO compounds were achieved by comparing their Kovats indexes from the original literature [39–41], retention time, and MS data with those of C_3-C_{24} *n*-alkanes, obtained from the NIST v.11 and Wiley v.07 libraries.

2.5. Dose–Mortality Relationship

The *O. vulgare* EO was prepared in 2 mL acetone to obtain a stock suspension and was tested on weevils using identical procedures to that which were outlined for topical application bioassays [11,13,18]. Six dilutions (0.75, 1.5, 2.5, 5, 10, and 20 μg $insect^{-1}$) besides the control (acetone) were utilized to determinate the lethality of this EO to *S. granarius* adults, set up the dose–response relation, and estimate the lethal doses (LD_{25}, LD_{50}, LD_{75}, and LD_{90}). For each EO dilution, one microliter (1 μL) was applied to the thorax of weevil adults, Ausing a Hamilton Model-7001 microsyringe. Subsequently, one adult exposed to the EO dilution was put into a glass tube (2.5 × 120 mm, covered with perforated lid) and fed wheat grain. Three replicates (50 weevils per replicate) were performed per dilution, and the dead weevils were quantified after exposure for 48 h to the essential oil.

2.6. Time–Mortality Relationship

The survival analysis for *S. granarius* obtained in the dose–mortality bioassay was evaluated. Solutions prepared with the estimated lethal doses (LD_{25}, LD_{50}, LD_{75}, and LD_{90}) were applied topically in weevils and air dried for 15 min. Acetone was utilized as the control. Weevil adults with the various lethal doses were individualized in glass tubes (2.5 × 120 mm), which were covered with perforated lids. Fifty weevils were employed for the lethal doses of *O. vulgare* EO, and each treatment was replicated three times. The live weevils were quantified each time at 6 h for 48 h.

2.7. Behavioral Response

The *Sitophilus granarius* adults were individually placed on a Petri dish arena (90 × 1.5 mm) with filter paper (Whatman No. 1, Merck KGaA, Darmstadt, Germany) at the bottom, and covered with Teflon® PTFE (E.I. Du Pont de Nemours & Co., Willmington, DE, USA). One half of the arena was impregnated with 250 μL of *O. vulgare* EO at the LD_{50} or LD_{90}, and the other half was treated with acetone. One *S. granarius* adult was released into the center of the arena and monitored for 10 min. Sixteen insects were used per treatment; for each repetition, the arena was changed. Behavioral responses were recorded using a camcorder, and videos were analyzed through the Videotrack computerized system (ViewPoint Behavior Technology, Lyon, France) to measure the distance walked, resting time, and velocity. Weevils that spent less than 1 min on the treated side of the arena were considered repelled, and those that spent less than 5 min were considered irritated [42,43].

2.8. Respiration Rate

The *S. granarius* respiration was assessed for 3 h after exposure to the *O. vulgare* EO (LD_{50} and LD_{90}) or the control, in accordance with the dose–mortality procedure. The CO_2 (carbon dioxide) evolution (μL of CO_2 h^{-1}/insect) was quantified with a respirometer of the CO_2 TR3C (Sable System Int., Las Vegas, NV, USA) type. One weevil adult was kept out in a glass chamber (25 mL) in a closed system. CO2 production was measured for 12 h at 27 ± 3 °C after weevil acclimatization. The O_2 (oxygen) molecules were infused in a glass chamber for 4 min at a flow of 125 mL min^{-1} to quantify the CO_2 exhaled in the chamber.

To measure the CO_2 exhaled by the insects in each chamber, an infrared reader attached to the system detected the CO_2 molecules during the passage of airflow. Fifteen weevils were employed for EO exposure (LD_{50} and LD_{90}, and the control).

2.9. Statistical Analysis

Probit analysis was performed on dose–mortality data to estimate the regression (intercept and slope) and lethal dose values with 95% confidence limits using SAS software (version 9.1.). The time–mortality data underwent Kaplan–Meier survival analysis using GraphPad Prism software (version 7.1.). The respiration rate data underwent a two-way ANOVA test and Tukey's HSD test. The behavioral response data were evaluated by one-way ANOVA and the means were compared with Tukey's test. Data analysis for the respiration and behavior was computerized with SAS software.

3. Results

3.1. Chemical *O. vulgare* EO Characterization

Twenty-five compounds were identified in *O. vulgare* EO, accounting for 98.31% of the total composition (Table 1).

Table 1. Chemical composition of *Origanum vulgare* essential oil.

Peaks	Compounds	Composition (%)
1	α-thujene	1.45 ± 0.01
2	α-pinene	2.74 ± 0.05
3	Camphene	1.99 ± 0.07
4	β-pinene	1.75 ± 0.02
5	β-myrcene	1.19 ± 0.01
6	α-phellandrene	1.91 ± 0.03
7	α-terpinene	1.25 ± 0.01
8	p-cymene	11.5 ± 0.11
9	Eucalyptol	2.98 ± 0.08
10	γ-terpinene	7.09 ± 0.15
11	Cis-sabinene hydrate	1.49 ± 0.01
12	Terpinolene	1.18 ± 0.01
13	Linalool	9.53 ± 0.22
14	Camphor	1.79 ± 0.01
15	Borneol	1.37 ± 0.01
16	Terpinen-4-ol	1.89 ± 0.02
17	α-terpineol	1.59 ± 0.01
18	Thymol methyl ether	1.91 ± 0.03
19	Carvacrol methyl ether	1.63 ± 0.01
20	Cuminaldehyde	1.22 ± 0.01
21	Thymol	7.51 ± 0.08
22	Carvacrol	25.4 ± 0.13
23	Aromandrene	1.39 ± 0.01
24	β-bisabolene	1.74 ± 0.01
25	Caryophyllene oxide	4.78 ± 0.09

3.2. Dose–Mortality Relationship

The dose–mortality data were suitable for a model probit fit ($p > 0.05$), demonstrating the lethality of *O. vulgare* EO to *S. granarius* (3.05 μg $insect^{-1}$), and allowing toxicological endpoints to be estimated (Table 2). Mortality remained at 1% in the control.

Table 2. Lethal doses of *Origanum vulgare* essential oil on *Sitophilus granarius* after 48 h exposure, obtained from probit analysis (df = 5, Slope ± SE = 2.481 ± 0.23, intercept = 1.949). The chi-square value refers to the goodness of fit test at $p > 0.05$.

N° Insects	Lethal Doses	Estimated Dose (μg Insect^{-1})	95% Confidence Interval (μg Insect^{-1})	χ^2 (p-Value)
150	LD_{25}	1.632	1.311–1.957	6.89 (0.14)
150	LD_{50}	3.053	2.576–3.645	
150	LD_{75}	5.709	4.691–7.331	
150	LD_{90}	10.02	7.748–14.27	

3.3. Time–Mortality Relationship

The survival rates of *S. granarius* revealed significant differences between the *O. vulgare* EO lethal doses (log-rank test, $\chi^2 = 19.41$; df = 4; $p < 0.0001$) (Figure 1). After 48 h, survival was 99.9% in the non-EO-exposed (the control) weevils, declining to 58.3% with LD_{25}, 44.9% with LD_{50}, 30.9% with LD_{75}, and 10.3% with LD_{90} of this EO.

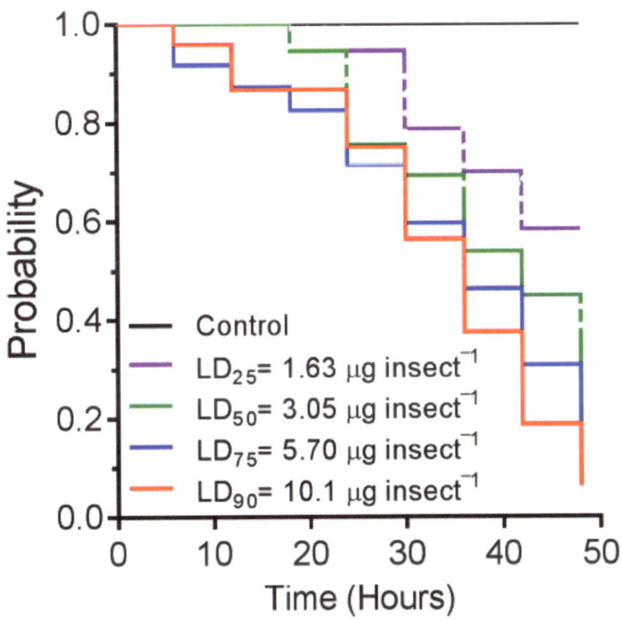

Figure 1. Survival curves of *Sitophilus granarius* exposed to different lethal doses of *Origanum vulgare* essential oil, subject to survival analysis using the Kaplan–Meier estimators (log-rank test $\chi^2 = 19.41$, DF = 4, $p < 0.001$).

3.4. Behavioral Response

Regarding *S. granarius* exposed to surfaces contaminated with *O. vulgare* EO, they gradually reduced the distance walked, their resting time, and the velocity, indicating repellency. *Sitophilus granarius* had a shorter walked distance in the half-arenas treated with *O. vulgare* EO (LD_{50} and LD_{90}) than in the control ($F_{2,15} = 26.57$, $p < 0.001$; Figure 2A). *Sitophilus granarius* had higher resting periods in the arenas exposed to lethal doses (LD_{50} and LD_{90}) than those exposed to the control ($F_{2,15} = 51.18$; $p < 0.001$; Figure 2B). The walking velocity of weevils was higher in the control than in the LD_{50} and LD_{90}-treated ones ($F_{2,15} = 12.53$; $p < 0.001$; Figure 2C).

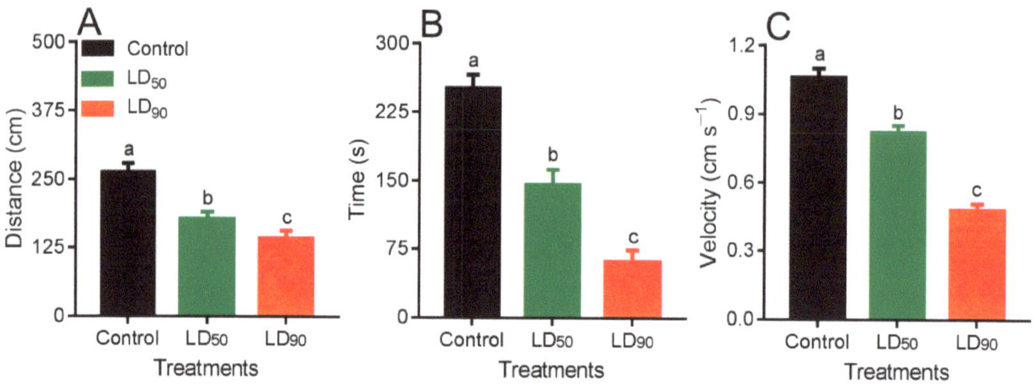

Figure 2. Behavioral response of *Sitophilus granarius* caused by *Origanum vulgare* essential oil. (**A**) Distance walked, (**B**) resting time, and (**C**) walking velocity of *S. granarius* subjected to essential oil (control, LD_{50}, and LD_{90} estimated values) for 10 min. Treatments (mean ± SEM) differ at $p < 0.05$ (Tukey's mean separation test).

3.5. Respiration Rate

The respiration of *S. granarius* was affected by exposure to *O. vulgare* EO at LD_{50} and LD_{90}. The respiration of weevils differed between the control (2.78 µL CO_2 h^{-1}), LD_{50} (2.36 µL CO_2 h^{-1}), and LD_{90} (1.68 µL CO_2 h^{-1}) 1 h after exposure; however, after 3 h, the respiration decreased to 2.53 µL CO_2 h^{-1} in the control group, followed by LD_{50} with 1.69 µL CO_2 h^{-1}, and LD_{90} with 1.25 µL CO_2 h^{-1} (Figure 3). The significant effect of treatments ($p < 0.0086$), time ($p < 0.0001$), and the interaction between treatments × time ($p < 0.0004$) were observed (Table 3).

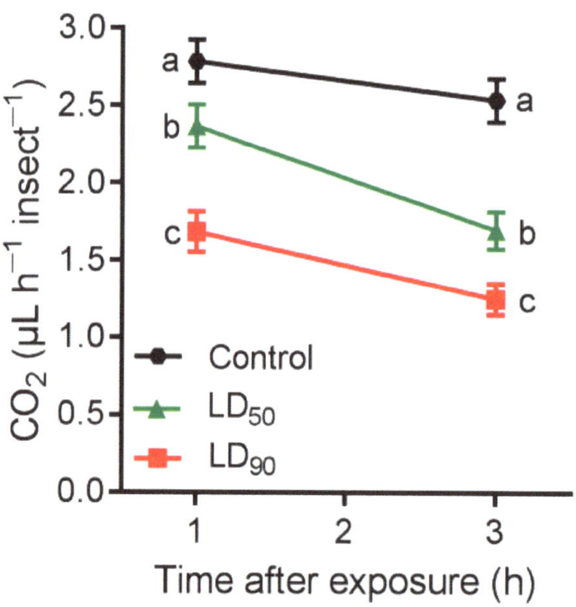

Figure 3. Respiration rate (mean ± SEM) of *Sitophilus granarius* exposure to *Origanum vulgare* essential oil (control, LD_{50} and LD_{90} estimated values) for 3 h. Treatments (mean ± SEM) differs at $p < 0.05$ (Tukey's mean separation test).

Table 3. Two-way ANOVA for respiration rate of *Sitophilus granarius* upon exposure to lethal doses (LD_{50} and LD_{90}) of *Origanum vulgare* essential oil. DF = degrees of freedom, SS = sum of squares, MS = mean square, n = numerator, d = denominator, *p* = probability of significance.

ANOVA Table	SS	DF	MS	F (DFn, DFd)	*p*-Value
Treatments	46.52	2	23.26	F (2,48) = 2.57	<0.0086
Time	138.1	1	138.1	F (1,48) = 15.2	<0.0001
Treatments × time	108.6	2	54.28	F (2,48) = 6.01	<0.0004
Residual	433.7	48	9.036		
Total	727	53			

4. Discussion

This work investigated the *O. vulgare* EO composition and evaluated the insecticidal properties caused of this EO on *S. granarius*. Twenty-five compounds were found: p-cymene, carvacrol, linalool, thymol, γ-terpinene, caryophyllene oxide, α-pinene, and eucalyptol were the principal compounds, in accordance with previous analytical chemical investigations on terpenoids of this EO [39–41]. Carvacrol, p-cymene, and thymol are highly toxic to insects and this action is due to the noncompetitive inhibition of acetylcholinesterase by interaction with nicotinic acetylcholine receptors [44]; meanwhile, linalool and γ-terpinene act upon the nervous system of insects by the reversible inhibition of acetylcholinesterase [45]. Specifically, a majority of *O. vulgare* EO compounds are terpenoids and can mediate herbivore–plant chemical communication, acting as allomones or kairomones [46]. Terpenoids are metabolites with various biochemical mechanisms [47], and they depict a crucial role in inducing defense responses against insects [48]. With respect to the mode of action, the evidence action of target proteins responsible for the bioactivity of *O. vulgare* EO is small, but it is probably its effect on the nervous system of *S. granarius*, owing to existence of terpenoids, which results in rapid mortality, as researched in other insects after EO exposure [17,31,49].

The lethality of the *O. vulgare* EO to *S. granarius* was assessed from the dose–mortality bioassay on topical application. This EO was lethal to adult *S. granarius* (LD_{50} = 3.05 μg $insect^{-1}$) and had an effect on cuticle contact. The *O. vulgare* EO induced dose–dependent lethality in *S. granarius*, as found in other insects after EO application [50–52]. Weevils treated to several doses of this EO showed altered locomotor and feeding activities. Some gradually lost mobility, followed by paralysis and death. These symptoms were consistent with the identifiable effect on the nervous system [19–21]. Different pest species, such as *Alphitobius diaperinus* Panzer (Coleoptera: Tenebrionidae) [37], *Nezara viridula* Linnaeus (Hemiptera: Pentatomidae) [53], and *Plutella xylostella* Linnaeus (Lepidoptera: Pyralidae), [54] were susceptible to EOs by contact exposure or fumigation, which caused irreversible effects in the neurons. This result demonstrated the strong neurotoxicity of *O. vulgare* EO in *S. granarius* when topically exposed, which can impair its populations.

High variability in *S. granarius* survival can be promoted when the *O. vulgare* EO interacts by contact exposure and penetration via the trachea, conducive to the suppression of nerve conduction. The reduced time exposures to the *O. vulgare* EO (24 to 48 h) were needed to induce lethality in this insect and were attributed with the quick action of this bioinsecticide. In this research, the comparative survival of weevils between lethal doses of this EO take place at various periods. These time differences were due to the ability of the EO to ingress via the insect's spiracles during respiration [23] and penetrate the integument cuticle layers [18], exerting its effect by acting as neurotoxin in insects [19]. EOs have been demonstrated to interrupt ion channel shutdown in neuronal axons and cause paralytic activity on *Acanthoscelides obtectus* Say (Coleoptera: Chrysomelidae) [55], *Anagasta khueniella* Zeller (Lepidoptera: Pyralidae) [50], and *Rhyzopertha dominica* Fabricius (Coleoptera: Bostrichidae) [56]. Low *S. granarius* survival prompts that this EO causes prejudicial effects on adults with quick exposure. Thus, *O. vulgare* EO may offer much protection against this insect in the management of stored products.

Alterations in the locomotion of *S. granarius* caused by *O. vulgare* EO are a result of the toxicant action of this biopesticide on the insect's neuronal receptors. Modifications in behavioral patterns have been observed in various insects after EO exposure [24,57,58], with serious consequences on orientation and olfactory responses [59–61]. In *S. granarius*, surfaces treated with the *O. vulgare* EO gradually reduced the walked distance of weevils and, subsequently, the resting time, suggesting repellency. Modifications in mobility with doses of this EO may be a result of its shutdown effect on neuron transmission during channel modulation, exerting a variation of action potentials along nerve axons and synapses [19,47,62]. The findings show that variations in the locomotor ability of *S. granarius* were dose-dependent on the *O. vulgare* EO, leading to repellency.

The *O. vulgare* EO compromises the *S. granarius* respiration, indicating physiological stress. In insects, inhaled EOs move into the respiratory system and can affect the gas exchange patterns [14,23,58]. The reduced respiration rate occurs during contact and EO exposure; consequently, this toxic event requires an energetic demand to lead the detoxification process [62]. An imbalance in respiration promotes a high fitness cost, and the energy utilized can be reused in other metabolic mechanisms [59]. A comparable reaction occurs in coleopteran pests, such as *Demotispa neivai* Bondar (Chrysomelidae) treated with neem EOs [30], *Ulomoides dermestoides* Fairmaire treated with lemongrass EOs [63], and *Tenebrio molitor* Linnaeus (Tenebrionidae) treated with cinnamon EOs [17], decreasing oxygen consumption and disrupting oxidative phosphorylation in respiration [24,31]. The findings obtained here show that *S. granarius* had a reduced respiration when exposed to the *O. vulgare* EO, with likely fitness costs and reallocated energy in other physiological processes.

5. Conclusions

Overall, the results indicate that *O. vulgare* EO has a significant range of prejudicial effects on *S. granarius*. This EO inflicts toxicity, low survival, altered behavioral response, and reduced respiration rate upon adults of this insect. The composition of this EO proves to be a blend that is abundant in terpenoids, actuating by contact or inhalation to exert neurotoxicity on *S. granarius*. Furthermore, this research provides data supporting *O. vulgare* EO as a potential source of natural insecticides, which might also be utilized as an innovative tool for the effective management of *S. granarius* populations.

Author Contributions: Conceptualization, A.P.-R., M.H.D.S., J.E.S. and L.C.M.; methodology, A.P.-R., M.H.D.S., J.E.S. and L.C.M.; validation, M.H.D.S. and L.C.M.; formal analysis, A.P.-R., M.H.D.S., J.E.S. and L.C.M.; investigation, A.P.-R., M.H.D.S., J.E.S. and L.C.M.; resources, M.H.D.S. and J.E.S.; writing—original draft preparation, A.P.-R., M.H.D.S., J.E.S. and L.C.M.; writing—review and editing, A.P.-R., M.H.D.S., J.E.S. and L.C.M.; project administration, M.H.D.S. and J.E.S. All authors have read and agreed to the published version of the manuscript.

Funding: This study was supported by Brazilian research agencies Conselho Nacional de Desenvolvimento Científico e Tecnológico (CNPq) and Coordenação de Aperfeiçoamento de Pessoal de Nível Superior (CAPES).

Institutional Review Board Statement: Not applicable for studies not involving humans or animals.

Informed Consent Statement: Not applicable for studies not involving humans.

Data Availability Statement: Not applicable.

Acknowledgments: We thank the Department of Chemistry, Entomology and General Biology of the "Universidade Federal de Viçosa" (Brazil) for technical support.

Conflicts of Interest: The authors declare no conflict of interest.

References

1. Flinn, P.W.; Hagstrum, D.W.; Reed, C.; Phillips, T.W. United States Department of Agriculture-Agricultural Research Service stored-grain area-wide integrated pest management program. *Pest Manag. Sci.* **2003**, *59*, 614–618. [CrossRef] [PubMed]
2. Neethirajan, S.; Karunakaran, C.; Jayas, D.S.; White, N.D.G. Detection techniques for stored-product insects in grain. *Food Control* **2007**, *18*, 157–162. [CrossRef]

3. Nesci, A.; Barra, P.; Etcheverry, M. Integrated management of insect vectors of *Aspergillus flavus* in stored maize, using synthetic antioxidants and natural phytochemicals. *J. Stored Prod. Res.* **2011**, *47*, 231–237. [CrossRef]
4. Boyer, S.; Zhang, H.; Lempérière, G. A review of control methods and resistance mechanisms in stored-product insects. *Bull. Entomol. Res.* **2012**, *102*, 213–229. [CrossRef] [PubMed]
5. Vincent, C.; Hallman, G.; Panneton, B.; Fleurat-Lessard, F. Management of agricultural insects with physical control methods. *Annu. Rev. Entomol.* **2003**, *48*, 261–281. [CrossRef]
6. Mohammed, M.E.; El-Shafie, H.A.; Alhajhoj, M.R. Design and efficacy evaluation of a modern automated controlled atmosphere system for pest management in stored dates. *J. Stored Prod. Res.* **2020**, *89*, 101719. [CrossRef]
7. Morrison, W.R., III; Scully, E.D.; Campbell, J.F. Towards developing areawide semiochemical-mediated, behaviorally-based integrated pest management programs for stored product insects. *Pest Manag. Sci.* **2021**, *77*, 2667–2682. [CrossRef]
8. Zettler, J.L.; Arthur, F.H. Chemical control of stored product insects with fumigants and residual treatments. *Crop Prot.* **2000**, *19*, 577–582. [CrossRef]
9. Sousa, A.; Faroni, L.; Pimentel, M.; Guedes, R. Developmental and population growth rates of phosphine-resistant and susceptible populations of stored product insect pests. *J. Stored Prod. Res.* **2009**, *45*, 241–246. [CrossRef]
10. Goulson, D. An overview of the environmental risks posed by neonicotinoid insecticides. *J. Appl. Ecol.* **2013**, *50*, 977–987. [CrossRef]
11. Plata-Rueda, A.; Campos, J.M.; da Silva Rolim, G.; Martínez, L.C.; Dos Santos, M.H.; Fernandes, F.L.; Serrão, J.E.; Zanuncio, J.C. Terpenoid constituents of cinnamon and clove essential oils cause toxic effects and behavior repellency response on granary weevil, *Sitophilus granarius*. *Ecotox. Environ. Saf.* **2018**, *156*, 263–270. [CrossRef] [PubMed]
12. Cossolin, J.F.S.; Pereira, M.J.; Martínez, L.C.; Turchen, L.M.; Fiaz, M.; Bozdoğan, H.; Serrão, J.E. Cytotoxicity of *Piper aduncum* (Piperaceae) essential oil in brown stink bug *Euchistus heros* (Heteroptera: Pentatomidae). *Ecotoxicology* **2019**, *28*, 63–770. [CrossRef] [PubMed]
13. Plata-Rueda, A.; Rolim, G.D.S.; Wilcken, C.F.; Zanuncio, J.C.; Serrão, J.E.; Martínez, L.C. Acute toxicity and sublethal effects of lemongrass essential oil and their components against the granary weevil, *Sitophilus granarius*. *Insects* **2020**, *11*, 379. [CrossRef] [PubMed]
14. Brügger, B.P.; Plata-Rueda, A.; Wilcken, C.F.; de Souza, L.S.A.; Serrão, J.E.; Carvalho, A.G.; Zanuncio, J.C.; Martínez, L.C. Exposure to lemongrass essential oil and its components causes behavior and respiratory disturbs in *Anticarsia gemmatalis*. *Int. J. Pest Manag.* **2021**, 1–11. [CrossRef]
15. Zanuncio, J.C.; Mourão, S.A.; Martínez, L.C.; Wilcken, C.F.; Ramalho, F.S.; Plata-Rueda, A.; Soares, M.A.; Serrão, J.E. Toxic effects of the neem oil (*Azadirachta indica*) formulation on the stink bug predator, *Podisus nigrispinus* (Heteroptera: Pentatomidae). *Sci. Rep.* **2016**, *6*, 30261. [CrossRef]
16. Amaral, K.D.; Martínez, L.C.; Lima, M.A.P.; Serrão, J.E.; Della Lucia, T.M.C. Azadirachtin impairs egg production in *Atta sexdens* leaf-cutting ant queens. *Environ. Pollut.* **2018**, *24*, 809–814. [CrossRef]
17. Martínez, L.C.; Plata-Rueda, A.; Colares, H.C.; Campos, J.M.; Dos Santos, M.H.; Fernandes, F.L.; Serrão, J.E.; Zanuncio, J.C. Toxic effects of two essential oils and their constituents on the mealworm beetle, *Tenebrio molitor*. *Bull. Entomol. Res.* **2018**, *108*, 716–725. [CrossRef]
18. Plata-Rueda, A.; Martínez, L.C.; Rolim, G.D.S.; Coelho, R.P.; Santos, M.H.; Tavares, W.de.S.; Zanuncio, J.C.; Serrão, J.E. Insecticidal and repellent activities of *Cymbopogon citratus* (Poaceae) essential oil and its terpenoids (citral and geranyl acetate) against *Ulomoides dermestoides*. *Crop Prot.* **2020**, *137*, 105299. [CrossRef]
19. Mukherjee, P.K.; Kumar, V.; Mal, M.; Houghton, P.J. Acetylcholinesterase inhibitors from plants. *Phytomedicine* **2007**, *14*, 289–300. [CrossRef]
20. Priestley, C.M.; Williamson, E.M.; Wafford, K.A.; Sattelle, D.B. Thymol, a constituent of thyme essential oil, is a positive allosteric modulator of human GABAA receptors and a homo-oligomeric GABA receptor from *Drosophila melanogaster*. *Braz. J. Pharmacol.* **2003**, *140*, 1363–1372. [CrossRef]
21. Kostyukovsky, M.; Rafaeli, A.; Gileadi, C.; Demchenko, N.; Shaaya, E. Activation of octopaminergic receptors by essential oil constituents isolated from aromatic plants: Possible mode of action against insect pests. *Pest Manag. Sci.* **2002**, *58*, 1101–1106. [CrossRef] [PubMed]
22. Fiaz, M.; Martínez, L.C.; da Silva Costa, M.; Plata-Rueda, A.; Gonçalves, W.G.; Sant'Ana, A.E.G.; Zanuncio, J.C.; Serrão, J.E. Squamocin induce histological and ultrastructural changes in the midgut cells of *Anticarsia gemmatalis* (Lepidoptera: Noctuidae). *Ecotox. Environ. Saf.* **2018**, *156*, 1–8. [CrossRef] [PubMed]
23. Farder-Gomes, C.; Saravanan, M.; Martínez, L.C.; Plata-Rueda, A.; Zanuncio, J.C.; Serrão, J.E. Azadirachtin-based biopesticide affects the respiration and digestion in *Anticarsia gemmatalis* caterpillars. *Toxin Rev.* **2022**, *41*, 466–475. [CrossRef]
24. Plata-Rueda, A.; Zanuncio, J.C.; Serrão, J.E.; Martínez, L.C. *Origanum vulgare* essential oil against *Tenebrio molitor* (Coleoptera: Tenebrionidae): Composition, insecticidal activity, and behavioral response. *Plants* **2021**, *10*, 2513. [CrossRef]
25. Olivero-Verbel, J.; Nerio, L.S.; Stashenko, E.E. Bioactivity against *Tribolium castaneum* Herbst (Coleoptera: Tenebrionidae) of *Cymbopogon citratus* and *Eucalyptus citriodora* essential oils grown in Colombia. *Pest Manag. Sci.* **2010**, *66*, 664–668. [CrossRef]
26. Basile, S.; Badalamenti, N.; Riccobono, O.; Guarino, S.; Ilardi, V.; Bruno, M.; Peri, E. Chemical composition and evaluation of insecticidal activity of *Calendula incana* subsp. *maritima* and *Laserpitium siler* subsp. *siculum* essential oils against stored products pests. *Molecules* **2022**, *27*, 588.
27. Kavallieratos, N.G.; Boukouvala, M.C.; Ntalli, N.; Skourti, A.; Karagianni, E.S.; Nika, E.P.; Kontodimas, D.C.; Cappellacci, L.; Petrelli, R.; Cianfaglione, K.; et al. Effectiveness of eight essential oils against two key stored-product beetles, *Prostephanus truncatus* (Horn) and *Trogoderma granarium* Everts. *Food Chem. Toxicol.* **2020**, *139*, 111255. [CrossRef]
28. Jayaram, C.S.; Chauhan, N.; Dolma, S.K.; Reddy, S.G.E. Chemical Composition and Insecticidal Activities of Essential Oils against the Pulse Beetle. *Molecules* **2022**, *27*, 568. [CrossRef]

29. Plata-Rueda, A.; Martínez, L.C.; Dos Santos, M.H.; Fernandes, F.L.; Wilcken, C.F.; Soares, M.A.; Serrão, J.E.; Zanuncio, J.C. Insecticidal activity of garlic essential oil and their constituents against the mealworm beetle, *Tenebrio molitor* Linnaeus (Coleoptera Tenebrionidae). *Sci. Rep.* **2017**, *7*, 46406. [CrossRef]
30. Martínez, L.C.; Plata-Rueda, A.; Zanuncio, J.C.; Serrão, J.E. Bioactivity of six plant extracts on adults of *Demotispa neivai* (Coleoptera: Chrysomelidae). *J. Insect Sci.* **2015**, *15*, 34. [CrossRef]
31. Brügger, B.P.; Martínez, L.C.; Plata-Rueda, A.; Castro, B.M.D.C.; Soares, M.A.; Wilcken, C.F.; Carvalho, A.G.; Serrão, J.E.; Zanuncio J.C. Bioactivity of the *Cymbopogon citratus* (Poaceae) essential oil and its terpenoid constituents on the predatory bug, *Podisus nigrispinus* (Heteroptera: Pentatomidae). *Sci. Rep.* **2019**, *9*, 8358. [CrossRef] [PubMed]
32. Fournomiti, M.; Kimbaris, A.; Mantzourani, I.; Plessas, S.; Theodoridou, I.; Papaemmanouil, V.; Kapsiotis, I.; Panopoulou, M. Stavropoulou, E.; Bezirtzoglou, E.E.; et al. Antimicrobial activity of essential oils of cultivated oregano (*Origanum vulgare*) sage (*Salvia officinalis*), and thyme (*Thymus vulgaris*) against clinical isolates of *Escherichia coli*, *Klebsiella oxytoca*, and *Klebsiella pneumoniae*. *Microb. Ecol. Health D* **2015**, *26*, 23289. [CrossRef] [PubMed]
33. Ibrahimand, F.A.; Al-Ebady, N. Evaluation of antifungal activity of some plant extracts and their applicability in extending the shelf life of stored tomato fruits. *J. Food Process Technol.* **2014**, *5*, 340.
34. Kim, S.-I.; Yoon, J.-S.; Jung, J.W.; Hong, K.-B.; Ahn, Y.-J.; Kwon, H.K. Toxicity and repellency of origanum essential oil and its components against *Tribolium castaneum* (Coleoptera: Tenebrionidae) adults. *J. Asia-Pac. Entomol.* **2010**, *13*, 369–373. [CrossRef]
35. Schillaci, D.; Napoli, E.M.; Cusimano, M.G.; Vitale, M.; Ruberto, A. *Origanum vulgare* subsp. hirtum essential oil prevented biofilm formation and showed antibacterial activity against planktonic and sessile bacterial cells. *J. Food Prot.* **2013**, *76*, 1747–1752. [CrossRef]
36. Papachristos, D.P.; Stamopoulos, D.C. Repellent, toxic and reproduction inhibitory effects of essential oil vapours on *Acanthoscelides obtectus* (Say) (Coleoptera: Bruchidae). *J. Stored Prod. Res.* **2002**, *38*, 117–128. [CrossRef]
37. Szczepanik, M.; Walczak, M.; Zawitowska, B.; Michalska-Sionkowska, M.; Szumny, A.; Wawrzeńczyk, C.; Brzezinska, M.S Chemical composition, antimicromicrobial activity and insecticidal activity against the lesser mealworm *Alphitobius diaperinus* (Panzer) (Coleoptera: Tenebrionidae) of *Origanum vulgare* L. ssp. *hirtum* (Link) and *Artemisia dracunculus* L. essential oils. *J. Sci. Food Agric.* **2018**, *98*, 767–774.
38. Papanikolaou, N.E.; Kavallieratos, N.G.; Iliopoulos, V.; Evergetis, E.; Skourti, A.; Nika, E.P.; Haroutounian, S.A. Essential oil coating: Mediterranean culinary plants as grain protectants against larvae and adults of *Tribolium castaneum* and *Trogoderma granarium*. *Insects* **2022**, *13*, 165. [CrossRef]
39. Kulisic, T.; Radonic, A.; Katalinic, V.; Milos, M. Use of different methods for testing antioxidative activity of oregano essential oil. *Food Chem.* **2004**, *85*, 633–640. [CrossRef]
40. Figiel, A.; Szumny, A.; Gutierrez-Ortiz, A.; Carbonell-Barrachina, A.A. Composition of oregano essential oil (*Origanum vulgare*) as affected by drying method. *J. Food Eng.* **2010**, *98*, 240–247. [CrossRef]
41. La Pergola, A.; Restuccia, C.; Napoli, E.; Bella, S.; Brighina, S.; Russo, A.; Suma, P. Commercial and wild Sicilian *Origanum vulgare* essential oils: Chemical composition, antimicrobial activity and repellent effects. *J. Essent. Oil Res.* **2017**, *29*, 451–460. [CrossRef]
42. Plata-Rueda, A.; Martínez, L.C.; Da Silva, B.K.R.; Zanuncio, J.C.; Fernandes, M.E.de.S.; Guedes, R.N.C.; Fernandes, F.L. Exposure to cyantraniliprole causes mortality and disturbs behavioral and respiratory response in the coffee berry borer (*Hypothenemus hampei*). *Pest Manag. Sci.* **2019**, *75*, 2236–2241. [CrossRef] [PubMed]
43. Silva, W.M.; Martínez, L.C.; Plata-Rueda, A.; Serrão, J.E.; Zanuncio, J.C. Respiration, predatory and prey consumption by *Podisus nigrispinus* (Heteroptera: Pentatomidae) nymphs exposed some to insecticides. *Chemosphere* **2020**, *261*, 127720. [CrossRef] [PubMed]
44. Campos, E.V.; Proença, P.L.; Oliveira, J.L.; Pereira, A.E.; de Morais Ribeiro, L.N.; Fernandes, F.O.; Gonçalves, K.C.; Polanczyk, R.A.; Pasquoto-Stigliani, T.; Lima, R.; et al. Carvacrol and linalool co-loaded in β-cyclodextrin-grafted chitosan nanoparticles as sustainable biopesticide aiming pest control. *Sci. Rep.* **2018**, *8*, 7623. [CrossRef] [PubMed]
45. Li, A.S.; Iijima, A.; Huang, J.; Li, Q.X.; Chen, Y. Putative mode of action of the monoterpenoids linalool, methyl eugenol, estragole, and citronellal on ligand-gated ion channels. *Engineering* **2020**, *6*, 541–545. [CrossRef]
46. War, A.R.; Paulraj, M.G.; Ahmad, T.; Buhroo, A.A.; Hussain, B.; Ignacimuthu, S.; Sharma, H.C. Mechanisms of plant defense against insect herbivores. *Plant Signal Behav.* **2012**, *7*, 1306–1320. [CrossRef]
47. Ibanez, S.; Gallet, C.; Després, L. Plant insecticidal toxins in ecological networks. *Toxins* **2012**, *4*, 228–243. [CrossRef]
48. Fürstenberg-Hägg, J.; Zagrobelny, M.; Bak, S. Plant defense against insect herbivores. *Int. J. Mol. Sci.* **2013**, *14*, 10242–10297. [CrossRef]
49. Tak, J.H.; Jovel, E.; Isman, M.B. Synergistic interactions among the major constituents of lemongrass essential oil against larvae and an ovarian cell line of the cabbage looper, *Trichoplusia ni*. *J. Pest Sci.* **2017**, *90*, 735–744. [CrossRef]
50. Aslan, I.; Çalmaşur, Ö.; Sahin, F.; Çağlar, Ö. Insecticidal effects of essential plant oils against *Ephestia kuehniella* (Zell.), *Lasioderma serricorne* (F.) and *Sitophilus granarius* (L.). *J. Plant Dis. Prot.* **2005**, *112*, 257–267.
51. Bertoli, A.; Conti, B.; Mazzoni, V.; Meini, L.; Pistelli, L. Volatile chemical composition and bioactivity of six essential oils against the stored food insect *Sitophilus zeamais* Motsch. (Coleoptera Dryophthoridae). *Nat. Prod. Res.* **2012**, *26*, 2063–2071. [PubMed]
52. Xie, Y.; Huang, Q.; Rao, Y.; Hong, L.; Zhang, D. Efficacy of *Origanum vulgare* essential oil and carvacrol against the housefly, *Musca domestica* L. (Diptera: Muscidae). *Environ. Sci. Pollut. Res.* **2019**, *26*, 23824–23831. [CrossRef] [PubMed]
53. Werdin González, J.O.; Gutiérrez, M.M.; Murray, A.P.; Ferrero, A.A. Composition and biological activity of essential oils from Labiatae against *Nezara viridula* (Hemiptera: Pentatomidae) soybean pest. *Pest Manag. Sci.* **2011**, *67*, 948–955. [CrossRef] [PubMed]

54. Nasr, M.; Sendi, J.J.; Moharramipourb, S.; Zibaee, A. Evaluation of *Origanum vulgare* L. essential oil as a source of toxicant and an inhibitor of physiological parameters in diamondback moth, *Plutella xylustella* L. (Lepidoptera: Pyralidae). *J. Saud. Soc. Agr. Sci.* **2017**, *16*, 184–190. [CrossRef]
55. Baricevic, D.; Milevoj, L.; Borstnik, J. Insecticidal effect of oregano (*Origanum vulgare* L. ssp. hirtum Ietswaart) on the dry bean weevil (*Acanthoscelides obtectus* Say). *Int. J. Hortic. Sci.* **2001**, *7*, 84–88. [CrossRef]
56. Alkan, M. Chemical composition and insecticidal potential of different *Origanum* spp. (Lamiaceae) essential oils against four stored product pests. *Turk J. Entomol.* **2020**, *44*, 149–163. [CrossRef]
57. Castro, B.M.C.; Martínez, L.; Plata-Rueda, A.; Soares, M.A.; Wilcken, C.F.; Zanuncio, A.J.V.; Fiaz, M.; Zanuncio, J.C.; Serrão, J.E. Exposure to chlorantraniliprole reduces locomotion, respiration, and causes histological changes in the midgut of velvetbean caterpillar *Anticarsia gemmatalis* (Lepidoptera: Noctuidae). *Chemosphere* **2021**, *263*, 128008. [CrossRef]
58. Vinha, G.L.; Plata-Rueda, A.; Soares, M.A.; Zanuncio, J.C.; Serrão, J.E.; Martínez, L.C. Deltamethrin-mediated effects on locomotion, respiration, feeding and histological changes in the midgut of *Spodoptera frugiperda* caterpillars. *Insects* **2021**, *12*, 483. [CrossRef]
59. Plata-Rueda, A.; Menezes, C.H.M.; Cunha, W.S.; Alvarenga, T.M.; Barbosa, B.F.; Zanuncio, J.C.; Martínez, L.C.; Serrão, J.E. Side-effects caused by chlorpyrifos in the velvetbean caterpillar *Anticarsia gemmatalis* (Lepidoptera: Noctuidae). *Chemosphere* **2020**, *259*, 127530. [CrossRef]
60. Fiaz, M.; Martínez, L.C.; Plata-Rueda, A.; Cossolin, J.F.S.; Serra, R.S.; Martins, G.F.; Serrão, J.E. Behavioral and ultrastructural effects of novaluron on *Aedes aegypti* larvae. *Infect. Genet. Evol.* **2021**, *93*, 104974. [CrossRef]
61. Lima, B.S.A.; Martínez, L.C.; Plata-Rueda, A.; dos Santos, M.H.; de Oliveira, E.E.; Zanuncio, J.C.; Serrão, J.E. Interaction between predatory and phytophagous stink bugs promoted by secretion of scent glands. *Chemoecology* **2021**, *31*, 209–219. [CrossRef]
62. Price, D.N.; Berry, M.S. Comparison of effects of octopamine and insecticidal essential oils on activity in the nerve cord, foregut, and dorsal unpaired median neurons of cockroaches. *J. Insect Physiol.* **2006**, *52*, 309–319. [CrossRef] [PubMed]
63. Plata-Rueda, A.; Fiaz, M.; Brügger, B.P.; Cañas, V.; Coelho, R.P.; Zanuncio, J.C.; Martínez, L.C.; Serrão, J.E. Lemongrass essential oil and its components cause effects on survival, locomotion, ingestion, and histological changes of the midgut in *Anticarsia gemmatalis* caterpillars. *Toxin Rev.* **2022**, *41*, 208–217. [CrossRef]

agronomy

MDPI

Article

Lamiaceae Plants in Bulgarian Rural Livelihoods—Diversity, Utilization, and Traditional Knowledge

Teodora Ivanova [1,*], Yulia Bosseva [1], Mihail Chervenkov [1,2] and Dessislava Dimitrova [1,*]

[1] Department of Plant and Fungal Diversity and Resources, Institute of Biodiversity and Ecosystem Research, Bulgarian Academy of Sciences, 1113 Sofia, Bulgaria; julibos@abv.bg (Y.B.); vdmchervenkov@abv.bg (M.C.)
[2] Faculty of Veterinary Medicine, University of Forestry, 1797 Sofia, Bulgaria
* Correspondence: tai@bio.bas.bg (T.I.); dessidim3010@gmail.com (D.D.)

Abstract: Lamiaceae comprises widely distributed medicinal and aromatic plants, many of which are traditionally used in European countries. The current study aimed to document Lamiaceae taxa used in rural Bulgaria (Southeast Europe) and to explore the related local knowledge and cultural practices that influence their utilization for various purposes. Field work included inventory of Lamiaceae diversity in home gardens and semi-structured interviews focused on the cultivation, collection, and utilization practices common among elderly inhabitants of 34 settlements in rural Bulgaria. We report the utilization of 27 Lamiaceae taxa, 9 of which were collected from the wild. Traditional and contemporary ways of utilizing Lamiaceae taxa as culinary and medicinal plants, in herbal teas, as repellents, ritual plants, etc., are presented. Recent knowledge on medicinal properties contributed to the introduction of new taxa in gardens (wild and cultivated), while traditional culinary practices were found to sustain the diversity of local forms (landraces).

Keywords: home gardens; culinary herbs; medicinal plants; ethnobotany; *Mentha*; *Satureja*

1. Introduction

Lamiaceae (Labiatae) is a cosmopolitan family and accounts for over 7000 species belonging to 245 genera [1]. Members of the family have long-lasting popularity for their diverse essential oil profiles, hence their multiple applications in medicine, cosmetics, gastronomy, etc. [2–4]. Many members of the Lamiaceae family used and cultivated since antiquity are still utilized [5–7]. Due to the broad spectrum of the biological activities (e.g., antimicrobial, antifungal, antioxidant, anti-inflammatory, anticancer, biocidal, etc.) of their secondary metabolites, Lamiaceae plants are used by local people based on their inherited empirical skills, or as a result of acquisition and exchange of traditional and/or modern knowledge [8–12]. Numerous ethnobotanical studies reported on the various past and current utilizations of different Lamiaceae taxa around the world and in Europe [4,13–21].

The status of Lamiaceae taxa as culinary and medicinal plants made them common in home gardens. Moreover, the growing urbanization and industrialization of agriculture turned several Lamiaceae taxa, e.g., *Lavandula*, *Ocimum*, *Origanum*, *Salvia*, etc., into important industrial crops that are grown around the world [22–24]. However, many of the traditionally used Lamiaceae plants are still harvested from the wild, often in an unsustainable way, which could be evaded through cultivation [25,26]. Global climate changes also threaten the crop yields and quality of production [26]. Large scale cultivation generally relies on limited biodiversity of crops which, in essential oil plantations, could result in a reduced range of bioactive constituents, flavors, and aromas, thus not meeting the expectations of different communities for their specific (traditional) medical practices, cuisines, and/or their cultural/spiritual needs. Conversely, small-scale farming offers opportunities to maintain a wider range of assorted crops, varieties, and/or landraces which not only provide for a more diverse diet and additional income for their owners, but can also assist the preservation of valuable genetic resources, like valuable Lamiaceae plants [27–29]. In this sense, small farms and home gardens, as compact, diverse, and multilayered agroforestry systems provide for an important multitude of services to their owners and to local communities [30–33]. These environmentally friendly and more sustainable systems also cater to the conservation of wild and agrobiodiversity [34–37]. The owners of small farms and larger home gardens are more likely to use traditional agricultural practices and the related traditional knowledge inherited from previous generations, which is regarded as an important stepping stone for the implementation of agroecological principles in practice [38–42].

Local plant genetic resources preserved by gardeners and farmers, as well as the traditional knowledge related to their use and cultivation, however, are threatened by the gradual urbanization of the industrialized societies and growing depopulation and ageing of the inhabitants of the rural areas [43,44]. Plant diversity in European home gardens, and especially in those of Eastern and Southeastern Europe, remains understudied in comparison to home gardens and homesteads in the tropics, mostly due to the specific socio-economic impact of the latter [30–33]. Bulgarian rural home gardens, which currently range in size between several square meters and half a hectare, were found to provide substantially for the family sustenance, harboring a relatively large number of annual and perennial crop species [45]. Lamiaceae was found to be the second most represented plant family in Bulgarian rural home gardens, after Rosaceae, the latter being represented mainly by singular trees and shrubs cultivated for their fruits or grown as ornamentals [45]. On the contrary, in home gardens across Europe and the East Mediterranean, members of Lamiaceae were found more scarcely [46–48]. Additionally, it was found that most of the gardening area was cultivated on an annual basis, while herbs/spices, ornamentals, and fruit trees bordering the plots were more permanent elements.

The aim of the current study is to assess the variety of Lamiaceae taxa cultivated in Bulgarian home gardens as a function of their traditional and modern uses and to evaluate the factors/drivers that maintain and/or change their taxonomic diversity and related knowledge. We present the case in the frame of the local tradition to use and cultivate plants of the Lamiaceae family as medicinal and aromatic plants all over Bulgaria.

2. Materials and Methods

2.1. Study Area and Data Collection

Bulgaria is located in the center of the Balkan Peninsula, in the most southeast corner of Europe. The territory of the country is in the transitional area between the temperate and the Mediterranean climatic zones, with slightly elevated average annual temperatures and lower precipitation rates in the last three decades: hardiness zones 7–8 [49]. Bulgarian vascular flora comprises 4064 species of spermatophytes affiliated with 921 genera and 159 families, of which nearly 150 taxa belong to Lamiaceae [50,51].

Representatives of 74 households took part in the field study (2017–2022). They were from settlements in eight Bulgarian provinces (Blagoevgrad, Haskovo, Plovdiv, and Smolyan on the southern site of the Balkan Mt. range, and Lovech, Montana, Pleven, and Vratsa on the northern site, Figure 1).

Participants were recruited directly using the snowball sampling approach. Assistance of local leaders (mayors, local cultural activists, etc.) was acquired so as to identify prominent gardeners, agronomists, and/or local healers, when needed. Formal information on the age, education and occupation of every participant was collected. Informed consent was verbally obtained from every participant prior to the interview. The guidelines prescribed in the Code of Ethics of the International Society of Ethnobiology [52] were followed during the field study, and their compliance was confirmed by the Scientific Council of the Institute of Biodiversity and Ecosystem Research, Bulgarian Academy of Sciences, acting as independent institutional Ethics Board (Decision No. 6/21/05/21).

Figure 1. Map of the studied provinces.

Participants were asked to share information about the ways they currently utilize plants of the Lamiaceae family, as well as the taxa they cultivate in their home gardens and those collected from the wild. Information on the source of different crop plants was collected together with data on the processing of the collected/cultivated plants, if any. Participants were invited to freely list the plants they utilized. Additional information was asked if some plants were present in the garden or stored on a visible place in the living premises. Guiding questions on popular culinary herbs and herbal teas were asked, if necessary.

Voucher specimens and/or image data were collected for identification purposes; herbarium specimens were deposited in the Herbarium of the Institute of Biodiversity and Ecosystem Research, Bulgarian Academy of Sciences (SOM). The identification of the plants was carried out at least to the genus and species taxonomical levels in accordance with the Handbook of Bulgarian vascular flora [53]. Plant names are in accordance with the Plant list (2013) [1].

Presented plants/herbal products were summarized in eight use categories:

- Culinary herbs (CUL)—used fresh and/or cooked in preparation of salads and/or dishes;
- Ornamental plants (ORN)—grown for their ornamental flowers and/or foliage;
- Medicinal plants (MED)—used for healing purposes incl. for preparation of recreational herbal teas;
- Aromatic plants (AROM)—used for aromatization of the garden and/or of the home (fresh or dry);
- Insect repellents (REP)—for deterring biting and vexing insects (mosquitos, flies), agricultural pests and clothes moths;
- Pollinator attraction (POLL)—plants planted and/or reported to be grown around other crops so to attract pollinators;
- Symbolic plants (SYM)—fresh/dry used traditionally for decoration of the home and other buildings or for personal decoration following religious or other rituals;
- Technical plant (TECH)—used for making of the household and other objects.

Importance of each taxon was assessed using the use reports (UR) of the participants [54]. We visualized the multitude of utilization via Venn diagrams and compared the use of the recorded taxa using the Jaccard (similarity) index (JI) for each pair of use records. Calculation of the JI was performed using the following formula:

$$JI\,(X, Y) = |X \cap Y| / |X \cup Y|,$$

where X and Y signify every two datasets. JI ranges from 0 (no similarity) to 1 (total equality) [55].

Sources of seeds and/or planting material were categorized as: local forms (landraces, purchased from the market/retail, introduced from the wild into the garden or collected from the wild for direct consumption and/or other uses).

2.2. Data Analysis and Statistics

Statistical association between nominal and ordinal variables was evaluated through chi-square tests (Fisher's exact test) and correlation analysis (Spearman rank-order correlation coefficient). All statistical tests were based on two-sided tests and with a significant level of at least $\alpha = 0.05$. Statistical analyses were performed using the SPSS statistical package (ver. 20.0, SPSS, IBM Inc., Armonk, NY, USA).

3. Results

Participants were primarily seniors, 90% over 56 years of age, actively engaged in home gardening (Table 1). The sets of single living participants and families were similar in number, 37 and 38, respectively. Education of the participants was prevalently secondary or higher. Half of the retired participants (79% of all participants) were professionally involved in industrial agriculture during their active years. However, even those who used to work or were still working in other professions were involved in agricultural activities throughout their lives. The retired family members were those responsible for decisions related to the composition of the gardens and spent most of their time in household organization and agricultural activities. Younger family members were those who were supplying new and foreign varieties for the gardens or helping in the processing of the garden yield, as well as introducing new recipes and knowledge.

Table 1. Characteristics of the participants.

Characteristic	Total
Sex	
Single female; N (%)	26 (34.7)
Single male; N (%)	11 (14.7)
Family; N (%)	38(50.7)
(N [1])	75
Participant age (years)	
Median	70.00
Mean ± SD	68.44 ± 10.19
Age range (years)	35–84
35–55 N (%)	7 (11.3)
56–75 N (%)	40 (64.5)
76–84 N (%)	16 (25.8)
Education	
Primary; N (%)	14 (18.7)
Secondary; N (%)	34 (45.3)
College/University; N (%)	19 (25.3)

[1] Sample sizes vary due to missing data in the different variables. N: Sample size.

Most of the participants (69%) used two to four taxa of Lamiaceae, while those interested in six or more were relatively few (Figure 2). The number of those growing only one taxon was fairly small (three). These were usually people with very small gardens, keen in ornamental gardening with little or no interest in herbal teas. Age, sex, and education were not significantly correlated with the number of cultivated/used taxa ($p > 0.2$). The highest number of taxa (nine) was recorded in a twin house with a common home garden managed by four people. There were no participants that would collect Lamiaceae plants only from the wild.

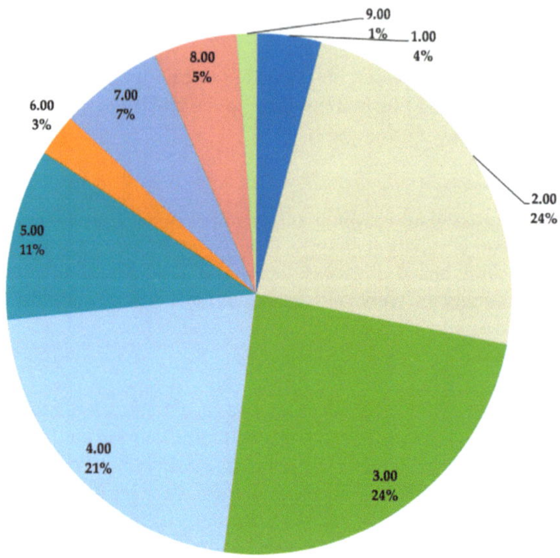

Figure 2. Number of used Lamiaceae taxa per participant (N, %).

Studied gardens were found to harbor small patches or separately planted individuals of 22 taxa of Lamiaceae, belonging to 15 genera (Table 2, Supplementary Table S1). Four of them were both cultivated and collected from the wild for personal/family use, while five species were collected by the participants only from the wild. *Salvia* was the most diverse genus with four cultivated species, followed by *Mentha* and *Satureja,* with three taxa each. The latter two genera concentrated most of the citations (76.5%). Third place was occupied by *Ocimum,* which was represented by two species, *Ocimum basilicum* L. and *O. minimum* L., the latter being rarely registered (28 and 4 mentions, respectively). *Clinopodium, Marrubium* and *Origanum* were presented by two taxa, one collected from the wild and one introduced into the gardens from the nearby populations and/or grown from seeds or planting material procured by informal exchange or from the market. Several participants demonstrated various *Thymus* sp. and *Mentha spicata* L. cultivars as well as *Origanum vulgare* subsp. *hirtum* (Link) Ietsw., imported from other European countries as culinary herbs. The remaining seven genera were represented by only one species.

Table 2. Cultivated and wild Lamiaceae taxa used in rural Bulgaria.

Genus	Taxon/Voucher Specimen Collection Reference	Origin	Source	Occurrence, % of Gardens (Landraces [3])	Provinces	UR	NU
Agastache	*Agastache foeniculum* (Pursh) Kuntze/BI300317_AF	I	C	2.7	Hs	5	4
Clinopodium	*Clinopodium dalmaticum* L./SOM177666; I040517_ClD	A	W	0.0	Sm	2	1
	Clinopodium vulgare L./B160617_ ClV	A	C, W	1.3	Pd	2	1
Lamium	*Lamium galeobdolon* (L.) Crantz/B140617_LG	I	C	1.3	Lv	1	1

Table 2. *Cont.*

Genus	Taxon/Voucher Specimen Collection Reference	Origin	Source	Occurrence, % of Gardens (Landraces [3])	Provinces	UR	NU
Lavandula	*Lavandula angustifolia* Mill./B300317_LAn;B140617_LAn;	I	C	13.3	Bl, Hs, Lv, Mo, Pv	19	4
Marrubium	*Marrubium peregrinum* L./BI300817_MP	A	W	0.0	Hs	1	1
	Marrubium vulgare L./B140617_MV	A	C	1.3	Bl	1	1
Melissa	*Melissa officinalis* L./BI140617_MO; BI300317_MO; B140617_MO; BY300717_MO; B160418_MO	A	C	24.0	Bl, Hs, Lv, Mo, Pv	26	4
Mentha	*Mentha pulegium* L./B230419_MPu	A	W	0.0	Bl	3	2
	Mentha spicata L./B260719_Ms; B030821_MS; B040821_MS; B300317_MS; B040517_MS; B130617_MS; B140617_MS; B280617_MS; B300817_MS; B100719_MS; BI260819_MS	A, I	C	98.7 (68)	All	118	5
	Mentha × *piperita* L./B300317_MxP; B040517_ MxP; B130617_ MxP; B140617_ MxP; B280617_ MxP; B300817_ MxP; B100719_ MxP; BI260819_ MxP	I	C	26.7	Bl, Hs, Lv, Mo, Pd, Pv, Sm	35	3
Ocimum	*Ocimum basilicum* L./SOM177657; SOM177659; B260819_OB; I101019_OB;	I	C	37.3 (16)	All	52	7
	Ocimum minimum L./SOM177655; SOM177660; B050717_OM; I110917_OM; I120917_OM	I	C	5.3 (1)	Hs, Sm, Vr	7	3
Origanum	*Origanum vulgare* L./B260819_OV; B280819_OV; B290819_OV,	A	W	0.0	Bl	3	2
	Origanum vulgare subsp. *hirtum* (Link) Ietsw./SOM177661; SOM177662	A, I [1]	C, W	21.3	Hs, Lv, Pd, Pv, Sm	25	5
Rosmarinus	*Rosmarinus officinalis* L./B300317_RO; B010417_RO	I	C	18.7	Bl, Hs, Mo, Pd, Pv	28	5
Salvia	*Salvia aethiopis* L./B290819_SAE	A	C	1.3	Bl	1	1
	Salvia officinalis L./BI160617_SO; I300817_SO; B260819_SO; B030821_SO	I	C	12.0	Bl, Hs, Lv, Pd, Sm	17	3
	Salvia splendens Sellow ex J.A. Schultes/B130617_SSp	I	C	1.3	Lv	1	1
Satureja	*Salvia viridis* L./BI120917_SV	I	C	1.3	Sm	1	1
	Satureja cuneifolia Ten./SOM177658	A	W	0.0	Bl	1	1
	Satureja hortensis L./SOM177656; BI150617_SH; BI260819_SH; I020821_SH; B030821_SH	I	C	74.7 (55)	All	57	2
	Satureja pilosa Velen./SOM177663 (c); SOM177664 (w); SOM177665 (w)	A	C, W	1.3	Hs	3	1
Sideritis	*Sideritis scardica* Gris./B310317_ SSc; B310317_SSc	A [2]	C	5.3	Hs	7	2
Stachys	*Stachys byzantina* K. Koch/B140617_StB	I	C	2.7	Bl, Mo	2	1
Thymus	*Thymus* sp./I040517_Th	A, I	C, W	2.7	Bl, Sm	5	3
	Thymus vulgaris L./B040517_TV; B280617_TV; B100719_TV; B030821_TV	I	C	5.3	Pd, Pv, Sm, Vr	6	3

[1] Cultivated plants were introduced from nearby wild populations or were grown from seeds/planting material procured from elsewhere; [2] cultivated plants were grown outside the natural distribution of the species; [3] number of local forms (landraces) if any; A: autochthonous; I: introduced; C: cultivated; W: wild; UR: use-reports; NU: number of uses; provinces (Bl: Blagoevgrad; Hs: Haskovo; Lv: Lovech; Mo: Montana; Pd: Plovdiv; Pv: Pleven; Sm: Smolyan; and Vr: Vratsa).

Altogether, ten taxa were accountable for the 88.1% of all mentions for utilization of Lamiaceae plants led by *Mentha spicata* L. and *Satureja hortensis* L. They also encompassed most of the local forms (landraces) that participants were maintaining in their gardens, 68 and 55, respectively. *Ocimum basilicum* and *O. minimum* were the other taxa of which local forms were preferred.

Each participant cited no more than four uses per taxon, which was related not only to the overall use of the taxon, but also to differences in utilization of varieties/local forms. *Ocimum basilicum* was the species with the highest number of uses (seven uses). Additionally, *O. basilicum* was, respectively, mentioned two and three times less frequently than the most popular—*M. spicata* and *S. hortensis*. More than half of the taxa (55.5%) had one or two uses of which 11 taxa had only one use. Culinary use was about twice as popular as the medicinal one.

Lamiaceae plants were used most often (61.4% of the use-reports) as culinary herbs, followed by those appreciated for their ornamental value (39.3%). The similarity between CUL and MED categories on a taxonomical level was high (JI = 0.83) (Figure 3). However, the varieties/local forms demonstrated as culinary herbs were rarely appreciated for other purposes, with negative correlation being significant (Spearman rank correlation $p < 0.01$, Table 3) for four out of seven categories (ORN, MED, AROM, and SYM).

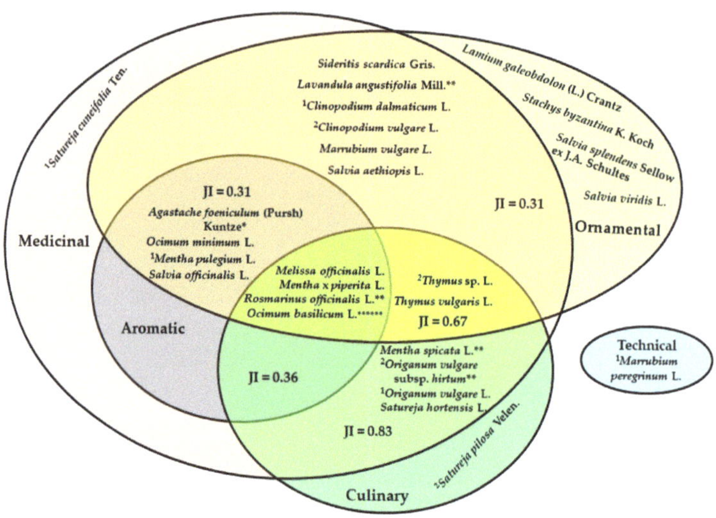

Figure 3. Multipurpose utilization of Lamiaceae representatives in rural Bulgaria. [1] Taxon collected only from the wild; [2] Taxon cultivated and collected from the wild; pollinator attraction; * repellent; ****** are the three pollinator attraction; * repellent; ** (*Ocimum basilicum* L. only) symbolic together; JI: Jaccard similarity index.

For example, *O. basilicum* was more frequently mentioned as an ornamental and aromatic plant than as a culinary one for which mostly foreign broad-leaf varieties were considered. On the other hand, only locally grown white flowering forms with small leaves were used for ritual purposes, but they were not consumed (Figure 4). *Ocimum basilicum* was the only Lamiaceae representative used symbolically in different rituals and for its protective (apotropaic) powers. Flowering *O. basilicum* branches were used by Ortodox Christians as church and home decoration and in various ceremonies as well as in rituals related to birth, weddings, and burials.

Table 3. Usage of Lamiaceae plants in rural Bulgaria and correlation between different usages.

Use	CUL [1]	MED	AROM	REP	POLL	TECH	SYM	Total Number of Use-Reports	Total Number of Taxa
ORN	−0.307 **	0.081	0.169 **	0.100	0.018	−0.048	0.082	112	17
CUL		−0.562 **	−0.331 **	−0.103	−0.106	−0.075	−0.185 **	175	11
MED			−0.03	0.003	0.036	−0.039	−0.043	86	21
AROM				−0.029	0.099	−0.022	0.172 **	34	10
REP					0.183 **	−0.013	0.085	13	5
POLL						−0.005	−0.012	2	2
TECH							−0.009	1	1
SYM								6	1

[1] Abbreviations—CUL: culinary herb: ORN: ornamental plant; MED: medicinal plant, incl. herbal teas; AROM: aromatic plant; REP: insect repellent; POLL: pollinator attraction; SYM: symbolic plant used ritually; TECH: technical plant (brooms). ** Spearman correlation coefficient is significant at the 0.01 level (2-tailed).

Figure 4. *Ocimum basilicum* L. in Bulgarian rural livelihoods.

Top (left to right): local landrace 'Zhenski bosilek' (women's basil) used as ritual and medicinal plant in Banichan village; *Ocimum basilicum* L. and *O. minimum* L. from retail seeds in Central Rhodope Mts.; and the drying of flowering whole plants of *O. basilicum*. Bottom (left to right): ornamental patch of *O. basilicum* and *Gomphrena globosa* L.; *O. basilicum* sown in front of greenhouse to attract pollinators; and the ritual decoration of *O. basilicum* for home protection.

Half of the studied taxa (11) were reported to be edible with only one collected exclusively from the wild (*Origanum vulgare* L.). The latter was also cited as spirit flavouring and as a medicinal plant. *Mentha spicata*, *S. hortensis*, and *O. vulgare* subsp. *hirtum* were most frequently used for preparation of traditional dishes. All culinary herbs were used dried, and only *M. spicata* (Figure 5) was also traditionally used fresh in the preparation of lamb meat, bean, and vegetable stews. Dried herbs were stocked in bunches or sealed in containers and bags in quantities matching the usual demands of the household until the next harvest. Fresh herbage or leaves of other taxa were used only for preparations of foreign dishes like *O. basilicum* for Italian dishes and salads, *Mentha* × *piperita* L. and *Melissa officinalis* L. for desserts and cocktails, as well as *Thymus* sp. and *Rosmarinus officinalis* L. for roasted meats. Interestingly, the latter was appreciated for its aromatic and decorative foliage rather than as a culinary herb, and some of our respondents shared that they are aware of its culinary use but never used it in their cooking.

Figure 5. Variety of *Mentha spicata* L. local forms (landraces) in Bulgarian rural home gardens.

Twenty-one taxa were used for their healing properties, with *M.* × *piperita*, *M. officinalis*, and *Salvia officinalis* being the most commonly mentioned. Still, only 30.2% of the reports were related to the healing properties of Lamiaceae taxa. For medicinal purposes, dried plants were used as infusions. Some of the participants consumed them daily as healthy herbal teas, while others associate herbal teas with childhood illnesses and reject to cultivate and/or collect medicinal plants. Taxa used for household aromatization (10 taxa), insect repellents (5 taxa), and as pollinator attraction/honey plants (2 taxa) overlapped completely with medicinal plant category. Dried *Lavandula angustifolia* Mill. and *Mentha spicata* bunches were reported as effective in the repelling of clothes moths. *Origanum vulgare* subsp. *hirtum* and *M. spicata* were demonstrated as pantry pest repellents, placed directly in the containers with legumes, grains, etc. *Ocimum balisicum* and *Rosmarinus officinalis* were grown in the gardens next to windows to repel mosquitoes.

Only one species was mentioned as a technical plant; *Marrubium peregrinum* L., which was used for garden brooms that were made of large, sturdy herbage collected towards the end of vegetation. (Figure 6). However, the local knowledge on broom making was fading away, demonstrated by only one participant. On the other hand, the use of Lamiaceae plants to attract pollinators in greenhouses so as to promote tomato pollination should be regarded as a relatively new practice, popular among gardeners interested in eco-friendly agriculture.

Figure 6. Home yard broom made of *Marrubium peregrinum* L.

The culinary use of some species resulted in the preservation of numerous local forms (Table 4). Contrastingly to the ornamental, aromatic and even medicinal plants that were sourced mainly from the market, seeds, rhizomes, and/or plantlets for culinary herbs were predominantly personally propagated. These were handed down through the generations and were replaced through exchange within the community only if inherited plants were lost. The practice of introducing plants from wild populations into home gardens was most popular for medicinal plants. Among these taxa, *Sideritis scardica* Gris., a species that became popular for its health benefits in recent years, was the only one that was transferred outside the area of its natural distribution. Cultivation of Lamiaceae species was preferred not only for the convenience and availability, but also due to the popularity of the protected status of some of the medicinal plants (e.g., *O. vulgare* ssp. *hirtum*, *S. scardica*). Additionally, our participants were more inclined to introduce wild-growing culinary herbs into their gardens than to lose time visiting natural populations.

Table 4. Sources of Lamiaceae taxa used in rural Bulgaria.

Source	ORN [1]	CUL	MED	AROM	REP	POLL	TECH	SYM
Local forms [2]	40	126	8	10	4	0	0	6
Introduced from the wild	6	7	20	3	1	0	0	0
Retail	51	29	38	13	5	2	0	0
Informal exchange	13	5	8	7	1	0	0	0
Collected from the wild [3]	0	5	9	1	0	0	1	0
Fisher exact test, significance (p)	<0.001	<0.001	<0.001	0.008	0.591	0.374	0.06	0.310

[1] Abbreviations—CUL: culinary herb: ORN: ornamental plant; MED: medicinal plant, incl. herbal teas; AROM: aromatic plant; REP: insect repellent; POLL: pollinator attraction; SYM: symbolic plant used ritually; TECH: technical plant (brooms); [2] data are presented as a number of reports; [3] plants collected regularly from the wild for direct consumption or other use.

Additionally to cultivated Lamiaceae taxa studied gardens harbored several wild Lamiaceae species (*Ballota nigra* L., *Lamium amplexicaule* L., *L. maculatum* L., *L. purpureum* L., *Leonurus cardiaca* L., *Glechoma hederacea* L., *Salvia nemorosa* L., *S. pratensis* L. *S. verticillata* L.) that were considered by the participants as weeds without practical application.

4. Discussion

The overall number of Lamiaceae family members (22) found cultivated in Bulgarian rural home gardens was similar to that reported for other European home gardens, however, none of the taxa mentioned in these studies were present even in half of the studied gardens [46,56]. In the current study, two species, namely *M. spicata* and *S. hortensis*, were almost compulsorily present in the home hardens, grown in 98% and 74% of the studied gardens, respectively. Although Lamiaceae taxa were grown in limited quantities in Bulgarian home gardens, it is noteworthy to mention that their occurrence was comparable with many of the garden vegetables that were found to be main crops in cultivation [45,57]. Even some Lamiaceae taxa, which are not so popular as culinary herbs in Bulgaria (e.g., *O. basilicum*, *R. officinalis*), were also found more frequently cultivated in the Bulgarian home gardens than in other European countries [29,48,58,59]. The observed high diversity of Lamiaceae taxa in Bulgarian rural home gardens should be attributed not only to the fact that nearly 70% of our participants cultivated two to four Lamiaceae taxa in their gardens, but also to the very high preference for local forms (landraces) of culinary herbs that were inherited from previous generations or obtained through informal exchange within their community. Utilization of local genetic resources for their local organoleptic perceptions and their perceived cultural value was previously shown to contribute to their in situ preservation, which creates additional opportunities for the development of local entrepreneurship [60,61]. Still, the relatively high age of the rural population in Bulgaria should be considered as an alarming factor in terms of the need for the development of targeted collection programs for the safeguarding of these resources [62].

Mentha spicata and *S. hortensis*, the culinary herbs that occupied the first two places among the reported taxa, also had the highest number of local forms. On the other hand, other, incl. wild-growing members of Lamiaceae, such as *Origanum*, *Thymus*, and *Mentha*, popularly used in different dishes, preserved and fermented foods in the Mediterranean area were far less consumed by Bulgarians [63–66]. Specific plant spices and their combinations play key roles in the local cuisines that underline their uniqueness [67–69]. Practicing of traditional (agro)ecological knowledge is of crucial importance, as demographic and socioeconomic changes in the rural areas, especially in industrialized societies, gradually diminish natural human connectedness [4,70,71]. While utilization of *M. spicata* was reported from all over the world [29,72–74], *S. hortensis* was found popular both as a culinary herb and a medicinal plant mainly in the Balkans, Iran, and Turkey [75–77]. In this sense traditional culinary use underpins the maintenance of local Lamiaceae diversity. Recent results from the Adriatic area show that traditional knowledge related to usage of certain species is not only diminishing, but also changes in agricultural practices and land use are making home gardens the main source of medicinal plants, rather than the wild [78]. Still, similarly to other studies, it is hard to designate every inherited form with a landrace status, as no genetic and/or phytochemical analyses were performed during the study [56]. Nonetheless, the high number of gardens in which *M. spicata* and *S. hortensis* were recorded highlights the importance of home gardens for the preservation of local plant genetic resources, and also urges for their more detailed characterization, given the numerous factors that cause genetic erosion and loss of crop diversity [59,79].

Traditionally, consumed plants that have two or more other applications and are relatively easy to grow would possibly attract more attention even from unexperienced gardeners, which would contribute to the broadening of the impact of home gardens for the safeguarding of local plant diversity [80–82]. In the current study, single use was mostly a signifier for recent introduction of taxa into the gardens (e.g., ornamental varieties) or of outdated practices such as the preparation of brooms from the herbage of *Marrubium peregrinum*. Others, such as *Clinopodium dalmaticum* L., have a limited distribution range in the country, which could be regarded as a restrictive factor for its usability [83]. Here, it is important to distinguish the *O. basilicum*, which was the only species with symbolic (ritual) use. Sweet basil is one of the most important ritual plants in Bulgarian Orthodox Christian traditions [84], and we found the local forms used for ritual purposes to be rarely consumed fresh or cooked. Similarly, *O. minimum* was perceived mostly as an aromatic and ornamental plant. The discerning and preservation of landraces and varieties of both *O. basilicum* and *O. minimum*, however, would be a complicated enterprise due to the high variability in genome size and chromosome number, plant morphology, essential oil profile, etc., that have created considerable taxonomical ambiguities throughout the years and were related also to the frequent intrageneric hybridization and extensive breeding of *Ocimum* [85,86].

While almost all of the reported taxa were known as medicinal plants, the number of use reports related to medicinal properties was less than 1/3. The number of taxa used in this category (21) was about twice as low when compared to ethnographical data previously reported for Bulgaria, however close to the number of Lamiaceae taxa used in other parts of the Balkans as medicinal plants [77,87,88]. Given the fact that in Bulgarian folk medicine several species of one taxa could be used for various ailments, sometimes interchangeably (e.g., *Lamium album* L., *L. maculatum*, *L. purpureum*), it is probable that knowledge of the uses of other members of Lamiaceae have remained preserved outside the assessed areas/among other communities [77]. Nonetheless, we should consider the transformative role of the publication of numerous phytotherapy books that provided the Bulgarian public with modern and/or foreign knowledge on the use of indigenous and introduced plant taxa since the 1950s [89–92]. Many of these books also provided information on the threats to natural populations of some medicinal plants and recommend more responsible utilization of the natural resources. The latter could also explain the low number of wild medicinal plants of the Lamiaceae family (8) used traditionally by Bulgarian farmers in the early 1990s [93]. Additionally, only five species of the Lamiaceae family, all of which were also shown here, were reported for recreational teas used in Bulgaria [94]. It is important to mention that Bulgarian folk medicine was, and in some places still is, typically practiced by skilled healers (called *bilkari*, *bilyari*, *znahari*, etc.) who collect, supply and often process medicinal plants that are further used by the patients. Hence, a substantial part of Bulgarian traditional knowledge related to medicinal plants was never largely available [77]. This could explain why the plants mentioned by our participants as medicinal were mainly sourced from the market and were exclusively consumed as infusions, most of which are taken as daily herbal teas. Common examples for such plants are *Thymus* spp., *Origanum vulgare*, and *Sideritis scardica*, which are popular winter teas sold allover Bulgaria. However, the traditional knowledge related to their use would be hard to trace. Complementary to the initial interest in the medicinal properties, we observed that the purchased plants were often appreciated as ornamental and aromatic plants. Still, utilization of Lamiaceae taxa for their healing properties was found to be an important factor for plant domestications in gardens which was found to be a useful approach to alleviate pressure on the natural populations of medicinal plants [33,95,96].

In conclusion, the cultivation of members of the Lamiaceae family in Bulgarian home gardens accommodates both the various needs and interests of their owners, as well as the preservation of wild and agrodiversity. While the medicinal properties of these taxa cater to higher diversity on a species level, local culinary practices were found to sustain the variety of local forms (landraces) that underline the role of home gardens as important pools of plant genetic resources that should be preserved and further explored in the frame of the multitude of benefits provided by these plants.

Supplementary Materials: The following supporting information can be downloaded at: https://www.mdpi.com/article/10.3390/agronomy12071631/s1; Table S1: Use reports and sources of the Lamiaceae taxa used in rural Bulgaria.

Author Contributions: Conceptualization, T.I. and D.D.; methodology, T.I. and D.D.; formal analysis, Y.B.; investigation, T.I., Y.B., M.C. and D.D.; data curation, Y.B.; writing—original draft preparation, T.I.; writing—review and editing, T.I., Y.B., M.C. and D.D. All authors have read and agreed to the published version of the manuscript.

Funding: The study is funded by the Bulgarian Ministry of Education and Science under the National Research Programme "Healthy Foods for a Strong Bio-Economy and Quality of Life" approved by DCM # 577/17.08.2018. Part of the field studies were supported under the project DN10/1/2016 "The Garden: Site of Biocultural Diversity and Interdisciplinary Junction" funded by the National Science Fund.

Institutional Review Board Statement: The study ethical compliance was confirmed by the Scientific Council of the Institute of Biodiversity and Ecosystem Research, Bulgarian Academy of Sciences, acting as independent institutional Ethics Board (Decision No. 6/21/05/21).

Informed Consent Statement: Informed consent was obtained from all participants involved in the study prior the interviews.

Acknowledgments: Authors are grateful to all participants for the shared knowledge and kind cooperation during the field study.

Conflicts of Interest: The authors declare no conflict of interest.

References

1. Royal Botanic Gardens Kew Garden; Missouri Botanical Garden The Plantlist Database. Available online: http://www.theplantlist.org (accessed on 26 May 2022).
2. Napoli, E.; Siracusa, L.; Ruberto, G. New Tricks for Old Guys: Recent Developments in the Chemistry, Biochemistry, Applications and Exploitation of Selected Species from the Lamiaceae Family. *Chem. Biodivers.* **2020**, *17*, e1900677. [CrossRef] [PubMed]
3. Waller, S.B.; Cleff, M.B.; Serra, E.F.; Silva, A.L.; Dos Reis Gomes, A.; de Mello, J.R.B.; de Faria, R.O.; Meireles, M.C.A. Plants from Lamiaceae family as source of antifungal molecules in humane and veterinary medicine. *Microb. Pathog.* **2017**, *104*, 232–237. [CrossRef] [PubMed]
4. Turner, N.J.; Luczaj, L.J.; Migliorini, P.; Pieroni, A.; Dreon, A.L.; Sacchetti, L.E.; Paoletti, M.G. Edible and tended wild plants, traditional ecological knowledge and Agroecology. *CRC Crit. Rev. Plant Sci.* **2011**, *30*, 198–225. [CrossRef]
5. Cilliers, L.; Retief, F.P. Horticulture in antiquity, with emphasis on the Graeco-Roman era | Akroterion. *Akroterion* **2009**, *54*, 1–10.
6. Hrisrova, I.; Atanassova, J.; Marinova, E. Plant economy and vegetation of the Iron Age in Bulgaria: Archaeobotanical evidence from pit deposits. *Archaeol. Anthropol. Sci.* **2017**, *9*, 1481–1494. [CrossRef]
7. Bosi, G.; Castiglioni, E.; Rinaldi, R.; Mazzanti, M.; Marchesini, M.; Rottoli, M. Archaeobotanical evidence of food plants in Northern Italy during the Roman period. *Veg. Hist. Archaeobot.* **2020**, *29*, 681–697. [CrossRef]
8. Ntalli, N.G.; Ozalexandridou, E.X.; Kasiotis, K.M.; Samara, M.; Golfinopoulos, S.K. Nematicidal Activity and Phytochemistry of Greek Lamiaceae Species. *Agronomy* **2020**, *10*, 1119. [CrossRef]
9. Do Nascimento, L.D.; de Moraes, A.A.B.; da Costa, K.S.; Galúcio, J.M.P.; Taube, P.S.; Costa, C.M.L.; Cruz, J.N.; de Aguiar Andrade, E.H.; de Faria, L.J.G. Bioactive natural compounds and antioxidant activity of essential oils from spice plants: New findings and potential applications. *Biomolecules* **2020**, *10*, 988. [CrossRef]
10. Ghalkhani, A.; Moradkhani, S.; Soleimani, M.; Dastan, D. Functional components, antibacterial, antioxidant, and cytotoxic activities of *Lamium garganicum* L. ssp. pictum as a novel natural agents from Lamiaceae family. *Food Biosci.* **2021**, *43*, 101265. [CrossRef]
11. Dzhambazov, B.; Daskalova, S.; Monteva, A.; Popov, N. In Vitro Screening for Antitumour Activity of *Clinopodium vulgare* L. (Lamiaceae) Extracts. *Biol. Pharm. Bull.* **2002**, *25*, 499–504. [CrossRef]
12. Mladenova, T.; Stoyanov, P.; Denev, P.; Dimitrova, S.; Katsarova, M.; Teneva, D.; Todorov, K.; Bivolarska, A. Phytochemical Composition, Antioxidant and Antimicrobial Activity of the Balkan Endemic *Micromeria frivaldszkyana* (Degen) Velen. (Lamiaceae). *Plants* **2021**, *10*, 710. [CrossRef] [PubMed]
13. Mustafa, B.; Hajdari, A.; Krasniqi, F.; Hoxha, E.; Ademi, H.; Quave, C.L.; Pieroni, A. Medical ethnobotany of the Albanian Alps in Kosovo. *J. Ethnobiol. Ethnomed.* **2012**, *8*, 6. [CrossRef] [PubMed]
14. Papp, N.; Birkás-Frendl, K.; Farkas, Á.; Pieroni, A. An ethnobotanical study on home gardens in a Transylvanian Hungarian Csángó village (Romania). *Genet. Resour. Crop Evol.* **2012**, *60*, 1423–1432. [CrossRef]
15. Mükemre, M.; Behçet, L.; Çakilcioğlu, U. Ethnobotanical study on medicinal plants in villages of Çatak (Van-Turkey). *J. Ethnopharmacol.* **2015**, *166*, 361–374. [CrossRef]

16. Sarri, M.; Boudjelal, A.; Hendel, N.; Sarri, D.; Benkhaled, A. Flora and ethnobotany of medicinal plants in the southeast of the capital of Hodna (Algeria). *Arab. J. Med. Aromat. Plants* **2015**, *1*, 24–30. [CrossRef]
17. Menković, N.; Šavikin, K.; Tasić, S.; Zdunić, G.; Stešević, D.; Milosavljević, S.; Vincek, D. Ethnobotanical study on traditional uses of wild medicinal plants in Prokletije Mountains (Montenegro). *J. Ethnopharmacol.* **2011**, *133*, 97–107. [CrossRef]
18. Šavikin, K.; Zdunić, G.; Menković, N.; Živković, J.; Ćujić, N.; Tereščenko, M.; Bigović, D. Ethnobotanical study on traditional use of medicinal plants in South-Western Serbia, Zlatibor district. *J. Ethnopharmacol.* **2013**, *146*, 803–810. [CrossRef]
19. Sõukand, R.; Pieroni, A. The importance of a border: Medical, veterinary, and wild food ethnobotany of the Hutsuls living on the Romanian and Ukrainian sides of Bukovina. *J. Ethnopharmacol.* **2016**, *185*, 17–40. [CrossRef]
20. Pieroni, A.; Quave, C.L.; Giusti, M.E.; Papp, N. "We Are Italians!": The Hybrid Ethnobotany of a Venetian Diaspora in Eastern Romania. *Hum. Ecol.* **2012**, *40*, 435–451. [CrossRef]
21. Pieroni, A.; Nedelcheva, A.; Dogan, Y. Local knowledge of medicinal plants and wild food plants among Tatars and Romanians in Dobruja (South-East Romania). *Genet. Resour. Crop Evol.* **2014**, *62*, 605–620. [CrossRef]
22. Nikolić, M.; Jovanović, K.K.; Marković, T.; Marković, D.; Gligorijević, N.; Radulović, S.; Soković, M. Chemical composition, antimicrobial, and cytotoxic properties of five *Lamiaceae essential* oils. *Ind. Crops Prod.* **2014**, *61*, 225–232. [CrossRef]
23. Chen, Y.; Liu, L.; Wang, H.; Ma, J.; Peng, W.; Li, X.; Lai, Y.; Zhang, B.; Zhang, D. Environmentally friendly plant essential oil: Liquid gold for human health. *Adv. Agron.* **2021**, *170*, 289–337. [CrossRef]
24. Kowalski, R.; Kowalska, G.; Jankowska, M.; Nawrocka, A.; Kałwa, K.; Pankiewicz, U.; Włodarczyk-Stasiak, M. Secretory structures and essential oil composition of selected industrial species of Lamiaceae. *Acta Sci. Pol. Hortorum Cultus* **2019**, *18*, 53–69. [CrossRef]
25. Schippmann, U.; Leaman, D.J.; Cunningham, A.B. Impact of Cultivation and Gathering of Medicinal Plants on Biodiversity: Global Trends and Issues. In *Biodiversity and the Ecosystem Approach in Agriculture, Forestry and Fisheries*; FAO: Rome, Italy, 2002; pp. 1–21. ISBN 92-5-104917-3.
26. Applequist, W.L.; Brinckmann, J.A.; Cunningham, A.B.; Hart, R.E.; Heinrich, M.; Katerere, D.R.; Van Andel, T. Scientists' Warning on Climate Change and Medicinal Plants. *Planta Med.* **2020**, *86*, 10–18. [CrossRef]
27. Ortiz-Sánchez, A.; Monroy-Ortiz, C.; Romero-Manzanares, A.; Luna-Cavazos, M.; Castillo-España, P. Multipurpose functions of home gardens for family subsistence. *Bot. Sci.* **2015**, *93*, 791–806. [CrossRef]
28. Rao, M.R.; Rao, B.R.R. Medicinal plants in tropical homegardens. In *Tropical Homegardens. Advances in Agroforestry*; Kumar, B.M., Nair, P.K.R., Eds.; Springer: Dordrecht, The Netherlands, 2006; pp. 205–232. ISBN 978-1-4020-4947-7.
29. Vogl, C.R.; Vogl-Lukasser, B. Tradition, Dynamics and Sustainability of Plant Species Composition and Management in Homegardens on Organic and Non-Organic Small Scale Farms in Alpine Eastern Tyrol, Austria. *Biol. Agric. Hortic.* **2003**, *21*, 349–366. [CrossRef]
30. Kumar, B.M.; Nair, P.K.R. The enigma of tropical homegardens. *Agrofor. Syst.* **2004**, *61*, 135–152.
31. Galluzzi, G.; Eyzaguirri, P.; Negri, V. Uncovering European home gardens: Their human and biological featurs and potential contribution to the conservation of agro-biodiversity. In *Crop Genetic Resources in European Home Gardens*; Bailey, A., Eyzaguirre, P., Maggioni, L., Eds.; Bioversity International: Rome, Italy, 2009; pp. 8–17.
32. Birol, E.; Bela, G.; Smale, M. The Role of Home Gardens in Promoting Multi-Functional Agriculture in Hungary. *EuroChoices* **2005**, *4*, 14–21. [CrossRef]
33. Calvet-Mir, L.; Gómez-Baggethun, E.; Reyes-García, V. Beyond food production: Ecosystem services provided by home gardens. A case study in Vall Fosca, Catalan Pyrenees, Northeastern Spain. *Ecol. Econ.* **2012**, *74*, 153–160. [CrossRef]
34. Calvet-Mir, L.; Riu-Bosoms, C.; González-Puente, M.; Ruiz-Mallén, I.; Reyes-García, V.; Molina, J.L. The Transmission of Home Garden Knowledge: Safeguarding Biocultural Diversity and Enhancing Social–Ecological Resilience. *Soc. Nat. Resour.* **2016**, *29*, 556–571. [CrossRef]
35. Raymond, C.M.; Diduck, A.P.; Buijs, A.; Boerchers, M.; Moquin, R. Exploring the co-benefits (and costs) of home gardening for biodiversity conservation. *Local Environ.* **2018**, *24*, 258–273. [CrossRef]
36. Ulug, C.; Horlings, L.G. Connecting resourcefulness and social innovation: Exploring conditions and processes in community gardens in the Netherlands. *Local Environ.* **2018**, *24*, 147–166. [CrossRef]
37. Ghosh, S. Measuring sustainability performance of local food production in home gardens. *Local Environ.* **2014**, *19*, 33–55. [CrossRef]
38. Altieri, M.A.; Nicholls, C.I. The adaptation and mitigation potential of traditional agriculture in a changing climate. *Clim. Chang.* **2013**, *140*, 33–45. [CrossRef]
39. Altieri, M.A.; Nicholls, C.I.; Henao, A.; Lana, M.A. Agroecology and the design of climate change-resilient farming systems. *Agron. Sustain. Dev.* **2015**, *35*, 869–890. [CrossRef]
40. Alzate, C.; Mertens, F.; Fillion, M.; Rozin, A. The study and use of traditional knowledge in agroecological contexts. *Rev. Fac. Cienc. Agrar. UNCuyo* **2019**, *51*, 337–350.
41. Gómez-Baggethun, E.; Reyes-García, V.; Olsson, P.; Montes, C. Traditional ecological knowledge and community resilience to environmental extremes: A case study in Doñana, SW Spain. *Glob. Environ. Chang.* **2012**, *22*, 640–650. [CrossRef]
42. Tataridas, A.; Kanatas, P.; Chatzigeorgiou, A.; Zannopoulos, S.; Travlos, I. Sustainable Crop and Weed Management in the Era of the EU Green Deal: A Survival Guide. *Agronomy* **2022**, *12*, 589. [CrossRef]

43. Burholt, V.; Dobbs, C. Research on rural ageing: Where have we got to and where are we going in Europe? *J. Rural Stud.* **2012**, *28* 432–446. [CrossRef]
44. Giannakis, E.; Bruggeman, A. Regional disparities in economic resilience in the European Union across the urban–rural divide *Reg. Stud.* **2020**, *54*, 1200–1213. [CrossRef]
45. Ivanova, T.; Bosseva, Y.; Chervenkov, M.; Dimitrova, D. Enough to Feed Ourselves!—Food Plants in Bulgarian Rural Home Gardens. *Plants* **2021**, *10*, 2520. [CrossRef] [PubMed]
46. Vogl-Lukasser, B.; Vogl, C.R. The changing face of farmers' home gardens: A diachronic analysis from Sillian (Eastern Tyrol Austria). *J. Ethnobiol. Ethnomed.* **2018**, *14*, 63. [CrossRef] [PubMed]
47. Bassullu, C.; Tolunay, A. General characteristics of traditional homegarden involving animal practices in rural areas of Isparta Region of Turkey. *J. Anim. Vet. Adv.* **2010**, *9*, 455–465. [CrossRef]
48. Reyes-García, V.; Aceituno, L.; Vila, S.; Calvet-Mir, L.; Garnatje, T.; Jesch, A.; Lastra, J.J.; Parada, M.; Rigat, M.; Vallès, J.; et al Home Gardens in Three Mountain Regions of the Iberian Peninsula: Description, Motivation for Gardening, and Gross Financial Benefits. *J. Sustain. Agric.* **2012**, *36*, 249–270. [CrossRef]
49. Bocheva, L.; Trifonova, L.; Marinova, T.; Malcheva, K. Climate profile of Bulgaria in the period 1988-2016 and brief climatic assessment of 2017. *Bulg. J. Meteorol. Hydrol.* **2017**, *22*, 2–15.
50. Petrova, A.; Vladimirov, V. Recent progress in floristic and taxonomic studies in Bulgaria. *Bot. Serbica* **2018**, *42*, 35–69. [CrossRef]
51. Stoyanov, S.; Marinov, Y. *Thymus jalasianus* (Lamiaceae), a New Species from the Serpentine Area of the Eastern Rhodope Mountains, Bulgaria. *Ann. Bot. Fenn.* **2020**, *57*, 163–172. [CrossRef]
52. International Society of Ethnobiology ISE Code of Ethics. Available online: http://ethnobiology.net/code-of-ethics/ (accessed on 21 June 2021).
53. Delipavlov, D.; Cheshmedzhiev, I.; Popova, M.; Terziyski, D.; Kovachev, I. *Handbook of Bulgarian Vascular Flora*; Delipavlov, D. Cheshmedzhiev, I., Eds.; Academic Publishing House of Agricultural University-Plovdiv: Plovdiv, Bulgaria, 2003.
54. Tardío, J.; Pardo-De-Santayana, M. Cultural Importance Indices: A Comparative Analysis Based on the Useful Wild Plants of Southern Cantabria (Northern Spain). *Econ. Bot.* **2008**, *62*, 24–39. [CrossRef]
55. Jaccard, P. The distribution of the flora in the alpine zone. *New Phytol.* **1912**, *11*, 37–50. [CrossRef]
56. Calvet-Mir, L.; Calvet-Mir, M.; Vaqué-Nuñez, L.; Reyes-García, V. Landraces in situ Conservation: A Case Study in High-Mountain Home Gardens in Vall Fosca, Catalan Pyrenees, Iberian Peninsula. *Econ. Bot.* **2011**, *65*, 146–157. [CrossRef]
57. Toleva-Novak, N. Spatial Aspects. Urban Gardening as Astrategy for Improvement of the Urban Community Well-Being Available online: https://www.youtube.com/watch?v=VKqEmkkElsg&t=1793s (accessed on 20 April 2021).
58. Grasser, S.; Schunko, C.; Vogl, C.R. Gathering "tea"—From necessity to connectedness with nature. Local knowledge about wild plant gathering in the Biosphere Reserve Grosses Walsertal (Austria). *J. Ethnobiol. Ethnomed.* **2012**, *8*, 31. [CrossRef] [PubMed]
59. Vitasović-Kosić, I.; Kaligarič, M.; Juračak, J. Divergence of Ethnobotanical Knowledge of Slovenians on the Edge of the Mediterranean as a Result of Historical, Geographical and Cultural Drivers. *Plants* **2021**, *10*, 2087. [CrossRef] [PubMed]
60. Riu-Bosoms, C.; Calvet-Mir, L.; Reyes-García, V. Factors Enhancing Landrace in Situ Conservation in Home Gardens and Fields in Vall de Gósol, Catalan Pyrenees, Iberian Peninsula. *J. Ethnobiol.* **2014**, *34*, 175–194. [CrossRef]
61. Thanopoulos, R.; Chatzigeorgiou, T.; Argyropoulou, K.; Kostouros, N.M.; Bebeli, P.J. State of Crop Landraces in Arcadia (Greece) and In-Situ Conservation Potential. *Diversity* **2021**, *13*, 558. [CrossRef]
62. Davidova, S.; Fredriksson, L.; Gorton, M.; Mishev, P.; Petrovici, D. Subsistence Farming, Incomes, and Agricultural Livelihoods in the New Member States of the European Union. *Environ. Plan. C* **2012**, *30*, 209–227. [CrossRef]
63. Hadjichambis, A.C.; Paraskeva-Hadjichambi, D.; Della, A.; Elena Giusti, M.; De Pasquale, C.; Lenzarini, C.; Censorii, E.; Reyes Gonzales-Tejero, M.; Patricia Sanchez-Rojas, C.; Ramiro-Gutierrez, J.; et al. Wild and semi-domesticated food plant consumption in seven circum-Mediterranean areas. *Int. J. Food Sci. Nutr.* **2008**, *59*, 383–414. [CrossRef]
64. Pieroni, A.; Nebel, S.; Santoro, R.F.; Heinrich, M. Food for two seasons: Culinary uses of non-cultivated local vegetables and mushrooms in a south Italian village. *Int. J. Food Sci. Nutr.* **2009**, *56*, 245–272. [CrossRef]
65. Tardío, J.; Pardo-de-Santayana, M.; Morales, M. Ethnobotanical review of wild edible plants in Spain. *Bot. J. Linn. Soc.* **2006**, *152*, 27–71. [CrossRef]
66. Łuczaj, Ł.; Jug-Dujaković, M.; Dolina, K.; Vitasović-Kosić, I. Plants in alcoholic beverages on the Croatian islands, with special reference to rakija travarica. *J. Ethnobiol. Ethnomed.* **2019**, *15*, 51. [CrossRef]
67. Chironi, S.; Bacarella, S.; Altamore, L.; Columba, P.; Ingrassia, M. Consumption of spices and ethnic contamination in the daily diet of Italians—Consumers' preferences and modification of eating habits. *J. Ethn. Foods* **2021**, *8*, 6. [CrossRef]
68. Verbeke, W.; López, G.P. Ethnic food attitudes and behaviour among Belgians and Hispanics living in Belgium. *Br. Food J.* **2005**, *107*, 823–840. [CrossRef]
69. Emmez, B.C. Plants and foods as identity markers: An ethnographic analysis on the case of the (Y)Ezidis from Batman (Turkey). *J. Ethn. Foods* **2020**, *7*, 5. [CrossRef]
70. Sõukand, R.; Kalle, R.; Pieroni, A. Homogenisation of Biocultural Diversity: Plant Ethnomedicine and Its Diachronic Change in Setomaa and Võromaa, Estonia, in the Last Century. *Biology* **2022**, *11*, 192. [CrossRef] [PubMed]
71. Łuczaj, Ł.; Pieroni, A.; Tardío, J.; Pardo-De-Santayana, M.; Sõukand, R.; Svanberg, I.; Kalle, R. Wild food plant use in 21st century Europe: The disappearance of old traditions and the search for new cuisines involving wild edibles. *Acta Soc. Bot. Pol.* **2012**, *81*, 359–370. [CrossRef]

72. Peroni, N.; Hanazaki, N.; Begossi, A.; Zuchiwschi, E.; Lacerda, V.D.; Miranda, T.M. Homegardens in a micro-regional scale: Contributions to agrobiodiversity conservation in an urban-rural context. *Ethnobiol. Conserv.* **2016**, *5*, 1–17. [CrossRef]
73. Sulaiman; Shah, S.M.; Sadaf; Amin, M.; Gul, B.; Begum, M. Ethnoecological, Elemental, and Phytochemical Evaluation of Five Plant Species of Lamiaceae in Peshawar, Pakistan. *Scientifica* **2020**, *2020*, 2982934. [CrossRef]
74. Motti, R. Wild Plants Used as Herbs and Spices in Italy: An Ethnobotanical Review. *Plants* **2021**, *10*, 563. [CrossRef]
75. Jarić, S.; MačUkanović-Jocić, M.; Djurdjević, L.; Mitrović, M.; Kostić, O.; Karadžić, B.; Pavlović, P. An ethnobotanical survey of traditionally used plants on Suva planina mountain (south-eastern Serbia). *J. Ethnopharmacol.* **2015**, *175*, 93–108. [CrossRef]
76. Fierascu, I.; Dinu-Pirvu, C.E.; Fierascu, R.C.; Velescu, B.S.; Anuta, V.; Ortan, A.; Jinga, V. Phytochemical Profile and Biological Activities of Satureja hortensis L.: A Review of the Last Decade. *Molecules* **2018**, *23*, 2458. [CrossRef]
77. Georgiev, M. *Bulgarian Folk Medicine*, 2nd ed.; Vasileva, M., Georgiev, M., Georgieva, I., Penchev, V., Popov, R., Simeonova, G., Troeva, E., Tsaneva, E., Eds.; Professor Marin Drinov Academic Publishing House: Sofia, Bulgaria, 2013; ISBN 978-954-322-542-8.
78. Łuczaj, Ł.; Jug-Dujaković, M.; Dolina, K.; Jeričević, M.; Vitasović-Kosić, I. Insular Pharmacopoeias: Ethnobotanical Characteristics of Medicinal Plants Used on the Adriatic Islands. *Front. Pharmacol.* **2021**, *12*, 597. [CrossRef]
79. Khoury, C.K.; Brush, S.; Costich, D.E.; Curry, H.A.; de Haan, S.; Engels, J.M.M.; Guarino, L.; Hoban, S.; Mercer, K.L.; Miller, A.J.; et al. Crop genetic erosion: Understanding and responding to loss of crop diversity. *New Phytol.* **2022**, *233*, 84–118. [CrossRef] [PubMed]
80. Galluzzi, G.; Eyzaguirre, P.; Negri, V. Home gardens: Neglected hotspots of agro-biodiversity and cultural diversity. *Biodivers. Conserv.* **2010**, *19*, 3635–3654. [CrossRef]
81. Huai, H.; Hamilton, A. Characteristics and functions of traditional homegardens: A review. *Front. Biol. China* **2008**, *4*, 151–157. [CrossRef]
82. Zhang, Y.; Yang, L.X.; Li, M.X.; Guo, Y.J.; Li, S.; Wang, Y.H. The best choices: The diversity and functions of the plants in the home gardens of the Tsang-la (Motuo Menba) communities in Yarlung Tsangpo Grand Canyon, Southwest China. *J. Ethnobiol. Ethnomed.* **2020**, *16*, 50. [CrossRef]
83. Mincheva, I.; Jordanova, M.; Benbassat, N.; Aneva, I.; Kozuharova, E. Ethnobotany and exploitation of medicinal plants in the Rhodope Mountains—Is there a hazard for Clinopodium dalmaticum? *Pharmacia* **2019**, *66*, 49–52. [CrossRef]
84. Dafni, A.; Petanidou, T.; Vallianatou, I.; Kozhuharova, E.; Blanché, C.; Pacini, E.; Peyman, M.; Dajić Stevanovic, Z.; Franchi, G.G.; Benítez, G. Myrtle, Basil, Rosemary, and Three-Lobed Sage as Ritual Plants in the Monotheistic Religions: An Historical–Ethnobotanical Comparison. *Econ. Bot.* **2019**, *74*, 330–355. [CrossRef]
85. Gupta, S.; Srivastava, A.; Shasany, A.K.; Gupta, A.K. Genetics, Cytogenetics, and Genetic Diversity in the Genus Ocimum. In *The Ocimum Genome*; Shasany, A.K., Kole, C., Eds.; Compendium of Plant Genomes; Springer International Publishing: Cham, Switzerland, 2018; pp. 73–78. ISBN 978-3-319-97429-3.
86. Naghibi, F.; Mosaddegh, M.; Motamed, S.M.; Ghorbani, A. Labiatae Family in folk Medicine in Iran: From Ethnobotany to Pharmacology. *Iran. J. Pharm. Res.* **2022**, *4*, 63–79. [CrossRef]
87. Zlatković, B.K.; Bogosavljević, S.S.; Radivojević, A.R.; Pavlović, M.A. Traditional use of the native medicinal plant resource of Mt. Rtanj (Eastern Serbia): Ethnobotanical evaluation and comparison. *J. Ethnopharmacol.* **2014**, *151*, 704–713. [CrossRef]
88. Mustafa, B.; Hajdari, A.; Pulaj, B.; Quave, C.L.; Pieroni, A. Medical and food ethnobotany among Albanians and Serbs living in the Shtërpcë/Štrpce area, South Kosovo. *J. Herb. Med.* **2020**, *22*, 100344. [CrossRef]
89. Jordanov, D.; Nikolov, P.; Boychinov, A. *Phytotherapy*, 3rd ed.; Medicina i Fizkultura State Publishing House: Sofia, Bulgaria, 1969.
90. Maleev, A.; Kuzmanov, B.; Petkov, V.; Vlahov, V.; Mitsov, V.; Furnadzhiev, G.; Zhelyazkov, D.; Minkov, E.; Petkova, E.; Krushkov, I.; et al. *Modern Phytotherapy*; Petkov, V., Ed.; Medicina i Fizkultura State Publishing House: Sofia, Bulgaria, 1982.
91. Nikolov, S.; Assenov, I.; Kuzmanov, B.; Marekov, N. *Specialized Encyclopedia of Medicinal Plants*; Nikolov, S., Assenov, I., Kuzmanov, B., Marekov, N., Eds.; Trud Publishing/Bulgarian Academy of Sciences: Sofia, Bulgaria, 2006; ISBN 978-954-528-555-4.
92. Asenov, I.; Gussev, C.; Kitanov, B.; Nikolov, S.; Petkov, T. *Guide to Medicinal Plant Collection*, 2nd ed.; GeoFar-2003: Sofia, Bulgaria, 2003; ISBN 978-619-91383-0-4.
93. Ivancheva, S.; Stantcheva, B. Ethnobotanical inventory of medicinal plants in Bulgaria. *J. Ethnopharmacol.* **2000**, *69*, 165–172. [CrossRef]
94. Sõukand, R.; Quave, C.L.; Pieroni, A.; Pardo-de-Santayana, M.; Tardío, J.; Kalle, R.; Łuczaj, Ł.; Svanberg, I.; Kolosova, V.; Aceituno-Mata, L.; et al. Plants used for making recreational tea in Europe: A review based on specific research sites. *J. Ethnobiol. Ethnomed.* **2013**, *9*, 58. [CrossRef] [PubMed]
95. Gautam, R.; Sunwar, S.; Subedi, A.; Shrestha, P.; Sthapit, B. Home Gardens and Their Roles in Domestication of Wild and Uncultivated Plant Genetic Resources in Nepal. *Acta Hortic.* **2009**, *806*, 677–684. [CrossRef]
96. Patel, S.K.; Sharma, A.; Singh, R.; Tiwari, A.K.; Singh, G.S. Diversity and Distribution of Traditional Home Gardens Along Different Disturbances in a Dry Tropical Region, India. *Front. For. Glob. Chang.* **2022**, *5*, 822320. [CrossRef]

agronomy

MDPI

Article

Joint Effects of Developmental Stage and Water Deficit on Essential Oil Traits (Content, Yield, Composition) and Related Gene Expression: A Case Study in Two *Thymus* Species

Kazem Yousefzadeh [1], Saadollah Houshmand [1], Behrouz Shiran [1,2], Sadegh Mousavi-Fard [3,*], Hossein Zeinali [4], Nikolaos Nikoloudakis [5], Mohammad Mehdi Gheisari [6] and Dimitrios Fanourakis [7]

1 Department of Plant Breeding and Biotechnology, Faculty of Agriculture, Shahrekord University, Shahrekord P.O. Box 115, Iran; ka_yousefzadeh@yahoo.com (K.Y.); houshmand@agr.sku.ac.ir (S.H.); beshiran45@gmail.com (B.S.)
2 Institute of Biotechnology, Shahrekord University, Shahrekord P.O. Box 115, Iran
3 Department of Horticultural Science, Faculty of Agriculture, Lorestan University, Khorramabad P.O. Box 465, Iran
4 Natural Science Department, Isfahan Agricultural and Natural Resources, Research and Education Center, AREEO, Isfahan P.O. Box 81785199, Iran; hoszeinali@yahoo.com
5 Department of Agricultural Sciences, Biotechnology and Food Science, Cyprus University of Technology, Limassol 3603, Cyprus; n.nikoloudakis@cut.ac.cy
6 Department of Science, Isfahan (Khorasgan) Branch, Islamic Azad University, Isfahan P.O. Box 81595-158, Iran; mgheisary@yahoo.com
7 Laboratory of Quality and Safety of Agricultural Products, Landscape and Environment, Department of Agriculture, School of Agricultural Sciences, Hellenic Mediterranean University, Estavromenos, 71004 Heraklion, Greece; dimitrios.fanourakis82@gmail.com
* Correspondence: mousavifard.s@lu.ac.ir or sadeghmosavifard@gmail.com

Abstract: In this study, the joint effects of developmental stage and soil water availability on biomass accumulation, harvest index, as well as essential oil content, yield and composition were investigated in *Thymus armeniacus*. For comparison purposes, *Thymus kotschyanus* was also considered. Plants were irrigated to either 75 or 50% of field capacity, and were sampled at 50 or 100% blooming. In both species, water deficit exerted limited effect on the time required to initiate or complete flowering. In most critical aspects of yield (harvestable organs dry weight, essential oil yield), *T. armeniacus* was found to be superior than *T. kotschyanus*. In these traits, however, *T. armeniacus* underwent a more drastic water deficit-induced decrease. Across treatments, metabolite levels fairly correlated to transcript accumulation profiles of terpene synthases and cytochrome P450 genes. Indices affiliated with reactive oxygen species were inter-correlated with the activity of five major antioxidant enzymes, while the same was noted between leaf water status and pigment content. Taken together, these results indicate that when water availability can be achieved, higher yields will be obtained by cultivation of *T. armeniacus*. Under water deficit conditions, instead, the more drought tolerant *T. kotschyanus* stands out as the primary choice.

Keywords: blooming; harvest index; terpene; *Thymus armeniacus*; *Thymus kotschyanus*; water deprivation

1. Introduction

In medicinal and aromatic plants, cultivated sources are increasingly gaining ground over wild populations [1]. This shift is underlain by many factors, including consistency of supply and alignment of product standards to market regulations and consumer preferences. Depending on the intended market, herbal material may be traded fresh or dried. Alternatively, the focus may be on the production of essential oils, and in this regard essential oil yield is of interest [2]. Since the composition of essential oil determines its properties, it also sets essential oil quality and market value [3]. Therefore, adjustments in cultivation protocols may be required to improve herbal material (fresh and dry) yield, essential oil yield or essential oil composition.

The dissemination of *Thymus* spp. (Lamiaceae) is centered in the Mediterranean area, where several species of the genus are also widely cultivated [4]. An extensive interspecific variation in every aspect of production has been documented. Although an important amount of information and knowledge has been gained, it is limited to certain species. Instead, data on other taxa is not publicly available for industrial use. For instance, both *Thymus kotschyanus* Boiss. and Hohen. and *Thymus armeniacus* Klokov and Des.-Shost. are widely cultivated in Iran. However, contrary to *T. kotschyanus* [5], data on biomass accumulation, essential oil yield and composition is essentially absent in *T. armeniacus*. In this perspective, there is a great demand of developing and establishing cultivation protocols.

The developmental stage at harvest is a factor affecting essential oil production. In some *Thymus* species, for instance, essential oil production is considerably affected by flowering induction, whereas flowering exerts a less pronounced effect in other taxa [6,7]. Therefore, the importance of harvesting time associated with essential oil features varies depending on the species of interest.

Although *Thymus* species are generally well adapted to water deficits, limited soil water availability may decrease productivity and alter essential oil composition [8]. These effects are underlain by a diverse range of processes. For instance, essential oil is highly concentrated in leaf trichomes, the density of which is affected by water deficit depending on the species. Another illustration is the water deficit-induced reduction in chlorophyll content, which impedes carbon assimilation. Furthermore, the water deficit elicits osmotic stress, impairing the activity of several enzymes and damaging the structure of critical macromolecules [2]. These effects may be effectively alleviated by the accumulation of specific osmolites (e.g., proline) [2]. In addition, plants under water deficit conditions generally experience an imbalance between the generation and scavenging of reactive oxygen species (ROS) such as hydrogen peroxide (H_2O_2). Ascorbate peroxidase (APX), catalase (CAT), guaiacol peroxidase (POX), polyphenol oxidase (PPO), and superoxide dismutase (SOD) are key ROS detoxification enzymes, while carotenoids and polyphenols are major non-enzymatic antioxidant metabolites [9–11]. Excessive ROS generation elicits a range of harmful effects, including lipid peroxidation [9,10,12].

The objectives of this study were to investigate for the first time the dual effects of harvesting time and soil water availability on herbal material marketable yield (leaves and flowers), as well as on essential oil yield and composition in *T. armeniacus*. For comparison purposes, *T. kotschyanus* was cultivated at the same stand and considered. To gain a more fundamental insight, several underlying processes were further investigated, including leaf trichome density, chlorophyll content, proline content, membrane lipid peroxidation and seven critical antioxidant defense elements. By studying the regulation of genes participating in the terpenoid biosynthetic pathway across developmental stages and species, the genetic-metabolic crosstalk was also elucidated.

2. Materials and Methods

2.1. Plant Material and Growth Conditions

Seeds of two *Thymus* species (*T. kotschyanus*, *T. armeniacus*) were sterilized, and then rinsed with distilled water. Seeds were then planted in a seedling tray (30 × 18 × 5 cm) filled with peat and perlite (9:1, *v*/*v*; Meegaa substrates BV, Rotterdam, The Netherlands). At the cotyledon stage, seedlings were transferred to 6 L pots (dimensions of 30 × 25 cm) containing the same mixture. Following sufficient growth (~six weeks), a single plant per species was selected as the mother plant. Experimental plants were further obtained by cutting propagation. For each species, branch cuttings were collected from the same (mother) plant, and were thus considered identical clones.

Following six weeks of cultivation, cuttings were fully rooted. Rooted cuttings with uniform height (~15 cm) and architecture were selected for potting. These were transplanted to 6 L pots (dimensions of 30 × 25 cm) containing a mixture of soil, sand, and animal manure (4:2:1, *v*/*v*/*v*). Pots were transferred in a multi-span plastic greenhouse, which was located to Shahrekord (32°19′32″ N 50°51′51″ E). A density of four plants m^{-2} was employed. The air temperature was set to 24 °C, and relative air humidity to 70%. Supplementary light was set at 250 μmol m^{-2} s^{-1} photosynthetic photon flux density at the top of the plant canopy for 16 h d^{-1}, and was provided by red (90%) and blue LED modules (Iraneon Co., Birjand, South Khorasan, Iran). Potting-media moisture was maintained by regular watering. With the dual purpose of inducing the same initial status among individuals and of ensuring that herbal material growth would take place under experimental conditions (described below), all plants were pruned at 5 cm above root-to-shoot interface.

Immediately afterwards, three factors [2 species (*T. kotschyanus*, *T. armeniacus*) × 2 irrigation levels (full, and deficit irrigation) × 2 sampling stages (50, and 100% blooming)] were applied as a factorial experiment based on a completely randomized design. Irrigation was adjusted daily to 75 and 50% of field capacity, representing full and deficit irrigation, respectively. Under these irrigation schemes, substrate volumetric water content corresponded to circa 8–10 and 5–7%, respectively. These irrigation levels were selected based on growth reduction in a preliminary study. Irrigation to 100% of field capacity was not performed in order to avoid potential adverse effects related to waterlogging. Given daily irrigation and the large substrate volume (6 L), day-to-day variation in substrate moisture content is expected to be rather minimal. The irrigation scheme was maintained until the final harvest, which was conducted at the 100% blooming developmental stage (15 August 2019). Experimental plants were collected for evaluation at two developmental stages (50 and 100% blooming), based on the percentage of branches bearing open flowers. When half of the branches had open flowers, it was considered as 50% blooming. Accordingly, when all branches were in that state, it was considered as 100% blooming.

Plants were cultivated for three months (15 May–15 August 2019). During this period, the mean air temperature was 23.8 ± 1.6 °C, while the mean relative air humidity was 66 ± 7%. The average daily light integral was 20.6 ± 0.3 mol m^{-2} d^{-1} (LI-250A, LI-COR, Lincoln, NE, USA).

Plant and leaf level measurements were conducted. For leaf-level measurements, leaves were selected from the upper (toward the apex) one-third of the leaf-bearing nodes. Sampled leaves had grown under direct light, and were fully expanded. In all evaluations, three replicates were assessed per treatment.

2.2. Growth, Flowering and Harvest Index

The time required for initiation of flowering, as well as for reaching the two developmental stages (50 and 100% blooming) under study was recorded. The inflorescence height and the number of flowers per inflorescence were also documented.

The length of five branches (from the root-to-shoot junction to the apical meristem) per plant was recorded and averaged to compute plant height.

By considering the overhead (top-view) 2D plant silhouette, a convex hull (the minimal polygon that encloses the entire silhouette perimeter) [13] was fitted. The maximum and the minimum distance spanned by pairs of points crossing the convex hull center were considered as plant length and width, respectively. Based on these data, canopy area ($\pi \times \frac{\text{plant length}}{2} \times \frac{\text{plant width}}{2}$) was computed.

For individual leaf trait assessment, leaves were scanned (HP Scanjet G4010, Irvine, CA, USA) and then evaluated by using the Digimizer software (version 4.1.1.0, MedCalc Software, Ostend, Belgium) [14]. Leaf length (midvein length; major axis), width (widest point perpendicular to the leaf major axis) and area (one-sided surface area) were digitally assessed [13,15].

Aerial plant and shoot (fresh and dry) masses were recorded. For measuring dry weight, samples were placed in a forced-air drying oven for 72 h at 80 °C. Economic yield was considered the combined mass of leaves and flowers (thus excluding shoots). Harvest index ($\frac{\text{economic yield}}{\text{plant dry weight}} \times 100\%$) was also computed.

2.3. Essential Oil Content

The essential oil content is as important as essential oil yield, since it determines the extraction cost and, therefore, the profitability of essential oil production [2]. Shade-dried samples were subjected to hydro-distillation using a Clevenger apparatus [3]. These (20 g) were added to a 1 L flask containing 200 mL of distilled water. The flask was then heated for 3.5 h. Essential oil content was recorded. The isolated essential oils were first dried over anhydrous sodium sulfate, and then kept in glass vials at −20 °C before further analysis [3]. Three replicates were assessed per sample.

2.4. Essential Oil Composition

Gas chromatography-mass spectrometry (GC–MS) analysis was conducted using a gas chromatograph (Model 7890A, Agilent, Palo Alto, CA, USA) coupled with a mass selective detector (Model 5673, Agilent, Palo Alto, CA, USA). A fused silica capillary column (30 m length × 0.25 mm i.d.; 0.25 μm BP-5 film thickness) was used to separate the oil compounds. The oven temperature was increased from 40 to 290 °C (at a rate of 7 °C min^{-1} for 40–200 °C, and of 40 °C min^{-1} for 200–290 °C), and finally held isothermal at 290 °C for 10 min. Ion source and transfer-line temperature was 290 °C. Ultra-pure helium was used as the carrier gas. Injector and interface temperatures were 290 °C and 280 °C, respectively. The mass spectrum was acquired over the mass range of 35–450 amu in full-scan acquisition mode. The split ratio was 1:50.

The GC-FID analysis of the essential oils was conducted using a Thermoquest Finnigan apparatus equipped with a flame ionization detector (FID) and a fused silica capillary column (30 m length ×0.25 mm i.d.; 0.25 μm BP-5 film thickness). The oven temperature was programmed as stated above. Injector and detector temperatures were 250 °C and 300 °C, respectively. Ultra-pure helium was used as the carrier gas with a flow rate of 2.1 mL min^{-1}. The split ratio was 1:10.

Retention indices (RI) of each compound were calculated using a homologous series of n-alkanes (C_6–C_{24}) injected to HP-5MS column in the same condition. Identification of oil constituents was performed by comparison of (1) their retention times with those of authentic standards, (2) their spectral mass with those of the internal reference mass spectra library (NIST08 and Wiley 9.0), and (3) their RI with those reported in the literature [16]. Quantification was conducted by the external standard method through calibration curves generated by running GC analysis of representative authentic compounds [3]. The relative percentage of each essential oil constituent was obtained according to the respective area under the curve by using the area normalization method and ignoring response factors [3].

2.5. Leaf Trichome Density

Glandular trichomes secrete essential oils. In this perspective, treatment effects on glandular trichome density of either leaf side were studied. The sampling area (1 × 1 cm) was located midway between the leaf base and tip, and between the midrib and lateral margin. Images were acquired using an optical microscope (Leitz Aristoplan; Ernst Leitz Wetzlar GmbH, Wetzlar, Germany) connected to a digital camera (Nikon DXM-1200; Nikon Corp., Tokyo, Japan). Glandular trichome density was counted on five non-overlapping fields of view per leaf (magnification × 100). Image processing was performed with ImageJ software (https://imagej.nih.gov/).

2.6. Leaf Water Status

Leaf water status was in situ assessed by measuring relative water content (RWC). Samples were collected 3 h following the onset of the photoperiod [17]. Following excision, fresh weight was gravimetrically obtained (±0.0001 g; Mettler AE 200, Giessen, Germany). Immediately after, samples were floated on distilled water inside a Petri dish covered with a lid. Following 24 h of incubation, the recorded weight was regarded as turgid (saturated). Then, dry weight (48 h at 80 °C) was determined. RWC was calculated according to Taheri-Garavand et al. [18].

2.7. Leaf Chlorophyll and Carotenoid Content

Leaf chlorophyll content is critical for photosynthesis, while carotenoids are important non-enzymatic antioxidants [1,11]. Samples were processed immediately after collection. Following fine chopping, portions weighing 0.1 g were homogenized with the addition of 10 mL of 100% acetone. The extract was then centrifuged (14,000× *g* for 20 min), and the supernatant was collected. Since chlorophyll is light sensitive, the extraction took place in a dark room [9,10]. The obtained extract was subjected to reading on a spectrophotometer (Mapada UV-1800; Shanghai Mapada Instruments Co., Ltd., Shanghai, China). Total chlorophyll and carotenoid contents were calculated according to Lichtenthaler and Wellburn [19].

2.8. Leaf Proline Content

Proline is actively involved in cell osmotic regulation via decreasing cell water potential, and in this way enzyme activity and the macromolecules' structure are protected [2]. In this perspective, the treatment effect on leaf proline content was assessed. Freshly cut leaf discs (0.5 g) were homogenized and then added in 10 mL of 3% (*w*/*v*) aqueous sulphosalycylic acid. The extract was filtered through Whatmann No. 2 filter paper, and 2 mL of the filtrate were mixed with 2 mL acid-ninhydrin and 2 mL of glacial acetic acid. The obtained solution was heated (100 °C for 1 h). The reaction mixture was extracted with 4 mL toluene, and the chromophore containing toluene was aspirated from liquid phase. After equilibration at 25 °C, the absorbance was measured at 520 nm with a spectrometer (Mapada UV-1800, Shanghai. Mapada Instruments Co., Ltd., Shanghai, China). Proline concentration was determined using a calibration curve [20].

2.9. Leaf Hydrogen Peroxide Content

H_2O_2 is a critical ROS which accumulates under adverse conditions [10]. Leaf H_2O_2 content was, therefore, assessed. The reaction mixture consisted of tissue extract supernatant, 0.5 mL 0.1% trichloroacetic acid, 0.5 mL of 0.1 M potassium-phosphate buffer (pH 7.0), and 1 mL of 1 M KI (*w*/*v*). Color developed for 45 min in darkness, and absorbance was then spectrophotometrically assessed at 390 nm (UV-1800, Shimadzu, Kyoto, Japan). H_2O_2 content was calculated by using a calibration curve prepared with eight known H_2O_2 concentrations.

2.10. Leaf Lipid Peroxidation

The treatment effects on the malondialdehyde (MDA) content, taken as an indication of lipid peroxidation level, were evaluated by employing the thiobarbituric acid reactive substance assay [2]. Freshly-cut leaf discs (0.1 g) were homogenized, and then added in 5 mL of 20% (*w*/*v*) trichloroacetic acid and 0.5% (*w*/*v*) thiobarbituric acid. The suspension was subsequently centrifuged (6000× *g* for 15 min). The obtained solution was heated (100 °C for 25 min). After equilibration at 25 °C, the precipitate was removed by centrifugation (6000× *g* for 5 min). The amount of MDA was calculated from the absorbance at 535 nm after subtracting the non-specific absorption at 450 and 600 nm (Mapada UV-1800; Shanghai Mapada Instruments Co., Ltd., Shanghai, China). The extinction coefficient 156 mmol MDA L^{-1} cm^{-1} was used. Four discs were assessed per replicate sample.

2.11. Antioxidant Defense Elements

2.11.1. Total Phenolic Content

As phenolics exhibit strong antioxidant properties, they can be beneficial for plant antioxidant defense [12]. Leaf total phenolic content was therefore assessed. Leaf samples (0.1 g) were ground with a mortar and pestle with liquid nitrogen, extracted with 1 mL of 80% aqueous methanol in an ultrasonic bath for 10 min, and were then centrifuged (15,000× *g* for 10 min). The content of total phenolics was determined by using the Folin-Ciocalteu assay [12]. The absorbance against prepared re-agent blank was determined using a microplate reader (Infinite 200 PRO, TECAN, Männedorf, Switzerland). For total phenolic content, gallic acid was used as the standard reference and gallic acid equivalent (GAE) was expressed as mg per g dry mass [12].

2.11.2. Activity of Five Antioxidant Enzymes

The treatment effects on the activity of five critical antioxidant enzymes (APX, CAT, POX, PPO, SOD) was assessed.

APX activity was assessed using the method described by Ahmadi-Majd et al. [9,10]. Fresh frozen leaf segments (0.1 g) were ground in liquid nitrogen, homogenized with 1 mL of 50 mM sodium phosphate buffer (pH 7.0) containing 2 mM EDTA and 1% polyvinylpyrrolidone (PVP), and centrifuged (14,000× *g* for 20 min) at 4 °C. APX activity in the supernatant was assessed by following the decrease in absorbance at 290 nm for 2 min (10 s intervals) in a reaction mixture containing sodium phosphate buffer, ascorbic acid, and H_2O_2. The extinction coefficient of 2.8 mM^{-1} cm^{-1} was used. APX activity was expressed as µmol of ascorbate oxidized min^{-1} g^{-1} tissue.

CAT activity was measured as described by Ahmadi-Majd et al. [9,10]. Fresh frozen leaf segments (0.3 g) were ground in liquid nitrogen, homogenized with 1.5 mL of K phosphate buffer (containing 1 mM EDTA and 2% PVP), and centrifuged (14,000× *g* for 20 min) at 4 °C. CAT activity in the supernatant was assessed by following the decrease in absorbance at 240 nm for 2 min (10 s intervals) in a reaction mixture containing potassium phosphate buffer and H_2O_2. The extinction coefficient of 39.4 M^{-1} cm^{-1} was used. CAT activity was expressed as µmol of H_2O_2 reduced min^{-1} g^{-1} tissue.

SOD activity was determined by the method of Ahmadi-Majd et al. [9,10] and was assayed by monitoring the inhibition of the photochemical reduction of nitro blue tetrazolium chloride (NBT). Fresh frozen leaf segments (0.5 g) were ground in liquid nitrogen, homogenized with 1 mL of 50 mM sodium phosphate buffer (pH 7.0) containing 2 mM EDTA and 1% PVP, and centrifuged (14,000× *g* for 20 min) at 4 °C. A reaction mixture of sodium phosphate buffer, methionine, NBT, EDTA, and riboflavin was used. The mixture was placed for 20 min at 25 °C under a fluorescent light (30 Watt). Absorbance at 560 nm was monitored using a spectrophotometer (Mapada UV-1800; Shanghai Mapada Instruments Co., Ltd., Shanghai, China). A SOD enzyme activity unit was considered as 50% of the NBT photoreduction and expressed as unit min^{-1} g^{-1} tissue.

PPO activity was determined according to Chen et al. [11], using a commercial assay kit (Nanjing Jiancheng Bioengineering Institute, Nanjing, China). Briefly, 0.15 mL supernatant was mixed with 0.6 mL phosphate buffer (25 mM) and 0.15 mL catechol (20 mM), and incubated for 10 min at 25 °C. It was then was immediately placed into boiling water for 10 min and cooled to room temperature (25 °C). Next it was centrifuged (10,000× *g* for 10 min) at 25 °C. Thereafter, using distilled water as a blank, the absorbance was measured at 410 nm with a microplate reader (Infinite 200 PRO, TECAN, Switzerland). One unit of PPO activity was defined as the amount of enzyme required to increase the absorbance by 0.01 min^{-1} under the assay conditions.

POX activity was performed according to the method of Narwal et al. [21]. The reaction mixture contained 1350 µL of 0.1 M potassium phosphate buffer (pH 6), 500 µL of 44 mM H_2O_2 solution, 100 µL of 45 mM guaiacol and 200 µL of extracted enzyme extract. Absorbance at 470 nm was monitored using a spectrophotometer (Mapada UV-1800; Shanghai Mapada Instruments Co., Ltd., Shanghai, China).

2.12. QPCR Expression Analysis

Total RNA from each sample was extracted using the total RNA isolation kit (S-1010-1, Dena Zist Asia, Mashhad, Iran). The genomic DNA was eliminated by DNAse I treatment according to the manufacturer's instructions (ThermoFisher, Waltham, MA, USA). Subsequently, first strand cDNA was amplified using M-MLV (Moloney Murine Leukemia Virus) Reverse Transcriptase and oligo (dT)18 primer, from 2 µg total RNA to 20 µL final volume (Cat. No. 18080-044; Invitrogen, Carlsbad, CA, USA).

All the RT-qPCR reactions were performed using RT-Master Mix (Cat. No. RR820L, Takara Bio, Shiga, Japan) containing SYBR Green on a Rotor Gene Q (Qiagen, Hilden, Germany). The following PCR profile was employed: (1) 95° C for 30 s (pre-denaturation step), (2) 95 °C for 15 s (denaturation step), (3) 56–57 °C for 20 s (annealing), and eventually (4) 72 °C for 25 s (extension). This sequence was repeated for 35 cycles. Then a melting curve (55–95 °C) was performed at the default ramp rate. The primers efficiency was acquired from calibration curves with 1:5 dilution series and at least three points fitted in a linear regression with R-square over 0.98.

The primer used in this study was designed based on gene sequences in the NCBI database and primers designed in previous reports (Crocell et al., 2010). The accuracy of sequences and size of primers were checked using the Primer-BLAST software of NCBI (Supplementary Table S1). The qRT-PCR analysis was carried out with elongation factor-1α (EF-1α) gene as the internal standard and was analyzed using the $2-\Delta\Delta CT$ method [22].

2.13. Statistical Analysis

Data analysis was performed using the SPSS software (version 23; SPSS Inc., Chicago, IL, USA). A three-way ANOVA was employed (species × irrigation level × sampling stage). The data were first tested for normality (Shapiro–Wilk test) and homogeneity of variances (Levene's test). The means were compared using Fisher's least significant difference test to be significant at $p \leq 0.05$.

For the four experimental units, eigenvalues were extracted and the most contributing variables for each dimension were computed and identified. The first two eigenvalues cumulated more than 77% of the total variance, and were retained to produce the principal components. A biplot principal component analysis (PCA) was produced to depict correlations across the traits, species and contributions to principal components. Individuals were grouped (by discrete color) and variables by their contribution to the principal components (gradient colors). The "FactoMineR", "factoextra" and "readxl" libraries were used under the R-studio integrated development environment (RStudio suite V 1.2.5033).

3. Results

3.1. Growth, Flowering and Harvest Index

The onset of flowering occurred much earlier in *T. armeniacus* as compared to *T. kotschyanus* (49.3 versus 79.3 d), though time to 100% blooming was rather similar among them (115–120 d; Table 1). *Thymus armeniacus* had shorter (47.6%) inflorescence length, and a lower (34.6%) number of flowers per inflorescence. Water deficit (50 as compared to 75% field capacity) reduced (~5%) time to 100% blooming similarly among the two species. In both species, the water deficit also decreased inflorescence length, and the number of flowers per inflorescence. The latter was more decreased in *T. armeniacus*, while the former in *T. kotschyanus*.

In both species, water deficit decreased plant height, length, and individual leaf dimensions (length, width, area) (Table 2). For plant height and individual leaf dimensions, this effect was less prominent in *T. armeniacus*.

Table 1. Effect of irrigation regime on time to flowering (initiation, 50 and 100%) and two key inflorescence characteristics of two *Thymus* species. The latter was evaluated at 100% flowering. Within each column, different letters indicate significant differences (n = 3).

Species	Irrigation (% of Field Capacity)	Time (d)			Inflorescence Length (cm)	Number of Flowers per Inflorescence
		Onset of Flowering	50% Flowering	100% Flowering		
T. armeniacus	75	49.3 c	63.0 c	120.3 a	2.31 b	50.22 c
	50	47.7 c	66.3 c	113.0 b	1.64 c	33.84 d
T. kotschyanus	75	79.3 b	88.0 b	114.7 ab	4.41 a	76.78 a
	50	88.0 a	96.0 a	108.7 b	2.34 b	63.07 b

Table 2. Effect of irrigation regime on plant and individual leaf dimensions of two *Thymus* species. Assessments were conducted at 100% flowering. Within each column, different letters indicate significant differences (n = 3).

Species	Irrigation (% of Field Capacity)	Plant			Canopy Area (cm^2)	Individual Leaf		
		Height (cm)	Length (cm)	Width (cm)		Length (mm)	Width (mm)	Area (mm^2)
T. armeniacus	75	14.6 bc	35.3 a	31.0 a	858 a	12.9 b	5.64 a	47.4 b
	50	13.1 c	28.8 bc	28.3 ab	640 b	11.2 c	4.89 b	41.9 c
T. kotschyanus	75	20.3 a	31.3 b	27.3 b	670 c	18. 5 a	4.71 b	57.1 a
	50	15.9 b	28.02 c	29.0 ab	637 d	14.0 b	3.74 c	38.1 c

Thymus armeniacus had higher plant fresh weight, dry weight, economic yield (dry weight of leaves and flowers), and harvest index (dry weight of leaves and flowers relative to plant dry weight) as compared to *T. kotschyanus* (111.6, 95.0, 122.7, and 14.3%, respectively; Table 3). In both species, water deficit decreased plant fresh weight, dry weight, economic yield, and harvest index. For plant fresh weight and economic yield, this effect was more pronounced in *T. armeniacus*.

Table 3. Effect of irrigation regime on plant (fresh and dry) weight, economic yield (dry mass of leaves and flowers) and harvest index ($\frac{\text{economic yield}}{\text{plant dry weight}} \times 100\%$) of two *Thymus* species. Assessments were conducted at 100% flowering. Within each column, different letters indicate significant differences (n = 3).

Species	Irrigation (% of Field Capacity)	Plant			Harvest Index (%)
		Fresh Weight (g)	Dry Weight (g)	Economic Yield (g)	
T. armeniacus	75	116.93 a	42.37 a	34.1 a	81.97 a
	50	83.78 b	32.2 b	23.71 b	76.63 b
T. kotschyanus	75	55.26 c	21.73 c	15.313 c	71.71 c
	50	42.81 d	16.413 d	12.167 c	63.99 d

3.2. Essential Oil Content, Yield and Composition

Thymus armeniacus had lower (49.9%) essential oil content, and higher (13.2%) essential oil yield as compared to *T. kotschyanus* (Table 4). The water deficit decreased (49.9%) the essential oil content in *T. kotschyanus*, whereas it exerted a minor effect (1.4%) in *T. armeniacus*. The water deficit decreased essential oil yield in both taxa, an effect that was more prominent in *T. armeniacus* (30.5 versus 24.6%).

Table 4. Effect of irrigation regime on essential oil content and yield, as well as on leaf trichome density of two *Thymus* species. Assessments were conducted at 100% flowering. Within each column, different letters indicate significant differences (n = 3).

Species	Irrigation (% of Field Capacity)	Essential Oil Content (%)	Essential Oil Yield (g)	Adaxial	Abaxial	Leaf (Adaxial + Abaxial)
				Trichome Density (mm^{-2})		
T. armeniacus	75	1.180 c	0.3870 a	4.75 c	5.640 c	10.38 c
	50	1.164 c	0.2689 b	18.1 a	12.74 a	30.84 a
T. kotschyanus	75	2.353 a	0.3420 a	8.88 b	10.32 b	19.2 b
	50	1.988 b	0.2578 b	10.0 b	11.65 ab	21.67 b

By using gas chromatography–mass spectrometry, essential oil composition was further analyzed (Table 5). For *T. armeniacus*, the most abundant metabolites were γ-Terpinene (16.65–23.58%), p-Cymene (6.70–16.56%) and α-Pinene (12.16–15.78%). The former two (i.e., γ-Terpinene, p-Cymene) were also the most abundant metabolites in *T. kotschyanus* (17.69–28.11 and 4.19–17.78%, respectively), whereas α-Pinene was not detected in this species. In *T. kotschyanus*, thymol was also a very abundant metabolite (11.41–15.73%).

The water deficit affected substance content depending on the metabolite and the species (Table 5; full range of documented compounds in Supplementary Table S2).

Hierarchical clustering of essential oil components revealed that these could be organized primarily based on species and secondarily on irrigation treatments (Figure 1). However, neither growth stage nor irrigation regime severely distorted essential oil fractions.

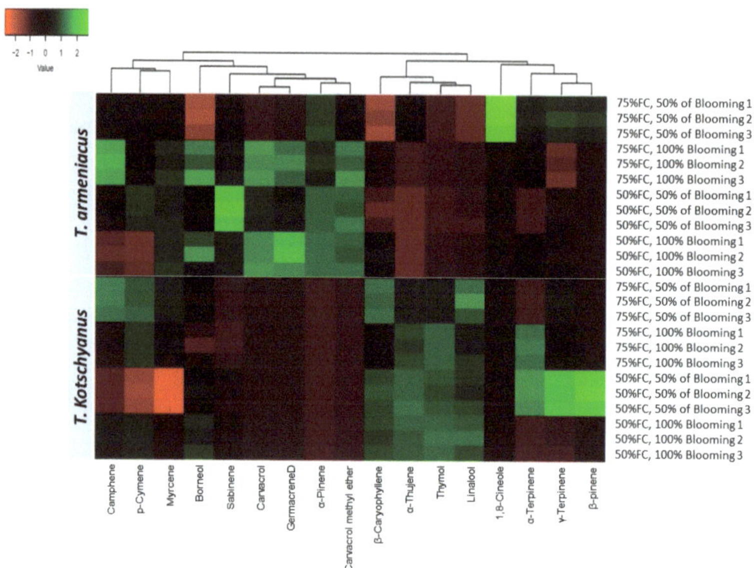

Figure 1. Heatmap of standardized values indicating percentages of essential oil components across treatments. The effect of irrigation regime (75 or 50% of field capacity) on essential oil composition of two *Thymus* species (*T. armeniacus, T. kotschyanus*) was investigated at two developmental stages (50 and 100% blooming). Essential oil composition is presented in Table 5. Color may indicate up- or downregulation (green and red, respectively). A scale of intensity is provided.

3.3. Principal Component Analysis in Plant Growth and Yield Traits

In order to identify and quantify the components that regulate the connections among the two *Thymus* species, as well as the effect of irrigation regime to the delineation of differential responses, a biplot PCA was conducted (Figure 2). In this analysis, all the growth traits were included (data in Tables 1–4 excluding trichome density). Eigenvalues were examined to determine the number of considered principal components (PC). The first two dimensions explained more than 77% of the total variance percentage (Supplementary Figure S1). The level of significant contribution of morphological traits to the PCA was estimated by using the cos2 index (Supplementary Figure S2). Among these descriptors, the plant fresh and dry weights as well as the economic yield and flowering/inflorescence indices had a profound impact for the categorization of species and treatments. The biplot PCA based on the first two components revealed the complex relationships among genotypes and treatments (Figure 2). The first axis revealed that *T. armeniacus* and *T. kotschyanus* can be clearly demarcated based on genotypic discrepancies under control conditions (75% of field capacity), while the second axis displays the discriminating effects imposed by water deficit within the same species. Furthermore, positive and negative correlations across morphological components were evident. Flowering and inflorescence indices were highly homogenous (indicating co-regulation) as were the plant fresh weight, dry weight and economic yield. Interestingly, the above-mentioned clusters had an inter-negative association showing that the earliness in blooming is an index of high biomass yield.

Table 5. Effect of irrigation regime on essential oil composition of two *Thymus* species at two developmental stages (50 and 100% blooming). Within each column, different letters indicate significant differences ($n = 3$). Retention indices (RI) were generated with a standard solution of n-alkanes (C_6–C_{24}) on the HP-5MS column. nd—no difference. The full range of documented compounds is provided in Supplementary Table S2.

Compound	RI	*T. armeniacus*				*T. kotschyanus*			
		75% Field Capacity		50% Field Capacity		75% Field Capacity		50% Field Capacity	
		50% Flowering	100% Flowering	50% Flowering	100% Flowering	50% Flowering	100% Flowering	50% Flowering	100% Flowering
α-Thujene	928	5.76 e	0.903 f	0 g	0 g	9.35 d	11.337 c	12.893 a	12.34 b
Camphene	934	3.62 d	8.12 a	3.13 e	1.34 g	7.19 b	2.82 f	1.307 g	4.883 c
α-Pinene	934	12.987 b	12.157 c	15.42 a	15.78 a	0 d	0 d	0 d	0 d
β-pinene	967	2.26 b	0 c	0 c	0 c	0 c	0 c	5.623 a	0 c
Sabinene	971	4.4067 d	6.42 c	12.43 a	6.873 b	2.347 f	2.137 f	3.71 e	4.06 de
Myrcene	988	6.1 d	8.483 ab	8.34 ab	8.623 a	8.13 b	7.22 c	0 e	5.73 d
α-Terpinene	1018	4.63667 c	3.53 d	0 f	2.623 e	0 f	7.1 b	7.797 a	0 f
p-Cymene	1034	12.067 e	13.507 d	16.56 b	6.703 f	17.783 a	16.63 b	4.187 g	15.267 c
1,8-Cineole	1037	12.53 a	0 b	0 b	0 b	0 b	0 b	0 b	0 b
γ-Terpinene	1073	23.583 b	16.65 h	19.21 f	19.84 e	21.93 c	20.74 d	28.11 a	17.69 g
linalool	1105	0 d	0.28 c	0.187 c	0.29 c	1.463 a	1.1 b	1.213 b	1.347 a
Borneol	1168	0.87 f	2.78 a	1.847 d	2.54 ab	1.927 d	1.48 e	2.067 cd	2.37 bc
Carvacrol methyl ether	1248	0.88 c	2.417 a	2.187 b	2.153 b	0	0	0	0
Thymol	1298	0.11 f	0.577 de	0.32 ef	0.66 d	11.41 c	15.66 a	14.337 b	15.73 a
Carvacrol	1313	0.827 d	11.483 a	7.657 b	11.34 a	1.45 c	1.653 c	1.52 c	1.78 c
β-Caryophyllene	1429	2.31333 f	5.22 d	3.803 e	6.693 c	8.917 a	6.52 c	8.3 b	8.193 b
Germacrene D	1484	nd	0.923 b	0.48 c	1.197 a	0 d	0 d	0 d	0 d

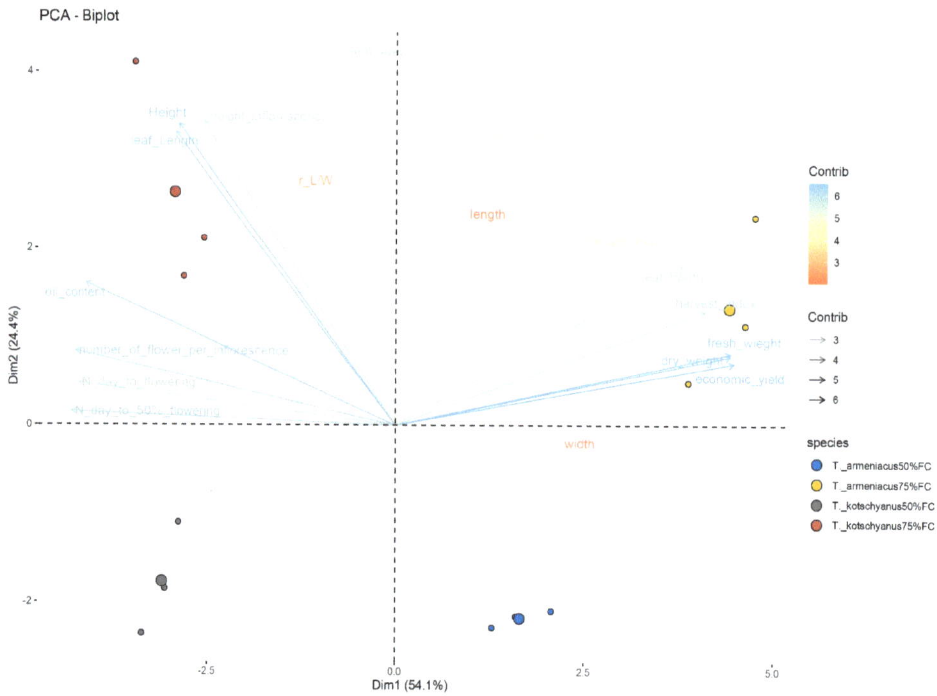

Figure 2. Principal Component Analysis (PCA) biplot depicting Principal Components (PC) scores of *Thymus* species and treatments (dots) and loadings of vector variables. The contribution of each morphological trait (data in Tables 1–4) in the two dimensions is indicated by a gradient scale and colour intensity (legend). Vectors near the plot center have lower cos2 values. Larger dots indicate mean values calculated from three discrete biological replications.

3.4. *Leaf Trichome Density*

Thymus armeniacus had lower (~45%) trichome density in either leaf side as compared to *T. kotschyanus* (Table 4). Most trichomes were generally situated on the abaxial leaf side as compared to the adaxial one, besides *T. armeniacus* cultivated under water deficit.

The water deficit increased trichome density in either leaf side of both species (Table 4). This increase was considerably more prominent in *T. armeniacus* as compared to *T. kotschyanus* (126–281% versus 13%).

3.5. *Leaf Water Status*

Leaf RWC was determined as an indication of hydration status. In both species, water deficit impaired leaf hydration status at 100% blooming (Table 6). This effect was more prominent in *T. kotschyanus*, as compared to *T. armeniacus*.

3.6. *Leaf Chlorophyll, and Proline Contents*

In *T. armeniacus*, the water deficit improved leaf chlorophyll content at either developmental stage (Table 6). In *T. kotschyanus*, the water deficit decreased (21.7%) leaf chlorophyll content at 50% blooming, but not at 100%.

At 100% blooming, water deficit induced a sizeable increase (~1500%) in leaf proline content (Table 6). This effect was comparable among the two species.

3.7. *Leaf Hydrogen Peroxide Content and Lipid Peroxidation Level*

H_2O_2 is a critical ROS. At either developmental stage, water deficit increased (~95%) leaf H_2O_2 content in both species (Table 6). This effect was comparable among the two species.

Table 6. The effect of irrigation regime on the content of chlorophyll, carotenoids, total phenolics, proline, malondialdehyde (MDA), and hydrogen peroxide (H_2O_2), as well as on relative water content and five antioxidant enzymes activity of two *Thymus* species at two developmental stages (50 and 100% blooming). Within each column, different letters indicate significant differences ($n = 3$). APX, ascorbate peroxidase; AsA, ascorbic acid; CAT, catalase; dw, dry weight; fw, fresh weight; GAE, gallic acid equivalent; POX, guaiacol peroxidase; PPO, polyphenol oxidase; SOD, superoxide dismutase; U, unit.

	Trait	*T. armeniacus*				*T. kotschyanus*			
		75% Field Capacity		50% Field Capacity		75% Field Capacity		50% Field Capacity	
		50% Flowering	100% Flowering	50% Flowering	100% Flowering	50% Flowering	100% Flowering	50% Flowering	100% Flowering
Chlorophyll content (mg g^{-1} FW)	a	0.261 b	0.31 a	0.22 c	0.23 c	0.278 b	0.28 b	0.226 c	0.269 b
	b	0.09 abc	0.099 a	0.082 bc	0.086 abc	0.086 bc	0.094 ab	0.058 d	0.078 c
	total	0.351 c	0.409 a	0.302 de	0.316 d	0.364 bc	0.374 b	0.285 e	0.348 c
Non-enzymatic	Carotenoid content (mg g^{-1} FW)	2.611 c	3.444 a	1.596 e	2.611 c	2.55 c	3.085 ab	2.171 d	2.769 bc
	Total phenol content (mg GAE g^{-1} DW)	-	31.422 b	-	26.513 c	-	34.718 a	-	31.108 b
	Relative water content (%)	74.97 b	68.794 c	66.394 c	55.621 d	76.681 b	83.144 a	74.467 b	57.048 d
Osmolite	Proline content (μmol g^{-1} FW)	0.515 cd	0.378 d	0.839 bcd	6.9 a	0.921 bc	0.491 cd	1.156 b	7.088 a
	MDA content (nmol g^{-1} FW)	10.938 f	12.874 e	12.018 ef	18.707 b	12.565 e	16.355 c	14.598 d	22.323 a
	H_2O_2 content (nmol g^{-1} FW)	478.57 d	587.52 c	663.09 b	1154.65 a	422.8 e	350.25 f	702.41 b	677.79 b
Enzymatic	APX activity (μmol AsA min^{-1} mg^{-1} protein)	0.4641 de	0.3968 e	0.5707 cd	0.6533 bc	0.7366 b	0.4886 de	0.9265 a	0.6066 c
	CAT activity (μmol H_2O_2 min^{-1} mg^{-1} protein)	5.446 f	6.576 de	9.805 a	8.43 b	7.105 cd	6.23 e	7.901 b	7.735 bc
	POX activity (U mg^{-1} protein min^{-1})	34.561 c	35.915 c	86.375 b	163.423 a	35.928 c	33.459 c	94.706 b	92.627 b
	PPO activity (U mg^{-1} protein min^{-1})	0.556 e	0.666 d	0.882 b	0.974 ab	0.771 c	0.678 cd	0.962 ab	0.988 a
	SOD activity (U mg^{-1} protein min^{-1})	11.9416 b	6.553 d	13.987 a	9.941 c	7.797 d	7.663 d	9.975 c	10.350 c

MDA content was assessed as an indication of lipid peroxidation level [2]. At 100% blooming, the water deficit induced an increase in leaf MDA content (Table 6). This increase was more prominent in *T. armeniacus* as compared to *T. kotschyanus* (45.3 versus 36.4%).

3.8. Enzymatic and Non-Enzymatic Antioxidant Defense Elements

Carotenoids and polyphenols are critical non-enzymatic antioxidants, while APX, CAT, POX, PPO, SOD are important antioxidant enzymes.

In *T. armeniacus*, water deficit improved leaf carotenoid content at either developmental stage (Table 6). In *T. kotschyanus*, instead, no effect on carotenoid content was noted.

Water deficit decreased leaf total phenolic content in both species (Table 6). This effect was more prominent in *T. armeniacus* as compared to *T. kotschyanus* (15.6 versus 10.4%).

Different effects of the water deficit were noted among enzymes, species and developmental stages (Table 6).

3.9. Interplay of Physiological Traits

In order to delineate the interaction responses of each species to the watering regime across the two developmental stages, a hierarchical clustering was performed (Figure 3). In this analysis, all the physiological traits were included (data in Table 6). Two major clusters were acknowledged. The first contained oxidative indices (H_2O_2, proline and MDA levels) affiliated with ROS induction, as well as antioxidant enzymes (APX, CAT, POX, PPO, SOD). Oxidative indices were inter-correlated with antioxidant enzymes, suggesting a joint regulatory biosynthetic pathway. Further correlations among MDA and proline contents, as well as among H_2O_2 content and POX enzyme activity, were determined, suggesting a related biochemical connection. The second cluster contained the RWC index as well as chlorophyll and carotenoid contents.

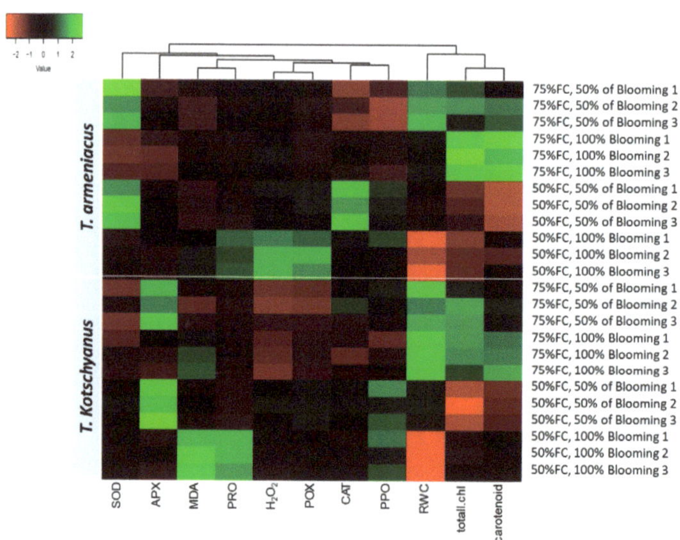

Figure 3. Heatmap of standardized values indicating physiological trait regulation across treatments. The effect of irrigation regime (75 or 50% of field capacity) on several physiological traits (data in Table 6) of two *Thymus* species (*T. armeniacus*, *T. kotschyanus*) was investigated at two developmental stages (50 and 100% blooming). Color may indicate up- or downregulation (green and red, respectively). A scale of intensity is provided.

Similar patterns across physiological traits were recorded among the two species under study (Figure 3). Specifically, a tight upregulation under water deficit that was mostly unaffected by the developmental stage, was demonstrated. Still, there were some instances where the species effect was more prominent and several discrepancies among the two species were recorded.

3.10. Relative Expression of Terpene Synthesis

The simultaneous analysis of transcript accumulation indicates that there are distinct developmentally regulated pathways as well as stress-induced responses (Figure 4). Moreover, gene regulation in the terpene pathway seems to follow different routes based primarily on the species. Hence, major discrepancies were detected among the two species under study. In *T. armeniacus*, a significant downregulation was evident for *TERPENE SYNTHASE 6* (*TPS6*) at the 100% blooming stage, which was not specifically correlated to the water deficit. In contrast, the same isoform seems unaffected in *T. kotschyanus*, having a rather conserved expression pattern. Genes encoding CYTOCHROME P450 (CYP) monooxygenases enzymes were also differentially regulated among the two species. *CYP71D179 182* seems to be linked to water availability, since in *T. kotschyanus* a significant upregulation was detected under water deficit regardless of the developmental stage. In *T. armeniacus*, transcription patterns fluctuated at a much lesser pace. By contrast, *TERPENE SYNTHASE 2* (*TPS2*) seems to correlate to water deficit only in *T. armeniacus*, since there was a significant upregulation under the water deficit. In *T. kotschyanus*, instead, a peak was established under adequate water supply (75% of field capacity) at the 100% blooming stage. *TERPENE SYNTHASE 3* (*TPS3*) had a complete opposite transcript accumulation among the two species. A clear developmental upregulation was established in *T. armeniacus*, while blooming progression was accompanied by a significant reduction in transcription of *T. kotschyanus*. Similar patterns were recorded during the water deficit, where in the former a notable increase of *TPS3* transcripts was established, as opposed to the latter. Nonetheless, *TERPENE SYNTHASE 4* (*TPS4*) was found to be equally responsive for both species and a general downregulation pattern was observed when compared to the adequate water supply and 50% blooming stage.

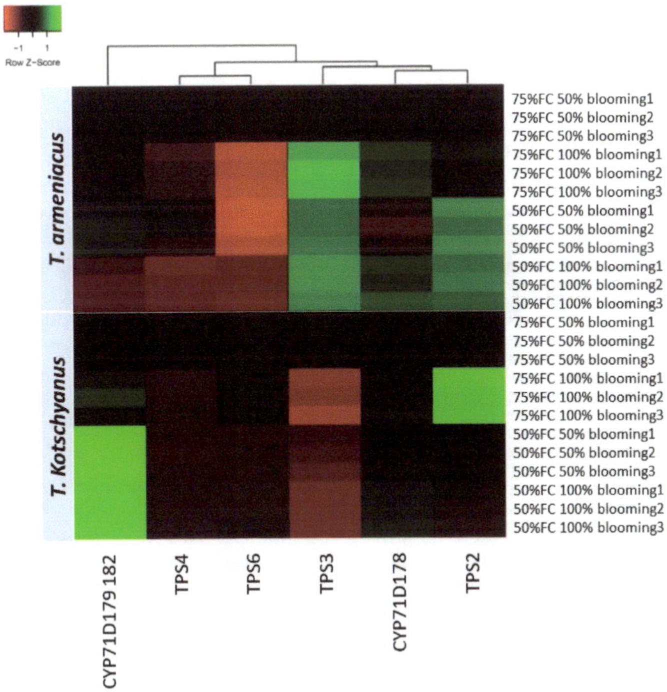

Figure 4. Heat map of relative transcript accumulation (log2) of the terpene biosynthetic pathway

across treatments. The effect of irrigation regime (75 or 50% of field capacity) on relative mRNA abundance of two *Thymus* species (*T. armeniacus, T. kotschyanus*) was investigated at two developmental stages (50 and 100% blooming). This was assessed using real-time RT-qPCR and three distinct biological repeats. Color may indicate up- or down regulation (green and red, respectively). A scale of intensity is provided.

4. Discussion

In Iran, *Thymus* spp. hold a wide distribution and long history of cultivation [4,23]. *Thymus kotschyanus* and *T. armeniacus* are not only highly popular amongst farmers, but they are also considered as greatly adapted for cultivation in Iran. Regardless, research has principally focused on the former [5], while information and knowledge on the yield traits of the latter is scarce. In this study and for the first time, the combined effects of developmental stage at harvest and soil water availability on several yield features were assessed in *T. armeniacus*. Evaluations included *T. kotschyanus* for comparison purposes.

Despite the little interspecies difference in the time required for full (100%) blooming, *T. armeniacus* underwent the beginning of flowering much sooner than *T. kotschyanus* (Table 1). Therefore, the overall period of blooming is considerably more extended in *T. armeniacus* as compared to *T. kotschyanus* (66 versus 41 d). Minor effects of soil water availability on flowering time were generally noted in both species under study (Table 1). Therefore, water deficit in the extent applied in the current study did not substantially affect the transition from the vegetative to the generative phase.

Thymus armeniacus developed shorter inflorescences with less flowers as compared to *T. kotschyanus* (Table 1). Both inflorescence traits were adversely affected by water deficit. Length was more decreased in *T. kotschyanus*, whereas the number of flowers per inflorescence was more impaired in *T. armeniacus* (Table 1). These results indicate that water deficit downgraded inflorescence features, though species differences were inconsistent among traits.

As discussed, yield may encompass fresh or dry herbal material, which in the case of *Thymus* spp. principally includes leaves and flowers. In all relevant aspects of yield, *T. armeniacus* was found to be superior than *T. kotschyanus* (Table 3). Nevertheless, the water deficit decreased plant fresh weight and economic yield more in *T. armeniacus* than *T. kotschyanus* (Table 3). In *T. kotschyanus*, a negative effect of drought on agronomic traits has also been earlier documented [24–26]. Complementarily or alternatively, the yield may be comprised of essential oil features [2,3]. Despite lower essential oil content, essential oil yield was higher in *T. armeniacus* than *T. kotschyanus* (Table 4). Similarly to plant fresh weight and economic yield, the water deficit decreased essential oil yield more in *T. armeniacus* than *T. kotschyanus* (Table 4). Taken together, these results indicate that when water availability can be secured, the cultivation of *T. armeniacus* will be associated with higher yield (thus returns), whereas under water deficit conditions the choice may be shifted to *T. kotschyanus*, which is more drought tolerant. In this perspective, *T. kotschyanus* bares a better potential for water limiting environments. Importantly, this potential is expected to increase in the near future, provided that global climate models project declining rainfall patterns allied with elevated temperatures [27].

Trichomes comprise highly dedicated secretory cells where most essential oils are anabolized and subsequently amassed in subcuticular cavities [28,29]. Although *T. armeniacus* had lower (46%) leaf trichome density as compared to *T. kotschyanus*, it underwent a much more drastic water stress-induced increase (197 vs. 13%; Table 4). Nonetheless, essential oil content was practically unaffected (1.33% reduction) in *T. armeniacus*, whereas a 15.48% reduction was noted in *T. kotschyanus* (Table 4). Still, under the molecular prism, a significant downregulation of terpene synthases and cytochromes was noted in *T. armeniacus* (Figure 4), which potentially reduces terpene precursors and thus leads to a decreased concentration of moieties in leaf trichomes.

Across *Thymus* taxa, over 360 discrete essential oil components have been characterized. An excessive part (>90%) of these moieties are attributable to the nature of monoterpenes. Among these derivatives, carvacrol and thymol are considered the most common [30]. In the current study, significant differences in terms of qualitative essential oil profiles across developmental stages at harvest and irrigation regimes were noted among the two species. *Thymus armeniacus'* terpene palette was mostly composed of terpinenes, although monoterpenes were positively regulated by the developmental stage, as well as the water deficit (Figure 1). In contrast, several pinene class terpenes were not detected in *T. kotschyanus*, revealing an inferior complexity of essential oil chemical markup. Nonetheless, several monoterpenes and monoterpene alcohols (thymol, linalool, a-Thujene) were found to be largely unaffected by the developmental stage at harvest or by soil water availability. Hence, at least in the case of *T. armeniacus* and *T. kotschyanus*, it seems that the main factor of chemical markup differences is predominantly species dependent.

In terms of physiological traits and ROS detoxifying responses, both species under study were affected by the water deficit (Table 6). Comparable patterns have been observed recently [31]. In some traits (MDA content, total phenolic content), *T. armeniacus* was more adversely affected than *T. kotschyanus*, while in others (chlorophyll content) the opposite was apparent. A comparable effect of water deficit among the two species was also documented on some traits (proline content, H_2O_2 content). Across treatments, metabolites affiliated with oxidative stress (H_2O_2, proline and MDA levels) were inter-correlated with antioxidant enzymes (APX, CAT, POX, PPO, SOD) (Figure 3). This correlation might be taken to indicate a combined regulatory biosynthetic pathway, which has also been earlier suggested [9–11]. Leaf hydration status (RWC) was also inter-correlated with leaf pigment (chlorophyll, carotenoid) content (Figure 3). Similar patterns have also previously recorded [8,31–33].

Since the present field study included a single growing season, additional research is evidently essential to reach solid results and conclusions. Nevertheless, the obtained findings serve both as a first step and a reference of how the cultivation protocol shapes marketable yield and the essential oil quality (composition) of a previously non-studied *Thymus* species.

5. Conclusions

For the first time, the dual effects of developmental stage at harvest and soil water availability on plant growth, marketable herbal material yield, as well as essential oil content, yield and composition were investigated in *Thymus armeniacus*. *Thymus kotschyanus* was also included for comparison purposes. Irrigation was adjusted to either 75 or 50% of field capacity, and plants were collected at 50 or 100% blooming. Water deficit did not substantially affect the time required for the transition from the vegetative to generative phase in either species. However, the water deficit decreased inflorescence length and the number of flowers per inflorescence. *Thymus armeniacus* was associated with enhanced yield (e.g., dry weight of harvestable organs, essential oil yield) as compared to *T. kotschyanus*, but was more affected by water deficit. Water deficit triggered an increase in leaf trichome density, an effect that was sizeable in *T. armeniacus*. Essential oil composition was largely unaffected by developmental stage at harvest or soil water availability. Oxidative stress indicators (proline, H_2O_2 and malondialdehyde levels) were inter-correlated with the activity of five key antioxidant enzymes, while a respective correlation was found for leaf water status and pigment (chlorophyll, carotenoid) accumulation. Collectively, these results denote that higher yields will be obtained by cultivating *T. armeniacus* when soil water is readily available. In contrast, under conditions of limited soil water availability, the choice shifts to the more drought tolerant *T. kotschyanus*.

Supplementary Materials: The following supporting information can be downloaded at: https://www.mdpi.com/article/10.3390/agronomy12051008/s1, Table S1. Primers used for the qPCR experiments. Table S2. Effect of irrigation regime on essential oil composition of two Thymus species at two developmental stages (50 and 100% blooming). Within each column, different letters indicate significant differences ($n = 3$). Retention indices (RI) were generated with a standard solution of n-alkanes (C_6–C_{24}) on the HP-5MS column. Figure S1. The first ten principal components and percentages of attributed variation. The first two eigenvalues were used to construct the principal component analysis biplot (accounting for the 77.4% of the cumulative percentage explained). Figure S2. Quality of representation (cos2) of the variables on factor map. variables on the first five dimensions are displayed. Size and colour intensity correlate to a better representation of specific morphological traits.

Author Contributions: Conceptualisation, S.H., B.S., S.M.-F. and D.F.; methodology, K.Y., S.M.-F., N.N. and D.F.; software, S.M.-F., N.N. and D.F.; validation, D.F.; formal analysis, N.N. and D.F.; resources, K.Y. and H.Z.; data curation, K.Y., S.M.-F., N.N. and D.F.; writing-original draft preparation, S.M.-F., K.Y. and D.F.; writing-review and editing, N.N. and D.F.; supervision, S.H. and B.S.; project administration, and funding acquisition, S.H., B.S. and M.M.G. All authors have read and agreed to the published version of the manuscript.

Funding: This research received no external funding.

Institutional Review Board Statement: Not applicable.

Informed Consent Statement: Not applicable.

Data Availability Statement: Raw data are available upon request from the corresponding author.

Acknowledgments: The authors gratefully acknowledge the laboratory staff for their contributions, continued diligence, and dedication to their craft. The valuable comments of the editor and two anonymous reviewers are greatly appreciated.

Conflicts of Interest: All authors declare that they have no conflict of interest.

Abbreviations

APX, ascorbate peroxidase; AsA, ascorbic acid; CAT, catalase enzyme; CYP, CYTOCHROME P450; dw, dry weight; fw, fresh weight; GAE, gallic acid equivalent; H_2O_2, hydrogen peroxide; MDA, malondialdehyde; NBT, nitro-blue tetrazolium chloride; PC, principal components; PCA, principal component analysis; POX, guaiacol peroxidase; PPO, polyphenol oxidase; PVP, polyvinylpyrrolidone; ROS, reactive oxygen species; RWC, relative water content; SOD, superoxide dismutase; *TPS2, TERPENE SYNTHASE 2; TPS3, TERPENE SYNTHASE 3; TPS4, TERPENE SYNTHASE 4; TPS6, TERPENE SYNTHASE 6*; u, unit.

References

1. Paschalidis, K.; Fanourakis, D.; Tsaniklidis, G.; Tzanakakis, V.A.; Bilias, F.; Samara, E.; Kalogiannakis, K.; Debouba, F.J.; Ipsilantis, I.; Tsoktouridis, G.; et al. Pilot cultivation of the vulnerable Cretan endemic *Verbascum arcturus* L. (Scrophulariaceae): Effect of fertilization on growth and quality features. *Sustainability* **2021**, *13*, 14030. [CrossRef]
2. Hassanvand, F.; Rezaei Nejad, A.; Fanourakis, D. Morphological and physiological components mediating the silicon-induced enhancement of geranium essential oil yield under saline conditions. *Ind. Crops Prod.* **2019**, *134*, 19–25. [CrossRef]
3. Taheri-Garavand, A.; Mumivand, H.; Fanourakis, D.; Fatahi, S.; Taghipour, S. An artificial neural network approach for non-invasive estimation of essential oil content and composition through considering drying processing factors: A case study in *Mentha aquatica*. *Ind. Crops Prod.* **2021**, *171*, 113985. [CrossRef]
4. Mumivand, H.; Shayganfar, A.; Tsaniklidis, G.; Emami Bistgani, Z.; Fanourakis, D.; Nicola, S. Pheno-Morphological and Essential Oil Composition Responses to UVA Radiation and Protectants: A Case Study in Three *Thymus* Species. *Horticulturae* **2022**, *8*, 31. [CrossRef]
5. Tohidi, B.; Rahimmalek, M.; Trindade, H. Review on essential oil, extracts composition, molecular and phytochemical properties of *Thymus* species in Iran. *Ind. Crops Prod.* **2019**, *134*, 89–99. [CrossRef]
6. Christensen, L.P.; Grevsen, K. Effect of development stage at harvest on the composition and yield of essential oils from thyme and oregano. *Dev. Food Sci.* **2006**, *43*, 261–264.

7. Abu-Darwish, M.S.; Alu'datt, M.H.; Al-Tawaha, A.R.; Ereifej, K.; Almajwal, A.; Odat, N.; Al Khateeb, W. Seasonal variation in essential oil yield and composition from *Thymus vulgaris* L. during different growth stages in the south of Jordan. *Nat. Prod. Res.* **2012**, *26*, 1310–1317. [CrossRef]
8. Amiri, H.; Dousty, B.; Hosseinzedeh, S.R. Water stress-induced changes of morphological, physiological and essential oil compounds in Thymus eriocalyx from Iran. *J. Essent. Oil-Bear. Plants.* **2018**, *21*, 1210–1223. [CrossRef]
9. Ahmadi-Majd, M.; Mousavi-Fard, S.; Rezaei Nejad, A.; Fanourakis, D. Carbon nanotubes in the holding solution stimulate flower opening and prolong vase life in carnation. *Chem. Biol. Technol. Agric.* **2022**, *9*, 15. [CrossRef]
10. Ahmadi-Majd, M.; Rezaei Nejad, A.; Mousavi-Fard, S.; Fanourakis, D. Postharvest application of single, multi-walled carbon nanotubes and nanographene oxide improves rose keeping quality. *J. Hortic. Sci. Biotechnol.* **2021**, 1–15. [CrossRef]
11. Chen, Y.; Fanourakis, D.; Tsaniklidis, G.; Aliniaeifard, S.; Yang, Q.; Li, T. Low UVA intensity during cultivation improves the lettuce shelf-life, an effect that is not sustained at higher intensity. *Postharvest Biol. Technol.* **2021**, *172*, 111376. [CrossRef]
12. Fanourakis, D.; Papadakis, V.M.; Psyllakis, E.; Tzanakakis, V.A.; Nektarios, P.A. The role of water relations and oxidative stress in the vase life response to prolonged storage: A case study in chrysanthemum. *Agriculture* **2022**, *12*, 185. [CrossRef]
13. Fanourakis, D.; Kazakos, P.; Nektarios, P.A. Allometric individual leaf area estimation in chrysanthemum. *Agronomy* **2021**, *11*, 795. [CrossRef]
14. Javadi Asayesh, E.; Aliniaeifard, S.; Askari, N.; Roozban, M.R.; Sobhani, M.; Tsaniklidis, G.; Woltering, E.J.; Fanourakis, D. Supplementary light with increased blue fraction accelerates emergence and improves development of the inflorescence in *Aechmea, Guzmania* and *Vriesea*. *Horticulturae* **2021**, *7*, 485. [CrossRef]
15. Fanourakis, D.; Papadopoulou, E.; Valla, A.; Tzanakakis, V.A.; Nektarios, P.A. Partitioning of transpiration to cut flower organs and its mediating role on vase life response to dry handling: A case study in chrysanthemum. *Postharvest Biol. Technol.* **2021**, *181*, 111636. [CrossRef]
16. Adams, R.P. *Identification of Essential oil Components by Gas Chromatography/Mass Spectrometry*, 4th ed.; Allured Publishing Corporation Carol Stream: Carol Stream, IL, USA, 2007.
17. Seif, M.; Aliniaeifard, S.; Arab, M.; Mehrjerdi, M.Z.; Shomali, A.; Fanourakis, D.; Li, T.; Woltering, E. Monochromatic red light during plant growth decreases the size and improves the functionality of stomata in chrysanthemum. *Funct. Plant Biol.* **2021**, *48*, 515–528. [CrossRef]
18. Taheri-Garavand, A.; Rezaei Nejad, A.; Fanourakis, D.; Fatahi, S.; Ahmadi-Majd, M. Employment of artificial neural networks for non-invasive estimation of leaf water status using color features: A case study in *Spathiphyllum wallisii*. *Acta Physiol. Plant.* **2021**, *43*, 78. [CrossRef]
19. Lichtenthaler, H.K.; Wellburn, A.R. Determinations of total carotenoids and chlorophylls a and b of leaf extracts in different solvents. *Biochem. Soc. Trans.* **1983**, *11*, 591–592. [CrossRef]
20. Bates, L.; Waldren, R.; Teare, I. Rapid determination of free proline for water-stress studies. *Plant Soil* **1973**, *39*, 205–207. [CrossRef]
21. Narwal, S.S.; Bogatek, R.; Zagdanska, B.M.; Sampietro, D.A.; Vattuone, M.A. *Plant Biochemistry*; Studium Press LLC: Houston, TX, USA, 2009.
22. Livak, K.J.; Schmittgen, T.D. Analysis of relative gene expression data using real-time quantitative PCR and the $2-\Delta\Delta CT$ method. *Methods* **2001**, *25*, 402–408. [CrossRef]
23. Bistgani, Z.E.; Hashemi, M.; DaCosta, M.; Craker, L.; Maggi, F.; Morshedloo, M.R. Effect of salinity stress on the physiological characteristics, phenolic compounds and antioxidant activity of *Thymus vulgaris* L. and *Thymus daenensis* Celak. *Ind. Crops Prod.* **2019**, *135*, 311–320. [CrossRef]
24. Asghari, B.; Mafakheri, S.; Rejali, F. Assessment of morphological, physiological, and biochemical characteristics of thymus kotschyanus bioss. And hohen under different bio and chemical fertilizers. *J. Agric. Sci. Technol.* **2021**, *23*, 1309–1325.
25. Moradi, P.; Ford-Lloyd, B.; Pritchard, J. Plant-water responses of different medicinal plant thyme (*Thymus* spp.) species to drought stress condition. *Aust. J. Crop Sci.* **2014**, *8*, 666.
26. Ashrafi, M.; Azimi-Moqadam, M.-R.; MohseniFard, E.; Shekari, F.; Jafary, H.; Moradi, P.; Pucci, M.; Abate, G.; Mastinu, A. Physiological and Molecular Aspects of Two *Thymus* Species Differently Sensitive to Drought Stress. *BioTech* **2022**, *11*, 8. [CrossRef]
27. Spinoni, J.; Barbosa, P.; Bucchignani, E.; Cassano, J.; Cavazos, T.; Christensen, J.H.; Christensen, O.B.; Coppola, E.; Evans, J.; Geyer, B.; et al. Future global meteorological drought hotspots: A study based on CORDEX data. *J. Clim.* **2020**, *33*, 3635–36561. [CrossRef]
28. Turner, G.; Gershenzon, J.; Nielson, E.E.; Froehlich, J.E.; Croteau, R. Limonene synthase, the enzyme responsible for monoterpene biosynthesis in peppermint, is localized to leucoplasts of oil gland secretory cells. *Plant Physiol.* **1999**, *120*, 879–886. [CrossRef]
29. Kowalski, R.; Kowalska, G.; Jankowska, M.; Nawrocka, A.; Kałwa, K.; Pankiewicz, U.; Włodarczyk-Stasiak, M. Secretory Structures and Essential Oil Composition of Selected Industrial Species of Lamiaceae. *Acta Sci. Pol. Hortorum Cultus* **2019**, *18*, 53–69. [CrossRef]
30. Wiese, N.; Fischer, J.; Heidler, J.; Lewkowski, O.; Degenhardt, J.; Erler, S. The terpenes of leaves, pollen, and nectar of thyme (*Thymus vulgaris*) inhibit growth of bee disease-associated microbes. *Sci. Rep.* **2018**, *8*, 14634. [CrossRef]
31. Seifikalhor, M.; Niknam, V.; Aliniaeifard, S.; Didaran, F.; Tsaniklidis, G.; Fanourakis, D.; Teymoorzadeh, M.; Mousavi, S.H.; Bustachi, M.; Li, T. The regulatory role of γ-Aminobutyric acid in chickpea plants depends on drought tolerance and water scarcity level. *Sci. Rep.* **2022**, *in press*. [CrossRef]

32. Safari, M.; Mousavi-Fard, S.; Rezaei Nejad, A.; Sorkheh, K.; Sofo, A. Exogenous salicylic acid positively affects morpho-physiological and molecular responses of *Impatiens walleriana* plants grown under drought stress. *Int. J. Environ. Sci. Technol.* **2022**, *19*, 969–984. [CrossRef]
33. Bijalwan, P.; Sharma, M.; Kaushik, P. Review of the Effects of Drought Stress on Plants: A Systematic Approach. *Preprints* **2022**, 2022020014. [CrossRef]

agronomy

MDPI

Article

Essential Oil Composition and Bioactive Properties of Lemon Balm Aerial Parts as Affected by Cropping System and Irrigation Regime

Antonios Chrysargyris [1,*], Spyridon A. Petropoulos [2] and Nikolaos Tzortzakis [1,*]

1 Department of Agricultural Sciences, Biotechnology and Food Science, Cyprus University of Technology, Lemesos 3603, Cyprus

2 Laboratory of Vegetable Production, University of Thessaly, 38446 Nea Ionia, Greece; spetropoulos@uth.gr

* Correspondence: a.chrysargyris@cut.ac.cy (A.C.); nikolaos.tzortzakis@cut.ac.cy (N.T.)

Abstract: The ongoing climate crisis necessitates the sustainable use of natural resources and the adoption of environmentally friendly agronomic practices. Deficit irrigation is an ecofriendly technique that allows for the improvement in the water use efficiency of crops. On the other hand, medicinal and aromatic crops, which usually have an innate tolerance to harsh conditions, are suitable candidates for cultivation under low-input cropping systems. In the present study, *Melissa officinalis* plants were cultivated under conventional and organic cropping systems, while in each system two irrigation regimes (full irrigation or deficit irrigation) were tested. The aerial parts of the plants were evaluated in terms of growth and physiological parameters, chemical composition, antioxidant activity, essential oil yield and essential oil composition. Our results indicate that prolonged water stress after two deficit irrigation cycles had detrimental effects on the plant growth and biomass production, whereas it significantly increased the essential oil yield, regardless of the cropping system (organic or conventional cultivation). The recorded physiological parameters are in agreement with morphological features, especially the stomatal conductance, which was significantly reduced under deficit irrigation for both cultivation systems, revealing that the growth inhibition was the result of stomatal closure and carbon dioxide deprivation. Deficit irrigation and organic cultivation also increased total phenol and total flavonoid content, especially in the second harvest, thus resulting in higher antioxidant activity assayed by the FRAP method. In contrast, DPPH and ABTS methods did not show any differences among the tested treatments in the second harvest, which suggests that other bioactive compounds are also involved in the overall antioxidant mechanism of lemon balm plants, as indicated by the increased ascorbic acid content. Regarding the essential oil composition, the major detected compounds were geranial and neral and, although they were both increased under the organic cropping in the first harvest, the same trend was not observed in the second harvest. Finally, a variable effect of cropping system and irrigation regime on minerals content was recorded. In conclusion, deficit irrigation is an ecofriendly practice that could be applied in conventional and organic cropping systems of lemon balm crops, aiming to reduce irrigation water consumption and compensate for reduced herb yields with increased essential oil yield and polyphenol content.

Keywords: *Melissa officinalis*; organic cultivation; water deficit; essential oils; volatile compounds; antioxidant activity; nutrient content

1. Introduction

Melissa officinalis (L.) or lemon balm is a widely consumed perennial medicinal herb belonging to the Lamiaceae family. It is native to the Mediterranean basin and central Asia; however nowadays it can be found throughout the world as a commonly cultivated herb [1]. Moreover, the distinctive aroma of its flowers make it suitable for the uses as a flavoring agent, while it is considered as a very important pollinator and it is commonly cultivated for ornamental purposes [2,3]. The aerial parts of the species are highly appreciated for their bioactive properties and are usually consumed in the form of infusions and herbal extracts in folk and traditional medicine [4,5]. The most common bioactive effects of these formulations include the treatment of headaches, gastrointestinal and neurodegenerative disorders, as well as liver and bladder diseases [2,6–8] and antimicrobial properties [9–11]. These effects are associated with various compounds, including various flavonoids, rosmarinic acid and lithospermic acid [1,12–17].

Apart from the herbal extracts and infusion which are intended for edible use, the aerial parts of the species contain essential oils which are also beneficial to human health due to their high content of monoterpenes (geranial, citronellal and neral) [18,19] and sesquiterpenes (β-caryophyllene) [16]. However, the composition of essential oils may be altered by growing conditions, especially when plants are subjected to abiotic stress such as water shortage [20], salinity [1], or nutrient imbalances [21], while the content of the individual compounds may vary depending on the harvesting date [22], post-harvesting treatments [23–25], genetic material [26–28] and plant part (leaves and flowers) [29]. Moreover, the cropping system (conventional vs. organic farming) may also affect the composition of essential oils [15], although there are studies in which the major compound content remained unaffected [19,30]. Finally, although lemon balm oil is generally regarded as safe (included in the GRAS list) [31], according to Stojanović et al. [32] the essential oil of the lemon balm, as well as the individual major compounds (citronella, geranial and neral) may exhibit moderate oral toxicity.

The environmental burden from fertilizer and agrochemical inputs in medicinal crops production, as well as the lack of certified pesticides and herbicides for various medicinal crops has increased the interest in adopting organic cropping systems [33,34]. Crop production following low-input cultivation strategies is suitable for medicinal and aromatic plants due to their low requirements for nutrients and their hardiness against biotic and abiotic stressors. Apart from environmental issues, organic cultivation of medicinal plants is associated with the higher quality of the final product compared to conventional practices, while the low use of chemical fertilizers and pesticides ensures the safety of the final products due to the lower content of heavy metals and chemical residuals [35,36]. Another aspect to be considered is that the commercial cultivation of medicinal and aromatic plants reduces the danger of genetic erosion due to the irrational harvest of wild plants, while it improves market availability throughout the year and allows the standardization of the final product following specific safety protocols [33,37]. So far, several reports have evaluated the possibility of growing lemon balm following organic farming practices with promising results [30,38,39]. Based on these reports, the lower fresh herb yields expected compared to conventional cropping methods could be compensated by the improved quality of the final product as well as by reduced production costs [39].

Deficit irrigation is another environmentally friendly agronomic practice that has been the focus of research in recent years due to the scarcity of water for agricultural uses and the climate crisis that necessitates the sustainable use of natural resources [40,41]. The main objective of deficit irrigation is to reduce water inputs and retain yield parameters as close as possible to the fully irrigated plants by adjusting the irrigation regime below plant requirements [42,43]. Therefore, any decrease in the obtained yields is compensated by the increased water use efficiency, and in several cases, by the improved quality of the final product. In particular, there are several studies in medicinal crops where deficit irrigation is reported to increase product quality (herbs or essential oils), since mild water stress conditions induce the biosynthesis of the secondary metabolites (phenolic compounds and antioxidants), thus increasing the bioactive properties of the products [3,18,44]. Moreover, deficit irrigation may result in higher essential oil yields per plant and increase the overall crop profitability [45–47]. However, the induction of abiotic stress conditions is also accompanied by changes in the volatile compound profile, which also should be considered in regards to the quality of the final product [17,44–46,48].

Considering the numerous factors that may affect essential oil and chemical composition of plant tissues, the aim of the present study was to evaluate the effect of environmentally friendly practices on the growth, yield and quality of lemon balm plants. For this purpose, an experiment was carried out where organic vs. conventional cropping systems, as well as full vs. deficit irrigation were evaluated. The parameters tested included plant growth and yield, as well as the chemical composition of plant tissues and essential oils of the aerial parts of the plant.

2. Materials and Methods

2.1. Plant Material and Experimental Conditions

Lemon balm (*Melissa officinalis*) seedlings were purchased from the Cypriot National Centre of Aromatic Plants in seeding trays, at the growth stage of 3–4 leaves and 4–5 cm height. The seedlings were established under field conditions, during the spring-summer growing period in a commercial organic farm, Limassol, Cyprus (34°38′ N, 32°56′ E, 7 m). The experimental field occupied approximately 350 m^2, and the soil properties were as follows: organic matter = 3.01%; available $CaCO_3$ = 21.23%; pH = 8.42; EC (electrical conductivity) = 0.78 mS/cm. The climate of the experimental location is dry with average midday temperature and air humidity during the summer months being ca. 34.2 °C and 59%, respectively.

2.2. Cultivation Practices

The seedlings were transplanted in soil and arranged in triple rows (rows were 0.2 m apart and plants were separated by 0.33 m within the same row) at a plant density of 56,818 plants/ha. The seedlings were grown for about four months. The experimental farm was divided into four treatments: (i) conventional cultivation with full irrigation (CFI); (ii) conventional cultivation with deficit irrigation (CDI); (iii) organic cultivation with full irrigation (OFI); and (iv) organic cultivation with deficit irrigation (ODI). Each treatment consisted of three plots (replicates) and each plot had 30 plants (360 plants were used in total). Registered organic or conventional fertilizers and pesticides were used according to the best practice guides for the species, with a detailed cultivation management being described in Table S1.

The amount of irrigation applied in each treatment was programmed based on the soil volumetric water content measured by field-scout TDR300 (Spectrum Technologies Inc., Aurora, IL, USA), equipped with 20 cm rods. The irrigation water was supplied approximately every 4–5 days. The soil water content measurements took place at intervals of 5 days. The plants were grown for two months under full irrigation. Then, the deficit irrigation (ca. 50% of the full irrigation treatment, namely 3.16 m^3 compared to 6.69 m^3 of the full irrigation regime) was applied for three weeks before the first and second harvest of the aerial parts. The first harvest took place at the early flowering stage of the plants (in May 2018), while the second harvest was performed in June 2018. Between the two harvests (four week period), the crop was irrigated normally and according to the plant water needs for one week after the first harvest in order to allow the recovery of the biomass production. Then, the second cycle of the deficit irrigation was applied for three weeks and until the second harvest.

2.3. Plant Growth, Physiology, and Minerals

Two weeks after the water stress initiation (one week before the first and second harvest), the physiological parameters were recorded, such as leaf stomatal conductance and chlorophyll fluorescence. The stomatal conductance measurements were carried out using a ΔT-Porometer AP4 (Delta-T Devices, Cambridge, UK). The leaf chlorophyll fluorescence levels (Chlorophyll fluoremeter, opti-sciences OS-30p, Hertfordshire, UK) and SPAD index values were measured in three fully expanded, sun-exposed leaves per plant [49]. The plant height was recorded in six plants per treatment before each harvesting. The plants were harvested at 3 cm above the soil in order to allow the recovering biomass production. Then, the upper fresh material was weighed (g) for fresh weight determination, while fresh samples were air-dried in a forced-air oven at 70 °C and until constant weight for dry matter content (%) calculation.

The chlorophyll a (Chl a), chlorophyll b (Chl b) and total chlorophyll (t-Chl) content were determined according to the protocol described by Richardson et al. [50]. Briefly, leaf disks (0.1 g consisted of a pool of two plants tissue) were incubated in a heat bath at 65 °C for 30 min, in the dark, with 10 mL dimethyl sulfoxide (DMSO, Sigma Aldrich, Taufkirchen, Germany) for the chlorophyll extraction. The absorbance of the extract was measured at 645 nm and 663 nm using a microplate spectrophotometer (Multiskan GO, Thermo Fischer Scientific, Massachusetts, MA, USA). The photosynthetic leaf pigments (Chl a, Chl b and t-Chl) content was calculated using the following equations: Chl a = $0.0127 \times A_{663} - 0.00269 \times A_{645}$; Chl b = $0.0229 \times A_{645} - 0.00468 \times A_{663}$; and t-Chl = $0.0202 \times A_{645} + 0.00802 \times A_{663}$. The results were expressed as mg of chlorophyll per g of fresh weight.

The mineral content in leaves was determined in three replications per treatment (three pooled plants per replication). The samples were dried to constant weight (at 65 °C for 4 d) and milled at <0.42 mm. Sub samples (~0.5 g) were burned to ash in a furnace (Carbolite, AAF 1100, GERO, Lilienthal, Germany) at 450 °C for 5 h and then were digested with acid (2 N HCl). The mineral assessment for potassium (K), sodium (Na), phosphorous (P) and nitrogen (N) was performed according to Chrysargyris et al. [51] and magnesium (Mg), calcium (Ca), copper (Cu), and zinc (Zn) by an atomic absorption spectrophotometer (PG Instruments AA500FG, Leicestershire, UK). The data were expressed in g/kg and mg/kg of dry weight, for macronutrients and micronutrients, respectively.

2.4. Essential Oil Extraction and Analysis

The aerial parts (leaves and flowers) of the lemon balm plants were harvested at the early flowering stage and three biological replicates (pooled samples of three individual plants per replicate) from each treatment were dried in a forced-air oven at 42 °C according to Calín-Sánchez et al. [52] and preliminary tests of our team. The dried plant material was chopped and hydro-distilled for 3 h, using a Clevenger apparatus for the essential oil (EO) extraction. The EO yield was calculated (in terms of % and L/ha) and oils were analyzed by gas chromatography-mass spectrometry (GC/MS- Shimadzu GC2010 gas chromatograph interfaced Shimadzu GC/MS QP2010plus mass spectrometer, Tokyo, Japan) and the constituents were determined [53].

2.5. Polyphenols, Flavonoids, Ascorbic Acid and Antioxidant Activity

Fresh samples (0.5 g) of the aerial parts (leaves and flowers) collected at the early flowering stage from four replicates (pooled by two individual plants/replicate) for each treatment were milled with 10 mL methanol (50%) and the extraction was assisted with ultrasounds. The antioxidant activity of the methanolic plant extracts was determined by using the assays of 2,2-diphenyl-1-picrylhydrazyl (DPPH) and ferric reducing antioxidant power (FRAP), as previously described by Chrysargyris et al. [54], while the 2,2′-azino-bis(3-ethylbenzothiazoline-6-sulphonic acid; ABTS) assay was implemented according to the methodology described by Woidjylo et al. [55]. The results were expressed as Trolox ((±)-6-Hydroxy-2,5,7,8-tetramethylchromane-2-carboxylic acid) equivalent (mg trolox/g of fresh weight). The total phenolic compound content was measured using the Folin-Ciocalteu method, and the results were expressed as gallic acid equivalents (μmol GAE/g of fresh weight (Fw)) as described previously by Tzortzakis et al. [56]. The total flavonoid content was determined according to the aluminium chloride colorimetric method [57] and expressed as rutin equivalents (mg rutin/g of Fw). The ascorbic acid (AA) content was quantified by titration with 2,6-dichlorophenol- indophenol [58] and the results were expressed as mg of AA per g of fresh weight.

2.6. Statistical Methods

The experiment was performed according to the split-split-plot design with three factors (cultivation practice, irrigation and harvesting period) and three replicates ($n = 3$) per treatment. The measurements were performed in three to six biological replications/treatments (each replication consisted of a poll of three individual measures/samples). Statistical analysis was performed using the SPSS statistical software (SPSS v.22; IBM, Armonk, NY, USA). The data means were also compared with one-way analysis of variance (ANOVA) and Duncan's multiple range test was used for the comparison of treatment means at $p < 0.05$.

3. Results and Discussion

The growth and essential oil yield parameters of lemon balm plants in relation to the cropping system and irrigation regime are presented in Table 1. In the first harvest, plant height and fresh weight showed decreasing trends under the deficit irrigation conditions, regardless of the cropping system, while the plants grown in organic farming had lower height and fresh weight than the conventionally grown ones. A similar trend was observed in the second harvest regarding the effect of the irrigation regime, whereas the organically grown and fully irrigated plants recorded the highest plant height. It is worth mentioning that the lowest overall values of plant height and fresh weight were observed in organically grown plants subjected to deficit irrigation. Moreover, dry matter content increased under deficit irrigation conditions for both cultivation systems, whereas in the second harvest significantly lower values were recorded in the organically grown and fully irrigated plants. Plant dry weight revealed similar trends as the plant fresh weight, especially in the second harvest, where the lowest values were recorded under deficit irrigation conditions, regardless of the cropping system. According to the literature, water stress may severely affect the growth of medicinal and aromatic plants such as lavender (*Lavandula angustifolia*) and Greek sage (*Salvia fruticosa*), while it increases dry matter content [53]. These effects are associated with reduced cell multiplication and expansion, which eventually reduce plant growth and biomass production [48]. Similar results were reported for lemon balm by Abbaszadeh et al. [44] and Ahmadi et al. [14], which indicate that water availability is a limiting factor for high biomass yield. Additionally, Németh-Zámbori et al. [46] evaluated the effect of water stress on biomass yield of four Lamiaceae species and suggested that lemon balm was the second most sensitive species after peppermint.

Table 1. Effect of cultivation (conventional-C or organic-O) and irrigation (full irrigation-FI or deficit irrigation-DI) practices on *Mellissa* plant growth and essential oil yields under two harvestings.

Treatment	Height (cm)	Fresh Weight (g)	Dry Matter Content (%)	Yield Dry Weight (ton/ha)	EO Yield (%)	EO Yield (L/ha)
1st harvest						
CFI	34.66 ± 0.61 a	132.13 ± 14.04 a	21.06 ± 0.72 c	1.34 ± 0.16 a	0.285 ± 0.026 a	3.94 ± 0.58 a
CDI	28.50 ± 1.83 b	86.02 ± 22.20 ab	23.21 ± 0.56 ab	1.14 ± 0.19 ab	0.305 ± 0.023 a	3.66 ± 0.81 a
OFI	28.16 ± 1.55 b	77.00 ± 15.57 b	21.63 ± 0.36 bc	0.94 ± 0.11 ab	0.328 ± 0.027 a	3.12 ± 0.47 a
ODI	26.25 ± 1.03 c	62.85 ± 2.73 c	24.48 ± 0.87 a	0.87 ± 0.04 b	0.335 ± 0.038 a	2.92 ± 0.22 a
2nd harvest						
CFI	29.25 ± 1.92 ab	155.23 ± 26.76 a	21.34 ± 0.68 a	1.87 ± 0.19 a	0.445 ± 0.047 b	8.45 ± 1.18 a
CDI	26.66 ± 1.69 bc	101.46 ± 17.03 ab	23.20 ± 1.03 a	1.34 ± 0.15 bc	0.511 ± 0.022 ab	6.92 ± 0.81 ab
OFI	32.50 ± 0.85 a	140.56 ± 16.55 a	19.16 ± 0.45 b	1.53 ± 0.12 ab	0.415 ± 0.030 b	6.53 ± 0.84 ab
ODI	24.83 ± 0.78 c	78.75 ± 0.84 b	22.71 ± 0.41 a	1.01 ± 0.02 c	0.580 ± 0.049 a	5.90 ± 0.33 b

Values ($n = 6$ for plant growth; $n = 4$ for oil yields) in column for each harvest followed by the same letter are not significantly different.

Similarly, García-Caparrós et al. [47] reported the varied response of six Lamiaceae species to water stress conditions in terms of biomass production and suggested a genotype dependent response. However, it has to be noted that deficit irrigation practices allow for the regulation of water availability according to the plant requirements and avoid the severe water stress conditions that may have a severe impact on plant growth. Finding the golden ratio between high plant growth and high water use efficiency is always the case in studies that evaluate the effects of deficit irrigation. According to the literature, the plant growth is negatively affected in lavender, Greek sage or peppermint when water availability is reduced by 50% [48,53,59], while the increase in the water stress intensity results in increasing yield losses [60]. In contrast, the moderate water stress (50% of field capacity) may increase the number of umbels and seed yield in cumin [61]. In the present study, the deficit irrigation practice was able to save more than 50% of water (3.16 m^3 at DI vs. 6.69 m^3 at FI) during the deficit irrigation application, which is translated to 499 m^3 vs. 1055 m^3 per hectare for the DI and FI, respectively. These amounts of water savings are of great importance in semi-arid and arid areas with water scarcity, such as the Mediterranean basin.

Therefore, the effectiveness of deficit irrigation practices depends on the marketable part of plants and the severity of water shortages, while stress duration is also a key factor in the plant response to drought [62]. The cropping system may also affect the plant growth of medicinal and aromatic species; according to Anwar et al. [63], the application of vermicompost or farmyard manure combined with inorganic fertilizers showed better results than inorganic fertilizers alone in terms of herb yield of French basil. However, varied results were reported for the effects of cropping systems on dry matter content of two medicinal species (peppermint and sage), where peppermint was not affected whereas sage dry matter content increased in the organically cultivated plants [64]. Moreover, the integration of vermicompost and chemical fertilizers resulted in higher dry matter content in French basil plants in comparison with the vermicompost, chemical fertilizers or the combination of farmyard manure and chemical fertilizers [63].

Apart from the biomass yield (fresh or dry), the medicinal plants are usually cultivated for essential oil production. Therefore, the effects of deficit irrigation and cropping system on essential oil yield are of great interest in order to evaluate the efficiency of these agronomic practices under the commercial farming conditions. In our study, the essential oil (EO) yield (%) was not affected by the tested factors in the first harvest, while it showed an increase for the deficit irrigation treatments of both cultivation systems. Interestingly, the EO yield per ha was not affected either by cropping system or the irrigation regime in the first harvest, whereas a significant decrease was recorded in organically grown plants under deficit irrigation only when compared to fully irrigated conventional ones. The lack of effect of cropping system and irrigation regime on oil yield (EO% or per hectare) in the first harvest could be associated with the low water requirements of the species at the first growth stages, since according to Ghamarnia et al. [65], the lemon balm plants need lower water amounts by approximately 23% when successive harvestings are implemented compared to the plants where a single harvest at flowering is implemented. Therefore, it seems that even the deficit irrigation levels applied in our study did not induce plant secondary metabolism for the biosynthesis of essential oils. In contrast, the prolonged water stress as indicated by the application of the second deficit irrigation cycle in our study resulted in stress conditions which consequently increased the essential oil biosynthesis, as expressed by the EO% values, although the reduction in biomass production resulted in a decrease of EO yield per harvested area.

The findings of our study are confirmed by Németh et al. [66], who suggested that short-term water shortages (up to three weeks) followed by rehydration does not affect the essential oil yield and only when water deficit is applied for six weeks are the effects on oil yield irreversible. In contrast to our study, Bonacina et al. [1] suggested that salinity stress decreased essential oil yield with increasing salinity levels, while it increased the number of detected volatile compounds. This disagreement indicates that salinity stress has more severe effects on lemon balm plants than water stress and the increasing salinity overcomes the defense mechanisms of the plants resulting in inhibited growth and limited essential oil yield. Ozturk et al. [67] further justifies this argument in a study where lemon balm plants were subjected to salinity and water stress, wherein despite the fact that plant growth was inhibited under both stress conditions, the essential oil yield decreased and increased under the increasing salinity and water stress, respectively. On top of that, Szabó et al. [68] highlighted the genotype effect of lemon balm species on plant response to water stress and reported significant differences in essential oil yield of five cultivars subjected to drought. Finally, Petropoulos et al. [69] suggested that the reduced biomass recorded in parsley plant grown under deficit irrigation allows the increase in plant density and eventually the increase of the EO yield per harvested area.

The effect of cropping system on essential oil yield was not significant in both harvests, regardless of the irrigation regime (Table 1). This finding is in agreement with the results of Seidler-Łożykowska et al. [30], who suggested that oil yield of lemon balm plants did not differ between the organic and conventional cultivation at three distinct locations in Poland. Similar results were reported by Sodré et al. [19], who compared the effect of different amounts of cattle manure with mineral fertilizers without recording significant differences in essential oil content of fresh or dried lemon balm leaves. In contrast to our study, Németh-Zámbori et al. [46] reported the decreased oil content in lemon balm leaves subjected to water stress. This difference could be due to different growing conditions between the study of Németh-Zámbori et al. and our study (pot cultivation vs. soil cultivation, respectively), different stress levels (40% vs. 50%) and stress duration (12 weeks vs. two cycles of three weeks).

The physiological parameters recorded during the growing period are presented in Figure 1 and Table 2. The SPAD index was the highest in fully irrigated and conventionally grown plants in both harvesting periods. Moreover, the plants subjected to deficit irrigation decreased their leaf stomatal conductance during the second harvesting period in order to maintain water storage in the leaves, whereas no significant differences were observed in the first harvesting season (Figure 1). Similarly, chlorophyll fluorescence (Fv/Fm) was not affected in the first deficit irrigation cycle, whereas the highest values were recorded for the fully irrigated and organically grown plants, being significantly different only from the conventionally grown and subjected to deficit irrigation plants. Regarding chlorophyll content, plants grown under the conventional conditions had the highest total chlorophyll and chlorophyll b content in the first harvest, while chlorophyll a was significantly higher in the conventionally grown and fully irrigated plants. In contrast, deficit irrigation resulted in a significant increase of individual and total chlorophyll content in the second harvest, regardless of the cropping system. The findings of our study indicate that the second deficit irrigation cycle subjected the lemon balm plants to water stress conditions as suggested by the decrease in stomatal conductance values, while the chlorophyll fluorescence values did not differ between the fully irrigated and water stressed plants, regardless of the cropping system. Similar results were reported by Chrysargyris et al. [54], who also recorded a decrease of stomatal conductance values in leaves of *Sideritis perfoliata* L. subsp. *perfoliata* plants subjected to experimental treatments similar to our study, while chlorophyll fluorescence values were not affected. Moreover, Marino et al. [70] recorded a 40% reduction in stomatal conductance of *Mentha spicata* plants subjected to water stress which was also associated with the reduced net photosynthesis and plant growth, while Bonacina et al. [1] reported similar results for the lemon balm plants grown under saline conditions. Stomatal closure is a protective mechanism that allows plants to retain water content, while at the same time it limits CO_2 absorptions, which results in reduced biosynthesis and plant growth [45]. In addition, Parkash and Singh [71] suggested that the protective mechanism of the stomatal closure is mostly effective under short term water shortages and it helps plants to retain leaf water potential. Apart from the stomatal closure, the degradation of chlorophyll is also a stress index which affects the photosynthetic activity and results in the inhibited plant growth [71]. However, the SPAD index values in our study do not indicate the decrease of chlorophyll content, which either increased in the case of the conventionally grown plants in the first harvest or remained unaffected in the organically grown plants in both harvests under deficit irrigation conditions. Moreover, the actual chlorophyll content values showed a significant increase under the prolonged water stress (see second harvest), regardless of the cropping system. These findings are of particular interest, since despite the stunted plant growth under deficit irrigation, it seems that the water stress does not have a negative effect on chlorophyll content, especially at the second harvest where total and individual chlorophyll content increases under the deficit irrigation, regardless of the cropping system. Therefore, it could be suggested that plant response of lemon balm to stressors is limited to the stomatal closure. The same results were reported by [72], who suggested that the deficit irrigation up to 50% of the daily pan evaporation did not affect the SPAD index values in lemon balm leaves.

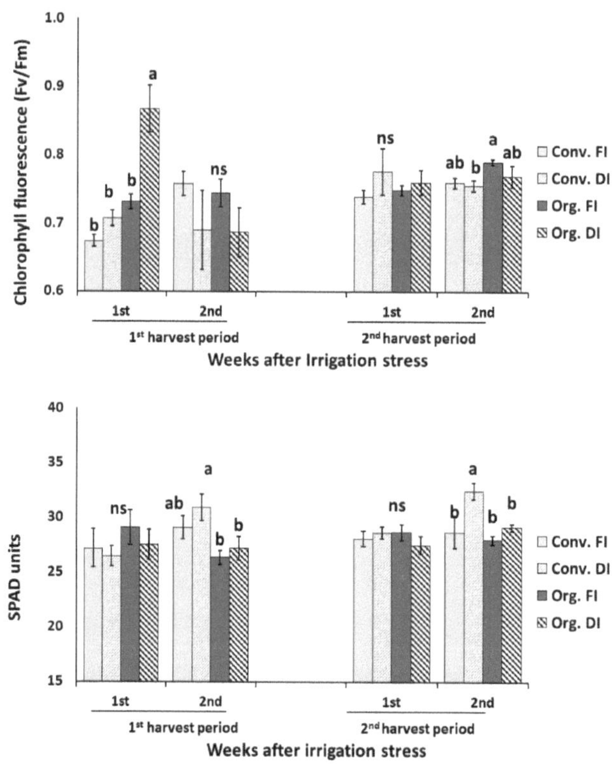

Figure 1. *Cont.*

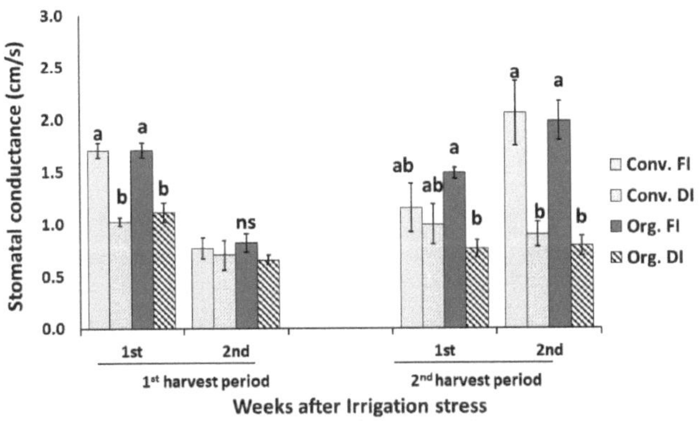

Figure 1. Effect of cultivation (conventional-C or organic-O) and irrigation (full irrigation-FI or deficit irrigation-DI) practices on Mellissa plants physiological parameters at two harvesting periods. Values represent mean (±SE) of measurements made on 6 independent replications per treatment. Mean values followed by the same letter do not differ significantly at $p \geq 0.05$ according to Duncan's MRT; ns: no significance.

Table 2. Effect of cultivation (conventional-C or organic-O) and irrigation (full irrigation-FI or deficit irrigation-DI) practices on *Mellissa* plants content (mg/g Fw) for chlorophyll a, chlorophyll b and total chlorophyll under two harvestings.

Treatment	Chlorophyll a	Chlorophyll b	Total Chlorophylls
		1st harvest	
CFI	1.22 ± 0.11 [a]	0.37 ± 0.04 [a]	1.59 ± 0.15 [a]
CDI	1.00 ± 0.02 [b]	0.36 ± 0.01 [a]	1.37 ± 0.03 [ab]
OFI	0.88 ± 0.04 [b]	0.24 ± 0.01 [b]	1.12 ± 0.06 [b]
ODI	0.98 ± 0.02 [b]	0.31 ± 0.01 [ab]	1.29 ± 0.03 [b]
		2nd harvest	
CFI	0.84 ± 0.03 [bc]	0.61 ± 0.02 [bc]	1.45 ± 0.06 [bc]
CDI	0.96 ± 0.03 [a]	0.70 ± 0.02 [a]	1.66 ± 0.05 [a]
OFI	0.78 ± 0.03 [c]	0.57 ± 0.02 [c]	1.36 ± 0.06 [c]
ODI	0.91 ± 0.02 [ab]	0.66 ± 0.01 [ab]	1.58 ± 0.03 [ab]

Values (n = 4) in column for each harvest followed by the same letter are not significantly different.

In contrast to our study, Golubkina et al. [73] suggested a concomitant decrease of total phenolic compounds and total chlorophyll content in *Artemisia dracunculus* and *Hyssopus officinalis* plants subjected to abiotic stress, Moreover, Hallmann and Sabała [35] reported that conventionally grown herbs contained higher amounts than the organic ones, a trend which was clearly observed only in the first harvest of our study. These contradictory results indicate that the lemon balm of our studies activated an effective protective mechanism under deficit irrigation through the induction of antioxidant compound biosynthesis (see the results of phenolic compounds and ascorbic acid content in Table 3), which increased their tolerance under abiotic stress conditions.

Table 3. Effect of cultivation (conventional-C or organic-O) and irrigation (full irrigation-FI or deficit irrigation-DI) practices on *Mellissa* plants, total phenolics (μmol GAE/g Fw), total flavonoids (mg rutin/g Fw), antioxidant status (ABTS, DPPH, FRAP; mg Trolox equivalents/g Fw) and ascorbic acid (AA; mg/g Fw) under two harvestings.

Treatment	Total Phenols	Total Flavonoids	ABTS	DPPH	FRAP	AA
			1st harvest			
CFI	225.53 ± 11.73 ^{b}	14.21 ± 1.56 ^{c}	27.95 ± 1.54 ^{b}	48.21 ± 3.07 ^{b}	88.36 ± 13.95 ^{b}	33.03 ± 2.01 ^{c}
CDI	205.64 ± 11.78 ^{b}	13.24 ± 0.29 ^{c}	28.52 ± 2.03 ^{b}	43.20 ± 2.03 ^{b}	83.82 ± 15.77 ^{b}	67.50 ± 5.50 ^{b}
OFI	350.94 ± 25.61 ^{a}	26.75 ± 1.19 ^{b}	36.58 ± 1.30 ^{a}	74.48 ± 3.41 ^{a}	135.34 ± 11.02 ^{a}	66.05 ± 2.60 ^{b}
ODI	365.28 ± 10.54 ^{a}	31.26 ± 1.36 ^{a}	43.08 ± 3.87 ^{a}	70.83 ± 3.77 ^{a}	150.70 ± 12.46 ^{a}	81.35 ± 2.85 ^{a}
			2nd harvest			
CFI	231.88 ± 20.38 ^{c}	29.91 ± 2.23 ^{b}	32.40 ± 0.96 ^{a}	78.85 ± 12.02 ^{a}	105.08 ± 8.88 ^{b}	73.75 ± 1.15 ^{c}
CDI	328.42 ± 23.18 ^{b}	35.84 ± 2.70 ^{ab}	42.02 ± 2.10 ^{a}	96.80 ± 10.38 ^{a}	127.82 ± 4.67 ^{b}	71.45 ± 3.05 ^{c}
OFI	271.32 ± 19.89 ^{bc}	32.09 ± 2.62 ^{ab}	36.79 ± 2.07 ^{a}	98.53 ± 5.08 ^{a}	123.90 ± 6.91 ^{b}	89.59 ± 3.01 ^{b}
ODI	410.43 ± 33.00 ^{a}	39.01 ± 2.31 ^{a}	45.57 ± 7.72 ^{a}	91.29 ± 9.49 ^{a}	55.96 ± 11.66 ^{a}	121.53 ± 2.45 ^{a}

Values (n = 4) in column for each harvest followed by the same letter are not significantly different.

The results related to the total phenol, ascorbic acid and total flavonoid content, as well as to the antioxidant activity assays performed are presented in Table 3. In the first harvest, total phenol content was higher in the organically grown plants by 55.6% and 77.6% under full irrigation and deficit irrigation, respectively, while total flavonoid content was the highest in organically grown plants subjected to deficit irrigation (31.26 mg rutin/g Fw), more than double the content of conventionally grown plants. Ascorbic acid content was also increased in organically grown plants with the highest values recorded in plants subjected to DI. Similarly, the antioxidant activity (ABTS, DPPH, FRAP) increased in the organically grown plants, regardless of the irrigation system. In the second harvest, the total phenol content and FRAP antioxidant activity were significantly higher in the organically grown plants subjected to water stress (410.43 μmol GAE/g Fw and 155.96 mg Trolox/g Fw, respectively), while no significant differences were observed between the tested treatments in the case of antioxidant activity determined with the ABTS and DPPH assays. Moreover, the total flavonoid content was the highest for the organically grown plants subjected to deficit irrigation (39.01 mg rutin/g Fw) without being significantly different from the normally irrigated plants of the same cropping system or from the conventionally grown plants subjected to deficit irrigation.

Similarly to our study, Németh-Zámbori et al. [18] reported an increase of the total phenolic compounds in the lemon balm shoots with the increasing water stress severity, whereas the opposite trend was recorded for root tissues. Moreover, in the same study total flavonoid content in shoots and roots was not affected by water stress levels [18]. Chrysargyris et al. [54] also reported an increase of the total phenol content in *S. perfoliata* L. subsp. *perfoliata* plants grown under deficit irrigation conditions, which was more profound in the first harvest compared to the second one. This difference with our study could probably be attributed to the severity of the water stress effects, since the growth inhibition in the study of Chrysargyris et al. [54] was less severe than that observed in our study, indicating two possible responses: firstly, the protective mechanisms of plants did not induce the biosynthesis of phenols because other antioxidant mechanisms were activated, or secondarily, the protective mechanism related to phenol biosynthesis collapsed and other antioxidant mechanisms were activated [18]. The total phenol content was also increased under moderate or severe water stress conditions in Greek sage plants, whereas no significant changes were recorded in lavender [53]. Chrysargyris et al. [54] observed the same trend in the total flavonoid content of *S. perfoliata* L. subsp. *perfoliata* between the two harvests, which also differed from our study in regards to total flavonoid content recorded in the second harvest. The explanation for this difference could be similar to that for the total phenol content and be due to the different plant species studied.

Regarding the effect of the organic cultivation practices, de Assis et al. [15] suggested that the combined application of organic manure and arbuscular mycorrhizal fungi on *Melissa officinalis* plants resulted in decreased content of the total phenols, whereas organic manure decreased total flavonoid content, regardless of arbuscular mycorrhizal fungi application. Similarly to our study, Hallmann and Sabała [35] evaluated for two consecutive years the quality of the organic and conventional herbs and reported that the organic cropping system significantly increased the total phenol and total flavonoid content. The same trend was recorded by Kazimierczak et al. [64], who suggested that the organic cultivation significantly increased both the total phenolic compounds and total flavonoid content in lemon balm. Therefore, it seems that the organic cultivation practices induce the biosynthesis of polyphenols, regardless of the presence of abiotic stressors or not. The reason for this increase could be associated with (a) the slower plant development under organic cultivation, which results in slower translocation and consumption of biosynthetic products and secondary metabolites in the plant growth, or (b) the presence of biotic stressors such as pests and pathogens, which induces the biosynthesis of protective compounds [74].

The results of our study indicate that the total phenols and flavonoids contributed to antioxidant activity, especially in the first harvest, where organic cultivation and/or deficit irrigation resulted in higher polyphenol content and antioxidant activity. However, in the second harvest this was the case only for the FRAP assay, since the detected Trolox values for ABTS and DPPH assays did not differ among the treatments. The positive correlation of total phenols and flavonoids with the antioxidant activity is well-documented in various crops, since the main role of these secondary metabolites is to protect the plants under the stressful conditions [75,76]. Moreover, the variable findings for the tested assays is very common in natural matrices, indicating the presence of other bioactive compounds not determined in our study (e.g., tocopherols, organic acids, essential oils) that could also contribute to the overall antioxidant activity of the species [77–79].

The essential oil composition of the aerial parts in relation to the cropping system and irrigation regime are presented in Table 4. The analysis of samples detected thirty individual compounds in the first harvest and twenty-two (excluding 2,4-heptadienal, (E,E)-, limonene, bergamal, cis rose oxide, carvone, geraniol, methyl citronellate and valeranone) compounds in the second harvest, representing $\geq$96.67% of the total oil profile in all the tested samples. It can be further noticed that the oxygenated monoterpene compounds were the most abundant class (89.91–94.53%) followed by sesquiterpene hydrocarbons (2.30–4.00%), followed by oxygenated sesquiterpenes (0.58–3.35%) and monoterpene hydrocarbons (0.07–0.20%). The most abundant compounds were geranial (47.27–54.6%) and neral (34.86–37.31%), followed by β-caryophyllene, *E*-isocitral, geranyl acetate, caryophyllene oxide and Z-isocitral. The same compounds were detected in similar amounts in the study of Németh-Zámbori et al. [46], who suggested geranial and neral as the major compounds, while they also detected citronellal, β-caryophyllene, geranyl acetate and caryophyllene oxide in lower amounts. Additionally, in a series of other studies, geranial and neral were also reported as the richest compounds in lemon balm essential oils; however, the detected amounts were lower than those in our study [11,24,80]. Bonacina et al. [1] suggested α-citral and neral as the major compounds of the lemon balm essential oil, while Khalid et al. [23] highlighted the presence of citronellal, citronellol and geranyl acetate. This variability in the literature reports could be attributed to genotypic differences, since Souihi et al. [27], who studied the essential composition of *Melissa officinalis* genotypes derived from Tunisia, Germany and France, observed great differences in individual compound content, especially in the major ones.

Table 4. Chemical composition (%) of essential oils of Melissa plants grown in conventional (C) or organic (O) cultivation systems and subjected to full (FI) or deficit (DI) irrigation.

		1st Harvest				2nd Harvest			
Compound	**RI**	**CFI**	**CDI**	**OFI**	**ODI**	**CFI**	**CDI**	**OFI**	**ODI**
1-Octen-3-ol	975	0.134 ^{a}	0.108 ^{b}	0.095 ^{c}	0.080 ^{d}	0.026 ^{b}	0.031 ^{ab}	0.038 ^{a}	0.027 ^{b}
5-Hepten-2-one,6-methyl	983	0.695 ^{a}	0.688 ^{a}	0.619 ^{ab}	0.578 ^{b}	0.313 ^{b}	0.438 ^{a}	0.308 ^{b}	0.451 ^{a}
β-Myrcene	989	0.117 ^{a}	0.089 ^{b}	0.091 ^{b}	0.093 ^{b}	0.071 ^{b}	0.093 ^{a}	0.095 ^{a}	0.096 ^{a}
2,4-Heptadienal, (E,E)-	1009	0.030 ^{a}	0.000 ^{b}	0.035 ^{a}	0.034 ^{a}				
Limonene	1028	0.091 ^{a}	0.059 ^{ab}	0.023 ^{bc}	0.000 ^{c}				
Benzene acetaldehyde	1041	0.149 ^{ab}	0.172 ^{a}	0.136 ^{b}	0.161 ^{ab}	0.128 ^{a}	0.112 ^{ab}	0.111 ^{ab}	0.070 ^{b}
trans β ocimene	1046	0.127 ^{a}	0.106 ^{b}	0.070 ^{c}	0.046 ^{d}	0.037 ^{a}	0.006 ^{b}	0.036 ^{a}	0.008 ^{b}
Bergamal	1050	0.035 ^{a}	0.011 ^{b}	0.031 ^{a}	0.032 ^{a}				
Linalool	1100	0.287 ^{b}	0.251 ^{c}	0.319 ^{a}	0.191 ^{d}	0.153 ^{a}	0.064 ^{b}	0.147 ^{a}	0.095 ^{ab}
cis Rose oxide	1109	0.000 ^{b}	0.000 ^{b}	0.000 ^{b}	0.021 ^{a}				
trans pinocarveol	1139	0.118 ^{b}	0.117 ^{b}	0.111 ^{b}	0.137 ^{a}	0.032 ^{d}	0.074 ^{b}	0.052 ^{c}	0.097 ^{a}
exo Isocitral	1142	0.257 ^{a}	0.197 ^{c}	0.223 ^{b}	0.230 ^{b}	0.181 ^{b}	0.216 ^{a}	0.183 ^{b}	0.221 ^{a}
neo Isopulegone	1148	0.329 ^{b}	0.323 ^{b}	0.320 ^{b}	0.430 ^{a}	0.200 ^{d}	0.342 ^{b}	0.261 ^{c}	0.413 ^{a}
Citronellal	1153	0.762 ^{ab}	0.690 ^{bc}	0.662 ^{c}	0.793 ^{a}	0.489 ^{a}	0.020 ^{b}	0.015 ^{b}	0.027 ^{b}
Z isocitral	1162	**1.288** ^{a}	**1.124** ^{b}	**1.245** ^{a}	**1.224** ^{a}	**0.590** ^{b}	**1.221** ^{a}	**1.112** ^{ab}	**1.188** ^{a}
Rosefuran epoxide	1173	0.354 ^{a}	0.301 ^{b}	0.282 ^{b}	0.304 ^{b}	0.056 ^{b}	0.097 ^{a}	0.048 ^{b}	0.086 ^{a}
E Isocitral	1180	**1.901** ^{a}	**1.723** ^{b}	**1.913** ^{a}	**1.946** ^{a}	**1.687** ^{a}	**1.879** ^{a}	**1.750** ^{a}	**1.803** ^{a}
Methyl salicylate	1192	0.024 ^{a}	0.028 ^{a}	0.023 ^{a}	0.027 ^{a}	0.000 ^{b}	0.023 ^{a}	0.000 ^{b}	0.000 ^{b}
Citronellol	1227	0.179 ^{a}	0.114 ^{a}	0.041 ^{b}	0.145 ^{a}	0.000 ^{b}	0.004 ^{a}	0.000 ^{b}	0.003 ^{a}
Neral	1242	**35.092** ^{bc}	**34.860** ^{c}	**35.690** ^{ab}	**36.243** ^{a}	**36.011** ^{b}	**36.966** ^{a}	**36.019** ^{b}	**37.307** ^{a}
Carvone	1244	**1.728** ^{b}	**2.524** ^{a}	**0.394** ^{c}	**0.041** ^{c}				
Geraniol	1253	0.330 ^{b}	0.423 ^{a}	0.116 ^{c}	0.077 ^{c}				
Methyl citronellate	1259	0.020 ^{a}	0.026 ^{a}	0.000 ^{b}	0.000 ^{b}				
Geranial	1271	**48.294** ^{b}	**47.266** ^{c}	**49.066** ^{ab}	**49.857** ^{a}	**54.601** ^{a}	**53.436** ^{b}	**53.571** ^{b}	**53.290** ^{b}
Methyl geranate	1321	0.358 ^{a}	0.333 ^{ab}	0.295 ^{b}	0.358 ^{a}	0.229 ^{bc}	0.273 ^{a}	0.209 ^{c}	0.246 ^{ab}
Geranyl acetate	1381	**1.847** ^{a}	**1.985** ^{a}	**1.614** ^{b}	**1.379** ^{c}	**0.968** ^{b}	**1.461** ^{a}	**1.185** ^{b}	**1.409** ^{a}
β caryophyllene	1425	**3.264** ^{ab}	**3.571** ^{a}	**3.854** ^{a}	**2.722** ^{b}	**3.375** ^{a}	**2.444** ^{b}	**3.777** ^{a}	**2.266** ^{b}
α Humulene	1462	0.121 ^{b}	0.141 ^{ab}	0.150 ^{a}	0.092 ^{c}	0.075 ^{a}	0.042 ^{b}	0.092 ^{a}	0.038 ^{b}
Caryophyllene oxide	1587	**1.741** ^{b}	**2.514** ^{a}	**2.283** ^{a}	**2.402** ^{a}	0.733 ^{a}	0.585 ^{a}	0.808 ^{a}	0.661 ^{a}
Valeranone	1673	0.000 ^{b}	0.000 ^{b}	0.057 ^{a}	0.036 ^{ab}				
Total Identified		99.678 ^{a}	99.748 ^{a}	99.765 ^{a}	99.703 ^{a}	99.958 ^{a}	99.833 ^{c}	99.932 ^{b}	99.806 ^{d}
Monoterpene hydrocarbons		0.208 ^{a}	0.148 ^{b}	0.115 ^{b}	0.093 ^{b}	0.071 ^{a}	0.093 ^{a}	0.095 ^{a}	0.096 ^{a}
Sesquiterpene hydrocarbons		3.385 ^{ab}	3.712 ^{a}	4.004 ^{a}	2.814 ^{b}	3.450 ^{a}	2.486 ^{b}	3.879 ^{a}	2.304 ^{b}
Oxygenated monoterpenes		90.922 ^{ab}	89.915 ^{c}	90.383 ^{bc}	91.660 ^{a}	94.003 ^{a}	94.322 ^{a}	93.160 ^{a}	94.532 ^{a}
Oxygenated sesquiterpenes		1.714 ^{b}	2.514 ^{a}	2.341 ^{a}	2.438 ^{a}	0.733 ^{a}	0.585 ^{b}	0.808 ^{a}	0.661 ^{b}
Others		3.293 ^{a}	3.352 ^{a}	2.851 ^{b}	2.651 ^{b}	1.663 ^{b}	2.339 ^{a}	1.952 ^{b}	2.205 ^{a}

Values ($n = 3$) in rows for each harvest followed by the same letter are not significantly different, $p \leq 0.05$. In bold indicated EO components > 1%.

Regarding the major compounds, geranial and neral content showed an increase in the plants grown organically and under the deficit irrigation in the first harvest, whereas in the second harvest the highest content of geranial compounds was recorded for the conventionally grown and fully irrigated plants. In contrast, neral content was beneficially affected by the deficit irrigation, especially in organic cropping systems. For the rest of the compounds, a variable response was observed in both harvesting periods. Apart from oil yield, the profile of essential oils of lemon balm was affected by the tested treatments and most of the identified compounds increased or decreased its content under the deficit irrigation conditions in both harvests. Similar results were recorded for the lemon balm and other aromatic plants, since the water stress conditions are associated with the increased density of oil glands in plant tissues due to reduced plant growth as well as with the increased production of terpenes, which resulted in enhanced oil yields [46,48,53,70].

The mineral content of the leaves is presented in Table 5, where a varied response to the cropping system and irrigation regime was recorded in both harvesting periods. In particular, N content was the highest in the organically grown plants in the first harvest, regardless of the irrigation regime, whereas in the second harvest the highest content was recorded in organically grown plants subjected to deficit irrigation. Similarly, K content in the first harvest was the highest for the fully irrigated and organically grown plants, while in the second harvest no differences between the irrigation treatments were observed in organically grown plants. Phosphorus content increased in organic cultivation and full irrigation in the first harvest, while no significant differences between the tested treatments were observed for the second harvest. The calcium content was decreased in organic cultivation and deficit irrigation in the first harvest, whereas the opposite trend was recorded in the second harvest. Regarding the Mg content, the organic cultivation and full irrigation resulted in the lowest overall content in the first harvest, whereas no significant differences among the studied treatments were observed in the second harvest. The sodium content was the highest in the conventional cropping system and deficit irrigation for both harvests, while Zn increased in organic system regardless of the irrigation regime. Finally, Cu content increased in organically grown plants in the first harvest, while no significant differences between the applied treatments were observed in the second harvest. According to Sussa et al. [81] the cultivation system may affect the mineral composition of the lemon balm leaves, while they also suggested a seasonal variation of the mineral content. The variability of the mineral composition recorded in our study could be associated with the involvement of minerals such Ca, Mg and Zn in terpene biosynthesis or with their participation in the mevalonic acid pathway [54]. However, no specific trends in the mineral composition of lemon balm leaves could be identified for the tested treatments and the applied harvests. Moreover, despite the literature reports, where it is suggested that the organic cultivation may result in increases in mineral content [64], this was not always the case in our study and a variable response to the cropping system was observed.

Table 5. Mineral composition of the aerial parts of *Melissa officinalis* plants grown in conventional (C) or organic (O) cultivation systems and subjected to full (FI) or deficit (DI) irrigation.

Treatment	N (g/kg)	K (g/kg)	P (g/kg)	Ca (g/kg)	Mg (g/kg)	Na (g/kg)	Zn (mg/kg)	Cu (mg/kg)
				1st harvest				
CFI	19.89 ± 0.22 [a]	32.74 ± 0.41 [a]	2.49 ± 0.03 [b]	23.42 ± 1.33 [a]	0.38 ± 0.03 [b]	0.35 ± 0.01 [c]	28.83 ± 0.82 [c]	125.89 ± 20.46 [b]
CDI	20.87 ± 0.47 [a]	29.19 ± 0.47 [bc]	2.51 ± 0.07 [b]	20.85 ± 0.57 [a]	0.48 ± 0.01 [a]	0.53 ± 0.01 [a]	37.0 ± 1.44 [b]	122.91 ± 14.18 [b]
OFI	18.26 ± 0.63 [b]	30.07 ± 0.14 [b]	3.35 ± 0.13 [a]	20.01 ± 1.88 [a]	0.43 ± 0.01 [ab]	0.28 ± 0.01 [d]	49.96 ± 3.08 [a]	266.43 ± 44.50 [a]
ODI	16..55 ± 0.29 [c]	28.36 ± 0.25 [c]	2.54 ± 0.01 [b]	14.87 ± 0.51 [b]	0.47 ± 0.01 [a]	0.47 ± 0.01 [b]	50.51 ± 0.95 [a]	290.70 ± 43.21 [a]
				2nd harvest				
CFI	19.17 ± 0.29 [b]	35.98 ± 0.50 [a]	2.52 ± 0.03 [a]	15.93 ± 0.74 [b]	0.36 ± 0.01 [a]	0.46 ± 0.01 [c]	37.41 ± 1.83 [b]	302.58 ± 41.10 [a]
CDI	22.21 ± 0.26 [a]	35.16 ± 0.34 [ab]	2.57 ± 0.24 [a]	20.23 ± 1.66 [a]	0.40 ± 0.01 [a]	0.69 ± 0.01 [a]	39.72 ± 1.07 [b]	249.18 ± 16.67 [a]
OFI	17.28 ± 0.49 [c]	33.78 ± 0.51 [bc]	2.35 ± 0.05 [a]	18.99 ± 1.40 [ab]	0.41 ± 0.01 [a]	0.48 ± 0.01 [c]	68.87 ± 2.08 [a]	335.77 ± 66.73 [a]
ODI	18.66 ± 0.55 [b]	32.32 ± 0.42 [c]	2.56 ± 0.16 [a]	22.41 ± 1.13 [a]	0.45 ± 0.08 [a]	0.59 ± 0.00 [b]	66.76 ± 3.24 [a]	254.08 ± 34.12 [a]

Values ($n = 3$) in rows for each harvest followed by the same letter are not significantly different, $p \leq 0.05$.

4. Conclusions

The ongoing climate crisis necessitates interventions that will reduce the environmental footprint of crops through reduced inputs and increased use efficiency of natural resources. Irrigation water is becoming more and more scarce, especially in the arid and semi-arid regions of the world. Therefore, deficit irrigation is an environmentally friendly agronomic practice that could help to mitigate the negative effects of the climate crisis. In the same context, the organic cultivation is closely connected with low input practices and is commonly applied in medicinal and aromatic plants. In our study, we evaluated the effect of both these practices on agronomic and quality features of the lemon balm plants. The obtained results were very promising, since despite the reduced plant growth and biomass production, the essential oil yield (%) increased while the composition of essential oils was not severely affected, especially regarding the two major compounds, i.e., geranial and neral. EO yield per harvested area showed a significant decrease in the organically grown plants under the prolonged water stress which could be compensated by the increased plant density when considering the lower biomass and shorter stature of plants grown under stress. Moreover, the total phenols and flavonoids increased under organic cultivation and deficit irrigation, thus increasing the bioactive potential of the obtained herb with the possible application in various sectors of the food industry. In conclusion, organic cultivation of lemon balm under deficit irrigation seems to be feasible; however, further studies are needed with more genotypes, while the application of other innovative practices such as biostimulants could also help to mitigate the negative effects of water stress on plant growth.

Supplementary Materials: The following supporting information can be downloaded at: https://www.mdpi.com/article/10.3390/agronomy12030649/s1, Table S1. Fertilizers and crop protection means applied during the experimental study.

Author Contributions: Conceptualization, N.T. and A.C.; methodology, N.T. and A.C.; software, A.C. and S.A.P.; validation, A.C., S.A.P. and N.T.; formal analysis, A.C., S.A.P. and N.T.; investigation, A.C. and N.T.; resources, N.T.; data curation, A.C., S.A.P. and N.T.; writing—original draft preparation, A.C. and S.A.P.; writing—review and editing, A.C., S.A.P. and N.T.; visualization, S.A.P. and N.T.; supervision, N.T.; project administration, N.T.; funding acquisition, N.T. All authors have read and agreed to the published version of the manuscript.

Funding: This research was co-financed by the project AgroLabs developed under the Programme Interreg V-B Balkan-Mediterranean 2014–2020 and co-funded by the European Union and National Funds of the participating countries.

Data Availability Statement: Not applicable.

Conflicts of Interest: The authors declare no conflict of interest.

References

1. Bonacina, C.; Trevizan, C.B.; Stracieri, J.; dos Santos, T.B.; Gonçalves, J.E.; Gazim, Z.C.; De Souza, S.G.H. Changes in growth, oxidative metabolism and essential oil composition of lemon balm (*Melissa officinalis* L.) subjected to salt stress. *Aust. J. Crop Sci* **2017**, *11*, 1665–1674. [CrossRef]
2. Lim, T.K. *Edible Medicinal and Non-Medicinal Plants*; Springer: Berlin/Heidelberg, Germany, 2016; Volume 10, ISBN 9789401772761
3. Moradkhani, H.; Sargsyan, E.; Bibak, H.; Naseri, B.; Sadat-Hosseini, M.; Fayazi-Barjin, A.; Meftahizade, H. *Melissa officinalis* L., a valuable medicine plant: A review. *J. Med. Plants Res.* **2010**, *4*, 2753–2759.
4. Barros, L.; Dueñas, M.; Dias, M.I.; Sousa, M.J.; Santos-Buelga, C.; Ferreira, I.C.F.R. Phenolic profiles of cultivated, in vitro cultured and commercial samples of *Melissa officinalis* L. infusions. *Food Chem.* **2013**, *136*, 1–8. [CrossRef] [PubMed]
5. Heghes, S.C.; Filip, L.; Vostinaru, O.; Mogosan, C.; Miere, D.; Iuga, C.A.; Moldovan, M. Essential Oil-Bearing Plants From Balkan Peninsula: Promising Sources for New Drug Candidates for the Prevention and Treatment of Diabetes Mellitus and Dyslipidemia *Front. Pharmacol.* **2020**, *11*, 989. [CrossRef]
6. Guran, K.; Buzatu, R.; Pinzaru, I.; Boruga, M.; Marcovici, I.; Coricovac, D.; Avram, S.; Poenaru, M.; Susan, M.; Susan, R.; et al In vitro pharmaco-toxicological characterization of melissa officinalis total extract using oral, pharynx and colorectal carcinoma cell lines. *Processes* **2021**, *9*, 850. [CrossRef]
7. Bukvicki, D.; Gottardi, D.; Prasad, S.; Novakovic, M.; Marin, D.P.; Tyagi, K.A. The Healing Effects of Spices in Chronic Diseases *Curr. Med. Chem.* **2020**, *27*, 4401–4420. [CrossRef]
8. Kim, M.; Yoo, G.; Randy, A.; Son, Y.J.; Hong, C.R.; Kim, S.M.; Nho, C.W. Lemon balm and its constituent, rosmarinic acid, alleviate liver damage in an animal model of nonalcoholic steatohepatitis. *Nutrients* **2020**, *12*, 1166. [CrossRef]
9. Mandras, N.; Roana, J.; Scalas, D.; Del Re, S.; Cavallo, L.; Ghisetti, V.; Tullio, V. The inhibition of non-albicans candida species and uncommon yeast pathogens by selected essential oils and their major compounds. *Molecules* **2021**, *26*, 4937. [CrossRef]
10. Abdellatif, F.; Akram, M.; Begaa, S.; Messaoudi, M.; Benarfa, A.; Egbuna, C.; Ouakouak, H.; Hassani, A.; Sawicka, B.; Elbossaty, W.F.M.; et al. Minerals, essential oils, and biological properties of *Melissa officinalis* L. *Plants* **2021**, *10*, 1066. [CrossRef]
11. Mimica-Dukic, N.; Bozin, B.; Sokovic, M.; Simin, N. Antimicrobial and Antioxidant Activities of *Melissa officinalis* L. (Lamiaceae) Essential Oil. *J. Agric. Food Chem.* **2004**, *52*, 2485–2489. [CrossRef]
12. Tmušić, N.; Ilić, Z.S.; Milenković, L.; Šunić, L.; Lalević, D.; Kevrešan, Ž.; Mastilović, J.; Stanojević, L.; Cvetković, D. Shading of medical plants affects the phytochemical quality of herbal extracts. *Horticulturae* **2021**, *7*, 437. [CrossRef]
13. Patora, J.; Majda, T.; Gora, J.; Klimek, B. Variability in the content and composition of essential oil from lemon balm. *Acta Pol. Pharm.-Drug Res.* **2003**, *60*, 395–400.

14. Ahmadi, T.; Shabani, L.; Sabzalian, M.R. Improvement in drought tolerance of lemon balm, *Melissa officinalis* L. under the pre-treatment of LED lighting. *Plant Physiol. Biochem.* **2019**, *139*, 548–557. [CrossRef] [PubMed]
15. De Assis, R.M.A.; Carneiro, J.J.; Medeiros, A.P.R.; de Carvalho, A.A.; da Cunha Honorato, A.; Carneiro, M.A.C.; Bertolucci, S.K.V.; Pinto, J.E.B.P. Arbuscular mycorrhizal fungi and organic manure enhance growth and accumulation of citral, total phenols, and flavonoids in *Melissa officinalis* L. *Ind. Crops Prod.* **2020**, *158*, 112981. [CrossRef]
16. Hawrylak-Nowak, B.; Dresler, S.; Stasińska-Jakubas, M.; Wójciak, M.; Sowa, I.; Matraszek-Gawron, R. NaCl-induced elicitation alters physiology and increases accumulation of phenolic compounds in *Melissa officinalis* L. *Int. J. Mol. Sci.* **2021**, *22*, 6844. [CrossRef]
17. Pistelli, L.; Tonelli, M.; Pellegrini, E.; Cotrozzi, L.; Pucciariello, C.; Trivellini, A.; Lorenzini, G.; Nali, C. Accumulation of rosmarinic acid and behaviour of ROS processing systems in *Melissa officinalis* L. under heat stress. *Ind. Crops Prod.* **2019**, *138*, 111469. [CrossRef]
18. Németh-Zámbori, É.; Pluhár, Z.; Sxabó, K.; Malekzadeh, M.; Radácsi, P.; Inotai, K.; Komáromi, B.; Seidler-Lozykowska, K. Effect of water supply on growth and polyphenols of lemon balm (*Melissa officinalis* L.) and thyme (*Thymus vulgaris* L.). *Acta Biol. Hung.* **2016**, *67*, 64–74. [CrossRef]
19. Sodré, A.C.B.; Luz, J.M.Q.; Haber, L.L.; Marques, M.O.M.; Rodrigues, C.R.; Blank, A.F. Organic and mineral fertilization and chemical composition of lemon balm (*Melissa officinalis*) essential oil. *Rev. Bras. Farmacogn.* **2011**, *22*, 40–44. [CrossRef]
20. Said-Al Ahl, H.; Omer, E.; Abdou, M. Effect of potassium fertilizer on lemon balm (*Melissa officinalis* L.) grown under water stress conditions. *Planta Med.* **2009**, *75*, 16–29. [CrossRef]
21. Silva, T.C.; Bertolucci, S.K.V.; Carvalho, A.A.; Tostes, W.N.; Alvarenga, I.C.A.; Pacheco, F.V.; de Assis, R.M.A.; da Honorato, A.C.; Pinto, J.E.B.P. Macroelement omission in hydroponic systems changes plant growth and chemical composition of *Melissa officinalis* L. essential oil. *J. Appl. Res. Med. Aromat. Plants* **2021**, *24*, 100297.
22. Said-Al Ahl, H.A.H.; Sabra, A.S.; Gendy, A.S.H.; Astatkie, T. Essential Oil Content and Concentration of Constituents of Lemon Balm (*Melissa officinalis* L.) at Different Harvest Dates. *J. Essent. Oil Bear. Plants* **2018**, *21*, 1410–1417. [CrossRef]
23. Khalid, K.A.; Hu, W.; Cai, W. The effects of harvesting and different drying methods on the essential oil composition of lemon balm (*Melissa officinalis* L.). *J. Essent. Oil-Bear. Plants* **2008**, *11*, 342–349. [CrossRef]
24. Argyropoulos, D.; Müller, J. Changes of essential oil content and composition during convective drying of lemon balm (*Melissa officinalis* L.). *Ind. Crops Prod.* **2014**, *52*, 118–124. [CrossRef]
25. Antal, T.; Chong, C.H.; Law, C.L.; Sikolya, L. Effects of freeze drying on retention of essential oils, changes in glandular trichomes of lemon balm leaves. *Int. Food Res. J.* **2014**, *21*, 387–394.
26. Adinee, J.; Piri, K.; Karami, O. Essential Oil Composition of Lemon Balm (*Melissa officinalis* L.) Leaves Grown in Hamadan Province, Iran. *Med. Aromat. Plant Sci. Biotechnol.* **2009**, *3*, 58–60.
27. Souihi, M.; Amri, I.; Souissi, A.; Hosni, K.; Brahim, N.B.; Annabi, M. Essential oil and fatty acid composition of *Melissa officinalis* L. *Prog. Nutr.* **2020**, *22*, 253–258.
28. Szabó, K.; Malekzadeh, M.; Radácsi, P.; Ladányi, M.; Rajhárt, P.; Inotai, K.; Tavaszi-Sárosi, S.; Németh, É. Could the variety influence the quantitative and qualitative outcome of lemon balm production? *Ind. Crops Prod.* **2016**, *83*, 710–716. [CrossRef]
29. Adinee, J.; Piri, K.; Karami, O. Essential oil component in flower of lemon balm (*Melissa officinalis* L.). *Am. J. Biochem. Biotechnol.* **2008**, *4*, 277–278. [CrossRef]
30. Seidler-Łożykowska, K.; Mordalski, R.; Kucharski, W.; Kędzia, E.; Nowosad, K.; Bocianowski, J. Effect of organic cultivation on yield and quality of lemon balm herb (*Melissa officinalis* L.). *Acta Sci. Pol. Hortorum Cultus* **2015**, *14*, 55–67.
31. Brendler, T.; Gruenwald, J.; Kligler, B.; Keifer, D.; Abrams, T.R.; Woods, J.; Kirkwood, C.D.; Basch, E.; Boon, H.; Lafferty, H.J.; et al. Lemon Balm (*Melissa officinalis L.*): An Evidence-Based Systematic Review by the Natural Standard Research Collaboration. *J. Herb. Pharmacother.* **2005**, *5*, 71–114. [CrossRef]
32. Stojanović, N.M.; Randjelović, P.J.; Mladenović, M.Z.; Ilić, I.R.; Petrović, V.; Stojiljković, N.; Ilić, S.; Radulović, N.S. Toxic essential oils, part VI: Acute oral toxicity of lemon balm (*Melissa officinalis L.*) essential oil in BALB/c mice. *Food Chem. Toxicol.* **2019**, *133*, 110794. [CrossRef] [PubMed]
33. Ahmad MALIK, A.; Suryapani, S.; Ahmad, J. Chemical vs. Organic Cultivation of Medicinal and Aromatic Plants: The choice is clear. *Int. J. Med. Arom. Plants* **2011**, *1*, 5–13.
34. Raei, Y.; Alami-milani, M. Organic cultivation of medicinal plants: A review. *J. Biodivers. Environ. Sci.* **2014**, *4*, 6–18.
35. Hallmann, E.; Sabała, P. Organic and conventional herbs quality reflected by their antioxidant compounds concentration. *Appl. Sci.* **2020**, *10*, 3468. [CrossRef]
36. Lv, J.; Huang, H.; Yu, L.; Whent, M.; Niu, Y.; Shi, H.; Wang, T.T.Y.; Luthria, D.; Charles, D.; Yu, L.L. Phenolic composition and nutraceutical properties of organic and conventional cinnamon and peppermint. *Food Chem.* **2012**, *132*, 1442–1450. [CrossRef]
37. Ahuja, M.; Mohan, S. *Genetic Diversity and Erosion in Plants-Case Histories Volume 2*; Ahuja, M.R., Jain, S.M., Eds.; Springer International Publishing: Basel, Switzerland, 2016; Volume 2, ISBN 9783319259543.
38. Ievinsh, G.; Andersone-Ozola, U.; Zeipiņa, S. Comparison of the effects of compost and vermicompost soil amendments in organic production of four herb species. *Biol. Agric. Hortic.* **2020**, *36*, 267–282. [CrossRef]
39. Mihajlov, L.; Ilieva, V.; Markova, N.; Zlatkovski, V. Organic Cultivation of Lemon Ballm (*Melissa officinalis*) in Macedonia. *J. Agric. Sci. Technol. B* **2013**, *3*, 769–775.

40. Mushtaq, S.; Moghaddasi, M. Evaluating the potentials of deficit irrigation as an adaptive response to climate change an environmental demand. *Environ. Sci. Policy* **2011**, *14*, 1139–1150. [CrossRef]
41. Ouda, S.; Zohry, A.E.H.; Noreldin, T. *Deficit Irrigation: A Remedy for Water Scarcity*; Springer: Berlin/Heidelberg, Germany, 202(ISBN 9783030355869.
42. Geerts, S.; Raes, D. Deficit irrigation as an on-farm strategy to maximize crop water productivity in dry areas. *Agric. Water Manag* **2009**, *96*, 1275–1284. [CrossRef]
43. Fereres, E.; Soriano, M.A. Deficit irrigation for reducing agricultural water use. *J. Exp. Bot.* **2007**, *58*, 147–159. [CrossRef]
44. Abbaszadeh, B.; Farahani, H.A.; Morteza, E. Effects of irrigation levels on essential oil of balm (*Melissa officinalis* L.). *Am. J. Sustain Agric.* **2009**, *3*, 53–56.
45. Farahani, H.A.; Valadabadi, S.A.; Daneshian, J.; Khalvati, M.A. Evaluation changing of essential oil of balm (*Melissa officinalis* L. under water deficit stress conditions. *J. Med. Plant Res.* **2009**, *3*, 329–333.
46. Németh-Zámbori, É.; Szabó, K.; Pluhár, Z.; Radácsi, P.; Inotai, K. Changes in biomass and essential oil profile of four Lamiaceae species due to different soil water levels. *J. Essent. Oil Res.* **2016**, *28*, 391–399. [CrossRef]
47. García-Caparrós, P.; Romero, M.J.; Llanderal, A.; Cermeño, P.; Lao, M.T.; Segura, M.L. Effects of drought stress on biomass essential oil content, nutritional parameters, and costs of production in six Lamiaceae species. *Water* **2019**, *11*, 573. [CrossRef]
48. Bettaieb, I.; Zakhama, N.; Wannes, W.A.; Kchouk, M.E.; Marzouk, B. Water deficit effects on Salvia officinalis fatty acids and essential oils composition. *Sci. Hortic.* **2009**, *120*, 271–275. [CrossRef]
49. Chrysargyris, A.; Drouza, C.; Tzortzakis, N. Optimization of potassium fertilization/nutrition for growth, physiological development, essential oil composition and antioxidant activity of *Lavandula angustifolia* Mill. *J. Soil Sci. Plant Nutr.* **2017**, *17*, 291–306 [CrossRef]
50. Richardson, A.D.; Duigan, S.P.; Berlyn, G.P. An evaluation of noninvasive methods to estimate foliar chlorophyll content. *New Phytol.* **2002**, *153*, 185–194. [CrossRef]
51. Chrysargyris, A.; Papakyriakou, E.; Petropoulos, S.A.; Tzortzakis, N. The combined and single effect of salinity and copper stress on growth and quality of Mentha spicata plants. *J. Hazard. Mater.* **2019**, *368*, 584–593. [CrossRef]
52. Calín-Sánchez, Á.; Figiel, A.; Lech, K.; Szumny, A.; Carbonell-Barrachina, Á.A. Effects of Drying Methods on the Composition of Thyme (*Thymus vulgaris* L.) Essential Oil. *Dry. Technol.* **2013**, *31*, 224–235. [CrossRef]
53. Chrysargyris, A.; Laoutari, S.; Litskas, V.D.; Stavrinides, M.C.; Tzortzakis, N. Effects of water stress on lavender and sage biomass production, essential oil composition and biocidal properties against Tetranychus urticae (Koch). *Sci. Hortic.* **2016**, *213*, 96–103 [CrossRef]
54. Chrysargyris, A.; Kloukina, C.; Vassiliou, R.; Tomou, E.-M.; Skaltsa, H.; Tzortzakis, N. Cultivation strategy to improve chemical profile and anti-oxidant activity of *Sideritis perfoliata* L. subsp. *perfoliata*. *Ind. Crops Prod.* **2019**, *140*, 111694. [CrossRef]
55. Wojdyło, A.; Oszmiański, J.; Czemerys, R. Antioxidant activity and phenolic compounds in 32 selected herbs. *Food Chem.* **2007** *105*, 940–949. [CrossRef]
56. Tzortzakis, N.G.; Tzanakaki, K.; Economakis, C.D. Effect of origanum oil and vinegar on the maintenance of postharvest quality of tomato. *Food Nutr. Sci.* **2011**, *2*, 974–982. [CrossRef]
57. Meyers, K.J.; Watkins, C.B.; Pritts, M.P.; Liu, R.H. Antioxidant and antiproliferative activities of strawberries. *J. Agric. Food Chem* **2003**, *51*, 6887–6892. [CrossRef] [PubMed]
58. AOAC Official. Methods of analysis of AOAC International. In *Official Methods of Analysis of AOAC International*; Horwitz, W. Latimer, G., Eds.; AOAC International: Gaithersburg, Maryland, 2016.
59. Okwany, R.O.; Peters, R.T.; Ringer, K.L.; Walsh, D.B. Sustained deficit irrigation effects on peppermint yield and oil quality in the semi-arid pacific northwest, USA. *Appl. Eng. Agric.* **2012**, *28*, 551–558. [CrossRef]
60. Okwany, R.O.; Peters, T.R.; Ringer, K.L.; Walsh, D.B.; Rubio, M. Impact of sustained deficit irrigation on spearmint (*Mentha spicata* L.) biomass production, oil yield, and oil quality. *Irrig. Sci.* **2012**, *30*, 213–219. [CrossRef]
61. Bettaieb Rebey, I.; Jabri-Karoui, I.; Hamrouni-Sellami, I.; Bourgou, S.; Limam, F.; Marzouk, B. Effect of drought on the biochemical composition and antioxidant activities of cumin (*Cuminum cyminum L.*) seeds. *Ind. Crops Prod.* **2012**, *36*, 238–245. [CrossRef]
62. Osakabe, Y.; Osakabe, K.; Shinozaki, K.; Tran, L.S.P. Response of plants to water stress. *Front. Plant Sci.* **2014**, *5*, 86. [CrossRef]
63. Anwar, M.; Patra, D.D.; Chand, S.; Alpesh, K.; Naqvi, A.A.; Khanuja, S.P.S. Effect of organic manures and inorganic fertilizer on growth, herb and oil yield, nutrient accumulation, and oil quality of French basil. *Commun. Soil Sci. Plant Anal.* **2005**, *36* 1737–1746. [CrossRef]
64. Kazimierczak, R.; Hallmann, E.; Rembiałkowska, E. Effects of organic and conventional production systems on the content of bioactive substances in four species of medicinal plants. *Biol. Agric. Hortic.* **2015**, *31*, 118–127. [CrossRef]
65. Ghamarnia, H.; Mousabeygi, F.; Arji, I. Lemon Balm (*Melissa officinalis* L.) Water Requirement, Crop Coefficients Determination and SIMDualKc Model Implementing. *Eur. J. Med. Plants* **2014**, *5*, 281–296. [CrossRef]
66. Németh, Z.; Radácsi, P.; Gosztola, B.; Rajhárt, P.; Szabó, K. Influence of water supply and fluctuations on yield and quality of lemon balm (*Melissa officinalis* L.). *Aust. J. Crop Sci.* **2017**, *11*, 1539–1546. [CrossRef]
67. Ozturk, A.; Unlukara, A.; Ipek, A.; Gurbuz, B. Effects of salt stress and water deficit on plant growth and essential oil content of lemon balm (*Melissa officinalis* L.). *Pakistan J. Bot.* **2004**, *36*, 787–792.
68. Szabó, K.; Radácsi, P.; Rajhárt, P.; Ladányi, M.; Németh, É. Stress-induced changes of growth, yield and bioactive compounds in lemon balm cultivars. *Plant Physiol. Biochem.* **2017**, *119*, 170–177. [CrossRef] [PubMed]

69. Petropoulos, S.A.; Daferera, D.; Polissiou, M.G.; Passam, H.C. The effect of water deficit stress on the growth, yield and composition of essential oils of parsley. *Sci. Hortic.* **2008**, *115*, 393–397. [CrossRef]
70. Marino, S.; Ahmad, U.; Ferreira, M.I. Evaluation of the effect of irrigation on biometric growth, physiological response, and essential oil of *Mentha spicata* (L.). *Water* **2019**, *11*, 2264. [CrossRef]
71. Parkash, V.; Singh, S. A review on potential plant-basedwater stress indicators for vegetable crops. *Sustainability* **2020**, *12*, 3945. [CrossRef]
72. Kokkinou, I.; Ntoulas, N.; Nektarios, P.A.; Varela, D. Response of native aromatic and medicinal plant species to water stress on adaptive green roof systems. *HortScience* **2016**, *51*, 608–614. [CrossRef]
73. Golubkina, N.; Logvinenko, L.; Novitsky, M.; Zamana, S.; Sokolov, S.; Molchanova, A.; Shevchuk, O.; Sekara, A.; Tallarita, A.; Caruso, G. Yield, essential oil and quality performances of *Artemisia dracunculus, Hyssopus officinalis* and *Lavandula angustifolia* as affected by arbuscular mycorrhizal fungi under organic management. *Plants* **2020**, *9*, 375. [CrossRef]
74. Faller, A.L.K.; Fialho, E. Polyphenol content and antioxidant capacity in organic and conventional plant foods. *J. Food Compos. Anal.* **2010**, *23*, 561–568. [CrossRef]
75. Mandim, F.; Petropoulos, S.A.; Pinela, J.; Dias, M.I.; Giannoulis, K.D.; Kostić, M.; Soković, M.; Queijo, B.; Santos-Buelga, C.; Ferreira, I.C.F.R.; et al. Chemical composition and biological activity of cardoon (*Cynara cardunculus* L. var. *altilis*) seeds harvested at different maturity stages. *Food Chem.* **2022**, *369*, 130875. [CrossRef] [PubMed]
76. Petropoulos, S.A.; Fernandes, Â.; Dias, M.I.; Pereira, C.; Calhelha, R.C.; Chrysargyris, A.; Tzortzakis, N.; Ivanov, M.; Sokovic, M.D.; Barros, L.; et al. Chemical composition and plant growth of *Centaurea raphanina* subsp. *mixta* plants cultivated under saline conditions. *Molecules* **2020**, *25*, 2204. [CrossRef] [PubMed]
77. Petropoulos, S.A.; Pereira, C.; Barros, L.; Ferreira, I.C.F.R. Leaf parts from Greek artichoke genotypes as a good source of bioactive compounds and antioxidants. *Food Funct.* **2017**, *8*, 2022–2029. [CrossRef]
78. Kollia, E.; Markaki, P.; Zoumpoulakis, P.; Proestos, C. Antioxidant activity of *Cynara scolymus* L. and *Cynara cardunculus* L. extracts obtained by different extraction techniques. *Nat. Prod. Res.* **2016**, *31*, 1163–1167. [CrossRef] [PubMed]
79. Stagos, D.; Portesis, N.; Spanou, C.; Mossialos, D.; Aligiannis, N.; Chaita, E.; Panagoulis, C.; Reri, E.; Skaltsounis, L.; Tsatsakis, A.M.; et al. Correlation of total polyphenolic content with antioxidant and antibacterial activity of 24 extracts from Greek domestic Lamiaceae species. *Food Chem. Toxicol.* **2012**, *50*, 4115–4124. [CrossRef]
80. Da Silva, S.; Sato, A.; Salgueiro Lage, C.L.; Da Silva San Gil, R.A.; Azevedo, D.D.A.; Esquibel, M.A. Essential oil composition of Melissa officinalis L. in vitro produced under the influence of growth regulators. *J. Braz. Chem. Soc.* **2005**, *16*, 1387–1390. [CrossRef]
81. Sussa, F.V.; Duarte, C.L.; Furlan, M.R.; da Silva, P.S.C. Agricultural management, season and trace elements effects on volatile oil production from *Melissa officinalis* L. (Lemon balm). *J. Radioanal. Nucl. Chem.* **2016**, *307*, 2365–2371. [CrossRef]

MDPI AG
Grosspeteranlage 5
4052 Basel
Switzerland
Tel.: +41 61 683 77 34

Agronomy Editorial Office
E-mail: agronomy@mdpi.com
www.mdpi.com/journal/agronomy

www.ingramcontent.com/pod-product-compliance
Lightning Source LLC
Chambersburg PA
CBHW041351210926
51850CB00013B/3049
* 9 7 8 3 7 2 5 8 4 4 3 3 3 *